History of Science, Technology, Environment, and Medicine in India

This volume studies the concept and relevance of HISTEM (History of Science, Technology, Environment, and Medicine) in shaping the histories of colonial and postcolonial South Asia. Tracing its evolution from the establishment of the East India Company through to the early decades after the Independence of India, it highlights the ways in which the discipline has changed over the years and examines the various influences that have shaped it. Drawing on extensive case studies, the book offers valuable insights into diverse themes such as the East–West encounter, appropriation of new knowledge, science in translation and communication, electricity and urbanization, the colonial context of engineering education, science of hydrology, oil and imperialism, epidemic and empire, vernacular medicine, gender and medicine, as well as environment and sustainable development in the colonial and postcolonial milieu.

An indispensable text on South Asia's experience of modernity in the nineteenth and twentieth centuries, this book will be of interest to scholars and researchers of modern South Asian studies, modern Indian history, sociology, history of science, cultural studies, colonialism, as well as studies on Science, Technology, and Society (STS).

Suvobrata Sarkar is Assistant Professor in the Department of History at the Rabindra Bharati University, Kolkata, India. His research explores history of technology in the context of nineteenth- and twentieth-century South Asia. His most recent publication is *Let There be Light: Engineering, Entrepreneurship and Electricity in Colonial Bengal, 1880–1945* (2020).

History of Science, Technology, Environment, and Medicine in India

Edited by Suvobrata Sarkar

Routledge
Taylor & Francis Group

LONDON AND NEW YORK

First published 2022
by Routledge
2 Park Square, Milton Park, Abingdon, Oxon OX14 4RN

and by Routledge
605 Third Avenue, New York, NY 10158

Routledge is an imprint of the Taylor & Francis Group, an informa business

British Library Cataloguing-in-Publication Data
A catalogue record for this book is available from the British Library

Library of Congress Cataloging-in-Publication Data
Names: Kumar, Deepak, 1952- honoree. | Sarkar, Suvobrata, editor.
Title: History of science, technology, environment, and medicine in India / edited by Suvobrata Sarkar.
Description: Abingdon, Oxon ; New York, NY : Routledge, 2022. | "Essays in honour of Professor Deepak Kumar." | Includes bibliographical references and index. |
Identifiers: LCCN 2021030009 (print) | LCCN 2021030010 (ebook) | ISBN 9781032148458 (hardback) | ISBN 9781032149691 (paperback) | ISBN 9781003241980 (ebook)
Subjects: LCSH: Kumar, Deepak, 1952- | Science--India--History. | Technology--India--History. | Environmental sciences--India--History. | Medicine--India--History. | Science--Historiography. | India--Historiography.
Classification: LCC Q127.I4 H597 2022 (print) | LCC Q127.I4 (ebook) | DDC 509.54--dc23
LC record available at https://lccn.loc.gov/2021030009
LC ebook record available at https://lccn.loc.gov/2021030010

ISBN: 978-1-032-14845-8 (hbk)
ISBN: 978-1-032-14969-1 (pbk)
ISBN: 978-1-003-24198-0 (ebk)

DOI: 10.4324/9781003241980

Typeset in Sabon
by SPi Technologies India Pvt Ltd (Straive)

Essays in Honour of Professor Deepak Kumar

Contents

SECTION II
Technology and culture

Contributors

Sahara Ahmed is Professor at the Department of History, Rabindra Bharati University, Kolkata, India. She received both her M.Phil. and Ph.D. degrees from the University of Calcutta, India. Her research focuses on the environmental history of Bengal, with particular references to aspects of forestry and related themes in colonial and postcolonial eras. She received the Charles Wallace India Trust Fellowship Award in 2017. Author of the book *Woods, Mines and Minds* (2019), she has several published chapters in national and international journals.

Jayanta Bhattacharya by training as a physician, did his Ph.D. on history of medicine. He is the author of *Biomediicne theke Najardari Medicine* (2017) and edited an international Bengali collection on history of medicine in India, *Bharatiya Patabhumite Cikitsa-Bijnaner Itihas: Ekti Sankshipta Paryalocana* (2009). His other significant publications include 'From Hospitals to Hospital Medicine: Epistemological Transformation of Medical Knowledge in India', in *Historia Hospitalium* (2012), 'Anatomical Knowledge and East-West Exchange', in *Medical Encounters in British India* (2012), and 'The Body: Epistemological Encounters in Colonial India', in *Making Sense of Health, Illness and Disease* (2004).

Santanu Chacraverti is an activist associated with Society for Direct Initiative for Social and Health Action and works on environmental and rights issues. His unpublished doctoral dissertation – *Ramendrasundar Trivedi and Bengal's Response to Modern Western Science* (Jadavpur University) – deals with Ramendra Sundar's epistemology of science in analytical detail. Further, his monograph on Bengal's *Subhankari*, published by the Asiatic Society, provides a detailed historical and mathematical treatment of the tradition. His book *The Sundarbans Fishers: Coping in an Overly Stressed Mangrove Estuary* (2014) has been published by the International Collective in Support of fish-workers.

Abhidha Dhumatkar is presently Head, Department of History, Sathaye College, University of Mumbai, Mumbai, India. She has several research publications to her credit in the reputed journals – *EPW*, *IJHS* – among others. She was recipient of the Senior Research Fellowship of the INSA, New

Delhi, India. She is the biographer of Balaji Prabhakar Modak, a forgotten science popularizer of Maharashtra. She is the first visually challenged woman from India to receive the Charles Wallace (India) Trust Research Fellowship.

Ch. Radha Gayathri works at Educational Records Research Unit (ERRU), JNU, New Delhi, India. She is author of the *Female Medical Education in Colonial India* (2008) and co-editor of *Education in Colonial India: Historical Insights* (2013). Many of her articles and chapters are published in reputed journals and books. She is currently working on the volume *Deciding the Destiny of Nation: Selections from Nationalist Ideas on Education in India 1920–1964*. Her areas of research and publications are history of education, medicine, healthcare systems which she primarily analyses from women's perspective.

Sarandha Jain is a doctoral and teaching fellow in the department of anthropology at Columbia University (USA). Her Ph.D. project is about the relationship between petroleum, the state, and citizens in contemporary India. Her previous work was on rivers, specifically a cultural study of the Yamuna: the relationship people share with this river, and how the state intervenes in it.

Prakash Kumar is Associate Professor of History and Asian Studies at Pennsylvania State University (USA). He received his Ph.D. from the Georgia Institute of Technology in 2004. He spent two years as a postdoc at Yale University's History Department and was Assistant and Associate Professor at Colorado State, before joining Penn State in 2014. He is the author of *Indigo Plantations and Science in Colonial India* (2012).

Nirmal Kumar Mahato is Associate Professor, Department of History, Vidyasagar University, Midnapore, West Bengal, India. He has been awarded Charles Wallace Fellowship, 2019–2020 in the UK. He is also a member, Calcutta Research Group, Kolkata. He is the author of *Sorrow Songs of Woods: Adivasi-Nature Relationship in the Anthropocene in Manbhum*(2020).

John Mathew is Associate Professor in the Divisions of the Humanities and Social Sciences and of Science, Krea University, Sri City, Andhra Pradesh. He has taught at Harvard University, Cambridge, Massachusetts, Duke University, Durham, North Carolina, USA, and the Indian Institute of Science Education and Research (IISER), Pune, Maharashtra, India.

B. Eswara Rao is Associate Professor at the Department of History, University of Hyderabad, Hyderabad, India. He is specialized in the history of medicine. His research interests include sanitation, hygienic practices, and policies in South India. His significant publications include 'From Rajayaks(h)ma (disease of kings) to "Blackman's plague": Perceptions on prevalence and aetiology of tuberculosis in the Madras Presidency,

1882–1947', *IESHR* (2006), and 'Taming "Liquid Gold" and Dam Technology: A Study of the Godavari Anicut', in Deepak Kumar *et al* (eds.), *The British Empire and the Natural World: Environmental Encounters in South Asia* (2011). He is also a recipient of the Travel Grant, Wellcome Trust for the History of Medicine, London.

Y. Srinivasa Rao did his M.A. and M.Phil. at the University of Hyderabad and Ph.D. at IIT Madras, Chennai, India. He is Associate Professor and Head, Department of History, Bharathidasan University, Tiruchirappalli, India. Though he is specialized on the history of science and technology of colonial India, his academic interest covers areas such as studies on subalterns, caste, development, social exclusion, and cinema. He was an International Scholar of the Society for the History of Technology (2008 and 2009).

Dhrub Kumar Singh is at the Department of History, Faculty of Social Sciences, Banaras Hindu University, Varanasi, India. His major publications include 'Choleraic Times and Mahendra Lal Sarkar: The Quest of Homoeopathy as "Cultivation of Science" in Nineteenth Century India', *Medizin, Gesellschaft und Geschichte*, No. 24, yearbook of the Institutfür Geschichte der Medizin der Robert Bosch Stiftung, Germany (2005), and, 'Cholera, Heroic Therapies, and Rise of Homoeopathy in 19th Century India', in Deepak Kumar and Raj Sekhar Basu (eds.), *Medical Encounters in British India* (2013), among others.

Kapil Subramanian is a historian of technology trained at Imperial College, London, and King's College, London, UK. His research has thus far focused on the Green Revolution. He is presently engaged in climate policy research.

Himanshu Upadhyaya is Assistant Professor at School of Development, Azim Premji University, Bengaluru, India. His doctoral research at the CSSP, JNU, New Delhi, India, explored the interrelations between crops and cattle and transformations brought about during the late-colonial and postcolonial period (1890–1980) in western India. He is interested in contributing to the areas of history of veterinary medicine in South Asia.

Shiju Sam Varughese is currently Assistant Professor at the Centre for Studies in Science, Technology and Innovation Policy, School of Social Sciences, Central University of Gujarat, Gandhinagar, India. His areas of interest include social history of science, public engagement with S&T, media and science communication, cultural studies of S&T, and regional modernities in South Asia. He is the author of *Contested Knowledge: Science, Media, and Democracy in Kerala* (2017). He has also co-edited *Kerala Modernity: Ideas, Spaces and Practices in Transition* (2015).

Foreword

Historical studies of science, technology, and medicine expanded in India in distinct phases. The first forays after independence were made by historians with Marxist leanings in the 1950s and 1960s. But these leads were not followed by any significant spurt in historical writings on the subject although historians remained broadly interested in questions of scientific and technological traditions and industrialization. A second round of disciplinary interest developed in the 1970s and 1980s. Deepak Kumar entered the realm in the second round of early stabilization of the field. He has remained a sentinel of the discipline as it expanded and therefore it is eminently suitable that an "Indian" history of STM should celebrate the scholarship of Professor Deepak Kumar.

In the middle of the 1970s when Deepak Kumar started exploring the fringes of history of science, the shape of the field was still malleable. A wide array of approaches filled the analytical space for exploring the nature of STM traditions in India's past. A bedrock of humanistic and social scientific analyses of Indian STM had slowly solidified. This corpus provided a rich if fissiparous ensemble of analytics, examining not any definitive set of core questions but rather clusters of themes that sought to track the antiquity of STM, embed them in India's religion and "culture," and find linkages between STM and India's hierarchical social structure. The world of academic communities and discourses on the history of STM that Deepak Kumar would have encountered was heterogeneous at best.

Deepak Kumar's scholarship became one with a cohort of other works in providing a definitive characteristic to the historiography of science, technology, and medicine. More than anything else, it was Kumar's unsullied commitment to the examination of the "colonial encounter" that imparted a unifying weave. From among a number of problematics that scholars were considering, Kumar chose to focus specifically on the "colonial question." This mirrored a broader trend in a rush to colonial archives to explore the history of STM and reflected a definitive turn in academic histories that declared a priority to examine the history of STM through an exhaustive treatment of the colonial context. The use of the vast resources of colonial archives pushed the existing historiography of STM towards a political economy analysis in the modern period.

Apparently, Kumar's individual journey into colonial criticism of STS developed over a long arc of time and in the face of significant headwinds. The argument that science was not "apolitical" drew upon well-known constructivist frameworks of the time, but the ingenuity came even before – in the very choice of "science" as an arena for investigation, given the modes of research and subject interests of modern India history scholars at the time. Significant works by pioneers aside, historians at large in the universities were still slow to warm up to the study of science, technology or medicine. As Deepak Kumar later wrote in a semi-autobiographical account, it took significant energy and commitment on his part to just zero in on "science" in the search for papers in the archives. In the indexes of archives, he had to run down a list of words starting with "A for Agriculture and B for Botany down to Z for Zoology" as he assembled materials to allow for studying the treatment of science under the colonial regime.[1] The conceptual framework for a critical analysis also emerged only slowly. There were defining moments in Kumar's intellectual trajectory that signalled these shifts. His Ph.D. theses was completed at the University of Delhi in 1984, titled *Science Policy of the Raj*. Alongside, he had been communicating a certain critique of STM that started to emerge at talks and symposia. He presented a paper specifically on "colonial science" at the Indian History Congress session at Hyderabad in 1976. This was later published in the *Indian Journal of History of Science* in 1980.[2]

The first, complete enunciation of Deepak Kumar's colonial science scholarly project came with the publication of *Science and the Raj* in 1995. It bears emphasizing that at this moment, Kumar's primary intellectual goal as a historian of STM was not in pressing the Kuhnian programme forward in the Indian historiography. Neither was his re-launch of historical studies of STM shaped by any enthusiastic embrace of the Social Construction of Technology formulations for an object-focused history of technology. Instead, Kumar's scholarship – as heralded by *Science and the Raj* – chose to align with the dominant Colonial Studies orientation of modern Indian history scholarship at the time. The book announced the core programme of Kumar's history of STM project, which was to explore the colonial underpinnings of STM. This quest was replicated in a number of edited volumes that seemed to cover the entire spectrum of STME: *Technology and the Raj* (1995); *Disease and Medicine in India: A Historical Overview* (2001); and *The British Empire and the Natural World: Environmental Encounters in South Asia* (2010). In books and articles, Kumar covered an exquisitely broad array of topics. The subjects of public health and tropical agriculture received further focused treatment in two edited books down the road: *Medical Encounter in British India* (2013) and *Tilling the Land: Agricultural Knowledge and Practices in Colonial India* (2016).

Kumar remained steadfast in his reliance on the formidable repositories of colonial state and its institutions for a recounting of science and technology histories. But through the three decades that were the prime period of Kumar's publishing career, South Asian historiography redefined its relationship with "colonial archives." A critique emerged that acknowledged the fact that colonial thought inhered in the archives created by the colonial state. In a note on bibliography for *Science and the Raj*, Kumar wrote about how frequently he was warned by his contemporaries against an over reliance on the archive. Kumar's resilience in the light of these attacks spoke of a certain conviction. It also spoke of a methodological choice in favour of archive and not theory as the ultimate arbiter of an historical account. His standard reply to the contrarian ripostes was to argue that internal debates and discussions in the official files, in conjunction with nationalist tracts and vernaculars, provided sufficient avenues to track the decolonized perspective.

That said, it would be far too simplistic to paint the entire scholarship of Deepak Kumar in one broad brush, as if that scholarship did not alter over the long span in which he has remained dominant in STM's historiography. There are two aspects that came to bear more emphasis in Kumar's later works. One of these was to move to the many institutions outside the state's immediate purview whose study enabled an exploration of how wider processes added to the colonial state's outreach. The other was an increased reliance on the vernacular archives. For the sake of convenience, we can call them, respectively, the *project of decentring* and the *project of vernacularizing*.

The *project of decentring* has paralleled trends within the South Asian historiography more generally. Within the scholarship on colonial South Asia, the early focus on the operations, institutions, and archives of the colonial state faded. The first act of decentring involved moving to locality and to "case studies" that were distant from the day-to-day imperatives of colonial policy and implementation. The second act involved moving away from a primary focus on the colonial state's institutionalized bureaucracy. The new points of focus in the later iterations invoked and considered activities of indigenous actors and recognized the importance of the concurrent emergence of new forms of institutions that developed outside the realm of the state. Their study allowed for a focus on micro-processes that no doubt reproduced state power in the final reckoning but were clearly outside the latter's direct influence. They did not reflect imperial thinking, but rather the conflict between norms and practice, and by that very fact also scientific and technological practice in small, relatively autonomous spaces.

The *project of vernacularizing* in Kumar's scholarship has also mimicked a trend within South Asian Studies. The pursuit of vernacular archives was a calculated search for a specific type of archive, revealing a deliberate effort towards studying more regions and languages and adding underlying layers of empiricism. The invoking of vernacular provided entry to looking at Indian actors and colonized elites and their engagement with colonialism. In the wider South Asian historiography, "vernacular" was an expansive descriptor for historical experiences related to the local and discordant, to kinships, and to identities, all of which has proved enormously useful in exploring alternate political imaginaries. But it bears mentioning that in Kumar's project, the vernacular was to be studied along with colonialism, not outside of it or independent of it.

These twin projects formed part of the kernel of Kumar's reenacted project of colonial STM. This reenactment came more or less around the time of publication of an important article in *Economic and Political Weekly* of 2003. Here, Kumar laid down his vision for the future direction of historiography of STM in India, which he fundamentally pinned on the exploration of local archives in their multitudes.[3] This was his call to develop an Indian perspective on STM by using empirical data in India in all of its regional, vernacular, and local contexts. His argument was simple: India had syncretic and multiple traditions of scientific knowledge and myriad engagements between local and western knowledge. STM in every region had "its own distinctive flavour and history"[4] and Indians had multiple visions for modernization. He wanted historians to map this internal heterogeneity through Indian case studies. The indigenous STM could best be explored by studying complex engagements in the Indian locale and societies "through their own literature and practitioners."[5] Kumar believed that more explorations of the Indian context would provide the rightful understanding of STM history in India, a task that "metropolitan theorisations"[6] were restricting.

The *EPW* article was also important in generally suggesting a subtle shift from the era of the first decade of publications. The first book, *Science and the Raj*, was fundamentally an economic and institutional critique of colonialism. The latter two decades of Kumar's scholarship summoned a much broader critical framework. The *EPW* article argued for the need "to move away from state-centric approaches and look to more complex engagements."[7] At this instance in this piece, there was not only a wide-ranging reference to ecological and environmental historians and to Bruno Latour's cultural critique on pasteurization of France, but also an explicit admission that the additional pivots of "nature" and "culture" must be examined. The later works of Kumar reflect this openness that seemed to mark a departure from those that determined the framework of *Science and the Raj*.

The current volume extends the goals of the *project of decentring* and the *project of vernacularizing* in Kumar's redefined agenda in three specific ways. One way the chapters implement the foregoing agenda is in their exploration of multiple institutions. Several chapters provide fleshed out accounts of what transpired in the making of institutions and in their operation. In this way, they give representation to institutional history outside of state and take analysis towards the sites of operation of STM in society. The discussions of education or engineering, or of electrification, street-lighting, and urban transport together give centrality to vignettes from Indian actors and recover their voices about questions on the ground and from their perspective. The focus on popular science-writing, on "popular science" or "science movements" serves the same purpose of drawing the argumentative axis away from the central concerns of the colonial state and instead attends to the Indian engagement with colonial imperatives. The focus on homeopathy or on curing and healing in the *zenana* shows that the institutions could be simultaneously complicit in the act of "delivering, disciplining, [while also standing in for the] disruptive" (p. 301).

The chapters execute simultaneous decentring from the colonial state and its networks on the one hand and from the single pivot of "nation" on the other hand by focusing on a number of "middle level" Indian actors and their writings. The middling positionality of these subjects are flagged as much by their distance from any subcontinent-wide common vision as by their spatial location in isolated places that are still significant locally. To the extent that they focus on colonized Indian elites, they reveal their multifaceted response to colonial conditions. They also underscore a "contested" colonial history which is narrativized from the perspective of the colonized. Their method of correlation between STM and nation is also nuanced. It is important to point out that in their zeal to avoid any single-minded focus on the nation the chapters do not avoid "nation" altogether. What comes out in their readings of colonized Indians is then a disaggregated notion of the nation and other solidarities. They show actors to be holding multiple conceptions of freedom and justice that shines through in the scientific and technological visions they espouse. The new emphasis seems to be on the "local" but also to speak to the "national" as an aggregate of multiple constituents. This conjuring of "Nation 2.0" may as well be a reaction against the extreme and comprehensive "anti-statism" and "anti-elitism" of specific writings.

The third major innovation comes in the use of vernaculars and in the emphasis on vernacular knowledge of STM. This trend is epitomized in the use of Indian language materials in the regional archives, in memoirs, tracts, and journals that are buried deep in local libraries across India which would stay unused unless they are utilized by language-proficient analysts who can spend the time in these archives. This treasure trove allows for placing Indians as knowledge-makers, including those who were not formally trained in the colonial institutions. The latter are likely to go amiss in historical accounts that only place reliance on the views and works of experts who were so recognized on account of their professional credentials. This unshackling from a formulaic view of expertise also opens the way for the study of Indian engagements with "skill" and "practice" that were not anointed by modern science. Such knowledge domains are more likely to be found in the vernacular materials. The vernacular as a historiographical stance pushes historians towards uncovering domains of knowledge forms, artisanship, and traditions of curing that would be missed if one is cloistered by the visions of state records alone. As recent works on Indian STM have highlighted, the vernacular itself was not static or in a state of changelessness. Indeed, it could sometimes grow using tools provided by colonial modernity while at other times it ran the risk of being swept aside by modernist forces. As these scholars set their minds on pursuing these new archives, they add new layers to the history of STM. Where they do not go the distance, they shine light on the path forward.

Prakash Kumar
Pennsylvania State University, University Park

Notes

1 Deepak Kumar, *Trishanku Nation*, New Delhi: OUP, 2016, p, 35.
2 Deepak Kumar, 'Patterns of Colonial Science in India', *IJHS*, Vol. 15, No. 1, 1980, pp. 105–113.
3 Deepak Kumar, 'Developing a History of Science and Technology in South Asia', *EPW*, Vol. 38, No. 23, 2003, pp. 2248–2251.
4 Ibid, p. 2248.
5 Ibid, p. 2250.
6 Ibid, p. 2250. A similar case for exploring multiple engagements in India with "western" and "Hindu" traditions of science or medicine, or about "disunity" in development discourse emerged at different points. Deepak Kumar, 'The "culture" of science and colonial culture, India 1820–1920', *BJHS*, Vol. 29, No. 2, June 1996, pp. 195–209; 'Reconstructing India: Disunity in the Science and Technology for Development Discourse,1900–1947', *Osiris*, Vol. 15, 2000, pp. 241–257.
7 Deepak Kumar, 'Developing a History of Science and Technology in South Asia', op. cit., p. 2251.

Acknowledgements

This is to bring to your kind attention the need to introduce courses in History of Science in university curriculum. History departments of various universities would, I think, be right vehicles for the dissemination of knowledge concerning developments in science and technology and its ramifications.

So wrote in 1983 a young history lecturer in a letter to Professor Madhuri Shah, the then Chairman of the University Grants Commission (UGC), requesting to introduce courses on history of science and technology. India's history students had to wait for another 15–20 years for such an opportunity! Probably in 2002–2003, UGC introduced a course on the theme as an optional chapter at the post-graduate level in history. By that time, the young lecturer has already pioneered a new genre of writing Indian history – in his own words – HISTEM (history of science, technology, environment, and medicine). The celebrated author of *Science and the Raj* (1st edition, 1995), Professor Deepak Kumar has tried to popularize HISTEM during his four decades of teaching and research in India and abroad. For a proper socio-economic development, he believes that HISTEM can be of great value to the civil society and the policy-makers. This anthology is for Professor Kumar, my teacher and mentor, a selfless man gifted with boundless generosity and inspiration.

The present volume reflects the diversity of the field and displays exciting new scholarship from early and mid-career researchers by bringing together several locations and dimensions of HISTEM and modernity in South Asia. The idea first grew sometime in 2013! Later, I approached some of my fellow colleagues and friends and they immediately accepted my invitation with great enthusiasm and thus eager to honour Professor Kumar for his enormous contribution to history of science. The original plan was to release the volume on 31 August, 2017 on the day of superannuation of Professor Kumar. However, things shaped slowly. I accept full responsibility for the delay in publication. My greatest debt is to my contributors for their participation, patience, and perseverance.

Dr Prakash Kumar has been very kind in his encouragement. He has been the guiding force throughout and set the tone of this festschrift with his foreword. Several senior scholars appreciated my labour – Professor Roy MacLeod, Professor David Arnold, Professor Robert Anderson, Professor Ross Bassett, Professor Smritikumar Sarkar, Professor Arnab Rai Choudhuri, Professor Chittabrata Palit, Professor Ranjan Chakrabarti, Professor Arun Bandyopadhyay, Professor Michael Mann, Professor Mahesh Rangarajan, Professor Arabinda Samanta, Professor Sujata Mukherjee, Professor Achintya Kumar Dutta, Professor Raj Sekhar Basu, Professor Sutapa Chatterjee Sarkar, Professor Bipasha Raha, Professor Mahua Sarkar, Professor Ishrat Alam, Professor Jagdish N. Sinha, Professor Madhumita Mazumdar, and of course the late Professor Srilata Chatterjee. I am also grateful to Dr Rohan D'Souza, Dr Aparajita Mukhopadhyay, Dr Sambit Mallick, and Dr Aparajith Ramnath for their support and encouragement. All of them, have enriched the theme – HISTEM – with their writings, and are very gratefully acknowledged here. I have also benefitted immensely from the interest, suggestions, and advice of Professor Suranjan Das and Professor Tirthankar Roy.

I take this opportunity to remember late Amitabha Ghosh (pen-name Siddhartha Ghosh), once a close friend of Professor Deepak Kumar, who had contributed immensely to the field of history of technology by emphasizing the relevance of Indian language sources in the writing of HISTEM. The response to modern science, technology, and medicine could be more interestingly traced in the vernacular press as it had greater reach among the general people. There indeed was a close relationship between the rise of the vernacular press and the development of engagement with modern science and technology in different sections of society.[1]

My thanks go to the anonymous reviewers of the manuscript. I would also like to express my gratitude to Shashank Sinha, Antara Ray Chaudhury, Anvitaa Bajaj, and others at Routledge for their help and encouragement.

We are glad to note that this festschrift comes out as Professor Deepak Kumar enters the seventh decade of his life.

This book is a collective intellectual product. I hope that readers will find this work useful.

Suvobrata Sarkar
Rabindra Bharati University
Kolkata

Note

1 Deepak Kumar and Bipasha Raha (eds.), 'Introduction', *Tilling the Land: Agricultural Knowledge and Practices in Colonial India*, Delhi: Primus, 2016, p. 8. Another recent example, Suvobrata Sarkar, *Let There Be Light: Engineering, Entrepreneurship and Electricity in Colonial Bengal, 1880–1945*, New Delhi: Cambridge University Press, 2020.

Abbreviations

BJHS	*The British Journal for the History of Science*
CUP	Cambridge University Press
EPW	*Economic and Political Weekly*
HISTEM	History of Science, Technology, Environment and Medicine
IESHR	*The Indian Economic and Social History Review*
IHR	*Indian Historical Review*
IISER	Indian Institute of Science Education and Research
IIT	Indian Institute of Technology
IJHS	*Indian Journal of History of Science*
INSA	Indian National Science Academy
MAS	*Modern Asian Studies*
MUP	Manchester University Press
NIT	National Institute of Technology
OUP	Oxford University Press
SHOT	Society for the History of Technology
SIH	*Studies in History*
STM	Science, Technology and Medicine
STS	Science, Technology and Society
T&C	*Technology and Culture*

Introduction

Suvobrata Sarkar

> The development of a history of science, technology, environment and medicine (HISTEM) in south Asia has not merely to draw on different discipline, but also has to shape its concerns from unique and divergent regional traditions and histories that prevail in the region. The south-Asian techno-scientific tradition has largely been a syncretic one, evolving as a result of socio-politico and cultural interactions through the ages; the colonial experience too played its part. The appeal of HISTEM is therefore wider, it belongs to the mainstream of social and cultural debates in history.
>
> Deepak Kumar (2003)[1]

The concept of 'HISTEM' is popularized in the academic arena by Deepak Kumar. Referring to William Jones, founder of the Asiatic Society (1784), he argues, taken together, man and nature (also society) constitute the basis for the history of science, technology, environment, and medicine (HISTEM). The most significant feature of HISTEM, according to him, lies in its necessarily interdisciplinary nature. 'There need not be one history, there can be many'. Technology, for example, can be seen from the perspectives of social history, economic history, even cultural history. Medical history can also be written in multiple directions. HISTEM, like any other historical project, 'involves a study of several cross-sections representing events and ideas that are inter-connected, and which exemplify the cause and effect relationship'.[2] It also affirms the complementary co-existence of the natural and social sciences (even engineering). But one should not make a fine historical analysis jargon-ridden by borrowing concepts from other disciplines frequently; theories and empirical study should go together, and HISTEM, according to Kumar, provides enough opportunities to do so.

However, Steven Shapin and Simon Schaffer, authors of one of the significant interdisciplinary studies of our time, *Leviathan and the Air-Pump* (first edition 1985), emphasize that 'The writing of history could be and should be governed by standards internal to the community of professional historians and not by standards circulating among the laity or among groups who spoke in the name of the practice – for the political history, present-day politicians; for the history of art, present-day artists and aestheticians; and,

DOI: 10.4324/9781003241980-1

for the history of science, present-day scientists'.[3] They intend to show that the creation of scientific knowledge is profoundly political and in a wider sense, science is part of the entire body politic.[4] There is no doubt that the western scientific discourse occupied an extremely important place in the colonization of India. The utility of the pioneering survey work of James Rennell, first surveyor-general of Bengal, was recognized by early colonial administrators like Clive and Hastings. Rennell's work, patronized by the East India Company and later the Royal Society, contributed simultaneously to the consolidation and expansion of British colonial power in India and to the emergent scientific discourse of geography and geology in eighteenth-century Europe.[5] Indeed, colonial encounter provides a good example of the mutually constitutive relationship between scientific knowledge and then-existing socio-political situation.

In the early modern period, it would be more accurate to use the term 'useful knowledge' than science. Savant institutions like the Royal Society and the trading companies including the British East India Company struggled to prove their usefulness in this period. Their mutual interests, according to Anna Winterbottom, encouraged the expansion of networks through which new knowledge could be transformed into 'useful and profitable knowledge'.[6] The knowledge-gathering–manufacturing process was heavily dependent on the assistance from the local informants and collaborators. Thus, the varieties of knowledge produced during early colonial rule included global and local realities. Over the past decades, Rohan Deb Roy affirms, these histories have exposed patterns of connection and correspondence between the colony and imperial state.[7] Thus, Eurocentric narratives of triumphalism, progress, and unilateral diffusion of scientific knowledge from Europe to the rest of the world have been questioned and rejected. The increasing emphasis on a variety of non-European actors and sources has added multiplicity to the histories of modern science. Recently, focussing on the long nineteenth century, Deb Roy explored how Malaria as a category reconsolidated the intellectual, cultural, and political histories of the British Empire and concurrently sustained as an object of natural knowledge and social control.[8]

In the context of nineteenth century, Guoyan Wang mentions that the Chinese translated science as 'Gezhi', which is knowledge acquired through experience. *Gezhi* was derived from Confucianism, and it meant understanding things and the ethics surrounding them; as well as emphasizing enlightenment as a part of knowing.[9] Equally important is the concept of scientific objectivity which emerged only in the mid-nineteenth century. Lorraine Daston and Peter Galison's path-breaking study reveals that modern objectivity 'mixes rather than integrate disparate components, which are historically and conceptually distinct'.[10] They argue each of these components has its own history, in addition to the collective history that pronounces objectivity as a multifarious, mutable concept – capable of new meanings.[11] Lorraine Daston has recently identified the history of knowledge as significant analytical strategies in the history of science: '…some version of the history of knowledge, of which the history of science is a part, is probably indispensable'. She emphasized that a comparative perspective is necessary to tackle the new resources now on offer in new areas of study previously marginal to the history of science, technology, and medicine.[12] The work of knowing and creating, according to Steven Epstein, has always been a central concern in the sociology of science and technology. One of the crucial moves in emergent schools such as the sociology of scientific knowledge is to assert that scientific precisions, like any form of knowledge that does serious work in a society, are basically social or cultural products.[13]

The very first issue of the *Calcutta Review* (1844) claimed, 'The history of science is almost a science; and one of the most interesting and important of them all'.[14] Sixty-five years later, an Indian philosophy professor at the Calcutta University, wrote the following (1909): 'My paper on the *Mechanical, Physical and Chemical Theories of the Ancient Hindus* is tended to be a synoptic view of the entire field of Hindu Physico-Chemical Science, so far as this reached the stage of positive Science as distinguished from the prior mythological and empirical stages'.[15] He was Brajendranath Seal (1864–1938), the celebrated author of *The Positive Sciences of the Ancient Hindus* (first published 1915). Written within an early Hindu nationalist framework, he was unaware that the subject would fascinate the academia as well even after hundred years or so.[16] Not only Seal but also his contemporary erudite class, starting from the men of science to literary giants, continuously expressed their admiration for the HISTEM.

Great geologist Pramatha Nath Bose (1855–1934) and internationally acclaimed chemist Prafulla Chandra Ray (1861–1944) both were science enthusiasts who later turned into excellent historians. Bose's quest for the distinct Indian tradition of science resulted in his monumental work, *A History of Hindu Civilization during British Rule*, in three volumes (1890s). Ray's *History of Hindu Chemistry*, published in two volumes in 1902 and 1908, firmly established the scientific credentials of the ancient Hindus and, by extension, their rightful place in the modern world of science. But, although they saw the history of ancient Indian science as a source of national pride and inspiration, they never considered it as something that could be revived as an alternative to modern science; the latter alone could bring India's modern nationhood. Swami Vivekananda (1863–1902) once met Jamsetji Nusserwanji Tata (1839–1904), an Indian pioneer industrialist, and requested him not to depend on foreign products but start manufacturing steel in India itself. Later, the monk even advised Tata to start a science-learning institute of excellence, and the Indian Institute of Science, Bangalore was born.[17] Rabindranath Tagore (1861–1941), the laureate, was also fascinated by the marvel of modern science and technology:

> Thus, through the help of science, as we come to know more of the laws of nature, we gain in power; we tend to attain a universal body. Our organ of sight, our organ of locomotion, our physical strength becomes world-wide; steam and electricity become our nerve and muscle...[18]

Interest in the development of science, technology, environment, and medicine in India under the British rule has grown in recent decades and has played an ever-increasing part in the reinterpretation of modern South Asian history. Jahnavi Phalkey has recently emphasized the significance of the history of science to understand the history of modern India and at the same time links science's advancement with the history of colonialism/imperialism interpreting the character of empire.[19] The discipline of the history of science established itself in academic departments, centres, and programmes in Europe and North America in the 1950s and 1960s. After initial hesitancy, such histories are now being taught at numerous universities and colleges of India. Not even that, the country's premier engineering and science institutes (IITs, NITs, and IISERs) give adequate emphasis on the theme in their under-graduate, post-graduate, and doctoral programmes.

The volume explores a variety of ways, sites, and confluences through which the concept 'HISTEM' has evolved. A wide and diverse array of case studies showcases the vibrancy of histories of science in South Asia and the range of topics that are now being taken up for research. Here, 'science' in a broad sense includes technology, medicine, and the environment. The volume as a whole is not intended as a critique of earlier perspectives; however, attempts have been made to examine official (colonial) and Indian language sources simultaneously to explore how 'HISTEM' has been constituted by actors and agencies in specific periods and settings. Taken together, the contributors of this volume reassert the significance of the theme in shaping the histories of colonial and post-colonial South Asia.

On methods, questions, and theory

An important question that has engaged historians of science in the last few decades is the relationship between science and European imperialism.[20] It is now apparent that science was influenced by the Europe's imperial experiences. Both modern science and colonialism grew together, and as Deepak Kumar argues 'hand-in-hand'. Hence, he talks about 'colonial science' to emphasize that coloniality was the most dominant feature of science in India during the period.[21] 'Science' was directly driven by imperial achievements and needs. Early map-making operations including the work of the Great Trigonometrical Survey of India came from the need of trade and military campaigns. The geological surveys were linked with intelligence gathering on minerals and local politics. Efforts to control various epidemic diseases led to attempts to regulate the habits, diets, and movements of colonial subjects. According to David Arnold, this was a political process (colonization of the body)[22], by which the imperialists converted medicine into a weapon to secure their rule. New technologies were also put to use expanding and consolidating the empire.[23]

Discussions on science and European imperialism inevitably open up the very question of the diffusion and institutionalization of Western science across the world. Most recently, Warwick Anderson argued, in response to

Basalla, it had become necessary to situate colonial science precisely in its local, political, economic, and cultural settings, to render it multi-centred.[24] Earlier, Palladino and Worboys claimed, 'Western methods and knowledge were not accepted passively, but were adapted and selectively absorbed in relation to existing traditions of natural knowledge and religion and other factors'.[25] Increasingly, we were exposed to concepts like contact zones of mobile knowledge practices, often using anthropological and post-colonial approaches. The post-colonial agenda in science, technology, and society studies (STS) has attempted to describe how formal knowledge and practice travel, and what happens to them at their arrival points, how they articulate across and within cultures. As Steven Shapin observes, 'we need to understand not only how knowledge is made in specific places but also how transactions occur between places'.[26]

As far as theoretical terms are concerned, it is the idea of networks, following the influential work of Bruno Latour, which has had tremendous impact in understanding the making of modern science. Latour probably first emphasized to look to the colonies, not the 'home country' to understand '...transformation of a society by a "science"'.[27] On the other hand, the popular term 'contact zones' comes from Mary Louise Pratt: 'By using the term "contact", I am to foreground the interactive, improvisational dimensions of colonial encounters so easily ignored or suppressed by diffusionist accounts of conquest and domination'.[28] Sandra Harding issued a plea to 'locate modern sciences on the more accurate historical and geographical maps produced by the postcolonial accounts'.[29] From a feminist perspective, she envisioned a multicultural science integrating the knowledge and practice of Third World peoples.[30] The flow of knowledge and practice from Europe, and into it, according to Warwick Anderson, began to seem more turbulent, no longer laminar; matters of local terrain and inescapable friction came to complex the entire picture. Problems of translation, mediation, transformation, as well as indifference and resistance, seemed ever more pressing.[31] Anderson firmly believes that post-colonial analysis offers a 'flexible and contingent framework for understanding contact zones of all sorts, for tracking unequal and messy translations and transactions that take place between different cultures and social positions, including different laboratories and disciplines even within Western Europe and North America'.[32]

Interest in indigenous responses to Western science, however, precedes current concerns with the nature of 'colonial modernity', as is evident, from the wide-ranging literature on South Asia. Several historians dwell upon India's rational approach to Western science, both pre-colonial and colonial times.[33] Sometimes, the indigenous revivalist attempts sought to regenerate India's techno-scientific traditions; whereas, the secular modernizers rejecting the past, completely embraced western rationality and science.[34] Most Indian responses to western science existed somewhere between these two extremes. A few of them, tried to accept western science, believing it as part of a universal knowledge tradition, to which they had earlier contributed; others attempted to search analogues of Western sciences in the ancient

Indian texts and culture.[35] By emphasizing the local condition (context), concepts like 'colonial science', 'Indian responses', etc., enhance complications to the idea of direct transplantation of western knowledge in the colony. Gyan Prakash argues that western science was reinterpreted in and for the colony by the Indian intellectuals. These science enthusiasts, in interpreting western scientific ideas into Indian languages, involved in a 'renegotiation of knowledge and power', and, thus became the champions of an 'Indian' modernity. Prakash views this as a form of 'hybridity'.[36] For Pratik Chakrabarti, the search for cultural legitimacy that characterized Indian science and technology, between the 1850s and 1900s, was displaced by an increasingly dominant discourse of scientific industrialism.[37]

David Arnold advocates that understanding the significance of 'science as modernity' is the best available alternative to diffusion theory. Thus, a need to construct a particularly Indian modernity has been seen as the driving force behind Indian elites' participation in the debates on science and technology. In the early twentieth century, techno-scientific ideas, closely associated with the ideas of modernity, became central to the discourse of Indian nationalism.[38] In this context, Kumar's interpretation of the 'disunity in the science and technology for the development discourse' leading up to independence is significant. He shows how Indian nationalist voices and the British government each articulating their own versions of what a modern (independent) India should look like. Between the two World Wars, the Indian National Congress had first adopted scientific planning as an ideal, whereas the colonial government also drafted its own agenda to upgrade the material life of the ordinary Indians, which according to Kumar, was a credible option to jeopardize Gandhi's call of 'Quit India'.[39]

These studies provide a further reminder of the limitations of the diffusionist conceptions of science. S. Irfan Habib and Dhruv Raina rightly observe that 'the standard tale of the assimilation of modern science as a Western cultural import was inadequate and missed out the multifarious nature of exchange between modern science and so-called traditional knowledge forms'.[40] Some scholars have emphasized the need to see beyond fixed centres and peripheries. Rather, each locality has the capacity to become central, to act as the point of a route of information.[41] As an alternative to Eurocentric diffusion theory, Kapil Raj proposes a 'circulatory' model for the spread of western science. He seeks to illuminate the 'co-production of the local and the global'.[42] As circulation can only be inferred from the 'intercultural encounter', he provides fascinating examples of the construction of scientific knowledge in the 'contact zone' – South Asia.[43]

Much of what we believe as western science was produced in the colonies, rather than being exported to them. The expansion of colonial power and the production of techno-scientific knowledge were closely related, and, in the process, India served as the arena for the construction of a large-scale scientific research system. For example, colonial expansion was crucial to the development of botany and geology – where the collection and comparison of specimens were dominant. Similarly, it provided a spur to the emergence of modern medicine and environmental thought.[44] An important point Zaheer Baber raises is the immanent connection between instruction on science in India and the emergence of the colonial capitalist state.[45] This required that the colonial state be innovative in the founding of formal technical institutions. Until the end of the nineteenth century, Britain had no formal institutions imparting technical education and engineers received their training as apprentices; the engineering colleges established in colonial India served as models for replication in Britain and the colonial encounter contributed to the development of technical education in the metropole.[46]

Recently, Sujit Sivasundaram explored the technicalities of researching and writing globally oriented histories of science. There should not be one history, but history in all possible dimensions – multiple histories. To understand colonial science, he argues, it is necessary to think beyond categories of colonized and colonial and to identify traditions of knowledge from all corners: 'Take, for instance, Mughal traditions, which were themselves part of a Persian world. If all of this is taken into account, the global history of science becomes the history of the shifts and reinventions of a variety of ways of doing science across the world. European imperialism becomes a chapter in this story ...'[47] The global historians of science should address the question of modernity as well. For Sivasundaram, to be modern in the twentieth century meant using techno-scientific knowledge for development. The journey towards modernity is not teleological or linear; a true historian should contextualize it in the broader history of movement and reappearance of traditions of knowledge. Thus, 'networks', 'contact zones', 'people-in-between', 'mobility', 'circulation', 'practice theory', among others, have emerged as crucial categories to identify the emergence of modern science. These recent works have continued to reveal how modern knowledge was built by the mechanism and need of imperialism.

To reject the core–periphery deliberation altogether, Kumar argues, would be like 'throwing the baby with the bathwater'.[48] Studies in HISTEM may help us get a better understanding of colonial modernity. Scholars have talked and written about how modernity reached India riding the colonial wave for long. The evolution of 'Indian modernity' is different from Europe. The traditional knowledge which they inherited played a significant role in the formation of 'Indian modernity', along with the western knowledge, after the colonial-watershed. Exploration of this complex yet dynamic relationship from multiple dimensions is one of the agendas of HISTEM.

Science and society

Science has been integral to modern Indian history. Until the end of the eighteenth century, European travellers, traders, bureaucrats, army engineers, and missionaries remained busy in building up the colonial project, and they emerged as the major informants on what was happening. The British colonial government established several scientific institutions and surveys primarily to serve the economic needs of the colonial state and exploit India's natural resources through scientific research. Later, Indians also became curious about what was happening around them. They were no longer passive recipients, and, on several occasions, they emerged as collaborators – they contributed to the knowledge-dissemination process. Generation was not easy those days, as for generation, they always looked to the metropole. However, generation in certain areas also took place in the colonies, for example, in Canada, Australia, and India, in the zoological and botanical sciences. In these branches of knowledge, the metropole learnt from the colonies. In the mid-nineteenth century, India's share went up – when educated Indians wanted to trace European scientific ideas and principles within Indian culture as well as to adapt them for the material development of their country. Thus, according to Pratik Chakrabarti, science had two broad meanings in modern Indian history. Firstly, it symbolized European Enlightenment, modernity, and westernization. Secondly, embracing the western scientific ideas enthused Indians to identify scientific and rational traditions within their past. The deep engagements with these two, Chakrabarti argues, led to both assimilations of 'modern science' in the Indian society and culture and its redefinition.[49]

The history of natural history from the 'Age of Discovery' has been inseparable from the history of imperialism. Two scholars have recently emphasized, 'The networks of empire were not composed of a centre and radiating spokes, as is often imagined; rather, they formed a complex of crisscrossing flows and routes spanning much of the world. Nevertheless, Britain was flooded with new information and objects, and together they helped to transform the metropole. Conversely, the possession of such knowledge and data could be translated into power and authority'.[50] India, with much left unexplored, attracted broad scientific interest in the flora and fauna from the early days of the English East India Company; the interest intensified during the nineteenth century due to growing accessibility to the interiors. The state scientists recognized the importance of biological surveys in empire-building. As India became part of the British Empire and much of the research on its flora and fauna and natural environment was under the tutelage of the colonial state, most of the scientists involved in botanical and zoological surveys, as classifies by John Mathew, were 'translocators'.[51] The professionalization of natural history in British India had largely been the creation of this group of European experts – administrators, doctors, military officers, and missionaries. Yet Indians, apart from serving as seasoned collectors, trackers, and draughtsmen, were absent from this enterprise.

However, Indians were increasingly entering the ranks of medical men, there was no such related increase in their numbers in the branch of natural history. An extremely interesting kind of evidence was provided by John Mathew in his essay on Soorjo Coomar Goodeve Chuckerbutty, one of the first Indians to be sent to England for a higher medical degree (1845). Soorjo Coomar won the gold medal for comparative anatomy at University College London and was trained by a former teacher of Charles Darwin, Robert Grant. Despite such exposure and intelligence, Soorjo Coomar did not continue his studies in natural history to any great extent, and upon his return to India, he became professor of 'Materia Medica' at the Calcutta Medical College. In the early nineteenth century, the newly emerged middle class looked to western knowledge with great admiration. They tried to articulate the new knowledge in terms of Indian traditions or requirements. These were torn people in the sense, they wanted to have the best of both worlds – Soorjo Coomar was not an exception.

The desire to understand the new knowledge and machines that the colonizers had brought, and to appropriate them, was there. The vernacular press reflected the ripples in the Indian minds, and a handful of intellectuals responded by publishing scientific books and journals, while some took to founding associations and institutions of scientific nature. The most important characteristic of mid-nineteenth-century-Indian thinking was an enormous emphasis on cultural synthesis. The idea of a cultural synthesis, Kumar argues, gave them the best of both worlds – first, it empowered them to absorb the cultural shock, and, second, promised a possible opportunity to overcome the obstacles imposed by the colonial state.[52] The local interlocutors adopted several strategies – imitation, translation, appreciation, and then assimilation, even sometimes boycott – but without much success. Two things are striking in this interpretation, popularization of science through translations, and constant reference of ancient tradition to validate the existence of a spiritual, although rational, scientific past. The new paradigms in science were quickly accepted and numerous popular articles traced the seeds of modern advancements in ancient texts. Akshay Kumar Dutt (1820–1886), a contemporary crusader, worked for 'Indianizing western science'. But the man, who gave a new meaning to science popularization, taking the discourse to a new professional height, was none other than Ramendra Sundar Trivedi (1864–1919).

Ramendra Sundar simplified some of the most complicated physical and astronomical phenomena, all in a highly readable and informative, popular style in Bengali; and drew references from mythology, folklore, and popular local traditions. Santanu Chacraverti deals with Ramendra Sundar Trivedi's personal intellectual inclinations and the specific philosophical trends that influenced him. Trivedi's academic training allowed him to compare the western and Indian philosophical traditions. While analysing Ramendra Sundar's epistemological inquiry into the nature of scientific knowledge, Chacraverti indicates to his patriotic attachment towards Indian tradition – a combination of Brahmanical and *Loka* or popular traditions. He believed in scientific pursuit for the sake of knowledge only. This has some contemporary relevance as well. The notions of science and its terminologies entered so deep in the cultural lexicon of the country that no Indian erudite could afford to ignore them.

The science enthusiasts and popularizers, mostly from Bengal, have attracted considerable scholarly attention. However, there also existed substantial indigenous initiatives, without much intellectual limelight, in the field of modern science in other parts of India too. Abhidha Dhumatkar in her pioneering research on Balaji Prabhakar Modak (1847–1906), an unsung public intellectual from western India, seeks to explore his ideas and works as a prism to understand the reception of modern science and subsequent cognitive movements in South Asia. Balaji Prabhakar tried to create a scientific attitude in Maharashtra by spreading scientific knowledge through his books in Marathi language, public lectures and demonstrations, and annual science exhibitions in Kolahpur. He viewed science, Dhumatkar argues, as a stimulus to the regeneration not only of the Indian industry and agriculture but also of the Indian nation. He was one of the first in western India to begin translation of important scientific works into simple Marathi so that it could percolate deep in the society. Unfortunately, Balaji Prabhakar left no loyal band of followers after him as his movement was restricted to the intellectual domain – not social and political, which according to Dhumatkar, explains his invisibility in the discussion of the formation of a distinct Indian modernity.

Science communication as a pedagogical project was put forward by James A. Secord. He wrote that the image of scientific ideas and technological projects as social processes rested on the inter-dependence of the two variables: In order to appreciate the 'social roots' and 'social impact' of scientific knowledge, it was essential to acknowledge that a wider society existed beyond the known community of experts. Their appreciation and appropriation of scientific knowledge were as important as their generation.[53] Later researches add several complexities in the process – 'public communication' is shaped by the cooperation and conflict of several interest groups involved in the process, including the public. Shiju Sam Varughese has identified that research on public engagement with science and technology (PEST) is in its 'infancy' as an academic field, although there is a growing interest in recent years.[54] He talks about a 'scientific-citizen public' and explores the significant social and historical processes that made possible its emergence with reference to Kerala.

Ramendra Sundar Trivedi, Balaji Prabhakar Modak, along with other science popularizers of the late nineteenth and early twentieth century, successfully created for vernaculars a linguistic space that could accommodate scientific, philosophical, and epistemological themes without much jargon. It was their object to bring the joys of science to the learned audience. However, the medium of science communication, mostly vernacular periodicals, monographs, and tracts was confined to the elite minority – the readership of popular science writing could hardly be the 'public'. In his deliberation, Shiju Sam Varughese deals with a new form of public engagement with science that emerged in the post-independence Kerala and the *Kerala Sasthra Sahitya Parishad* (KSSP) was the prominent medium in creating a 'scientific-citizen public'. The KSSP and similar organizations involved in the People's Science Movements (PSMs), in the early decades of independent India, provided the impetus to a new cultural trend: promotion of a serious engagement with science beyond scientism. These organizations aimed at the democratization of scientific decision-making beyond a mere popularization of science.

Technology and culture

As a discipline, the history of technology was institutionalized with the formation of the Society for the History of Technology and the creation of the journal *Technology & Culture* in 1958.[55] According to Melvin Kranzberg, all history is relevant, but the history of technology is the most relevant.[56] The history of technology is an effort to recount the history of all those things, those artefacts that we have produced over the years. The social history of technology goes one step further, integrating the history of technology with the rest of human history. Works on the social construction of technology virtually revolutionized the field of history of technology.[57] A social history of technology assumes a mutual relationship between society and technology; it also assumes that changes in one can, and have, induced changes in the other.[58]

Until recently, much academic endeavour has tended to view technology from an imperial, post-imperial, and global-capitalist perspective on indigenous societies. Recent discussion on technology in nineteenth- and twentieth-century Asian or African countries have begun to move away from earlier insistence on the centrality of imperial agency and instrumentality of the empire's 'tools' of conquest and exploitation.[59] Enquiry has in part shifted away from a diffusionist preoccupation with a system of one-way 'technology transfer' that privileged Euro-American innovation over local agency, and form seeing technology in terms of European representations of machines as the measure of the imperial self and colonized other.[60] A language of 'transfers', 'diffusions', and global 'commodities' tells us remarkably little, as recent studies argue, about the origin and growth of various modern technologies in the non-Western world.

David Arnold and Erich Dewald have argued that technology's social fashioning is a difficult concept to employ in an Asian colonial context than in relation to autonomous European and North American industrial societies. These were not designed or manufactured locally but in the West. The social construction of these technologies in a colonial or semi-colonial setting perhaps takes in a different form. How various technologies were locally accepted or rejected were depended on significant local emendation and reinvention to match the local cultural norms and social usages.[61] Another interesting dimension in this archive is that there has been a growing interest in small-scale 'everyday technologies', such as the sewing machine, wristwatch, radio, typewriter, and bicycle. According to David Arnold, colonial regimes were unable to monopolize or reluctant to control these small-scale technologies and they passed with relative ease into the work-regimes, recreational activities, social life, and cultural aspirations of colonized and post-colonial populations.[62] These small-scale 'everyday technologies', according to this new trend of research, can be found in the diaries, novels, journals, pamphlets written mostly in Indian languages – in the self-representation of the people – rather than in the official archives.

However, the emphasis on 'everyday technology' does not mean that large-scale technologies are not researched or written about, albeit no longer in a 'celebratory' way.[63] Recently, Smritikumar Sarkar demonstrated that it is possible to write a history of technology in modern India without differentiating the 'big' and 'small' technologies, where the voices of the British Raj can be analysed together with the ordinary Indians.[64] In this burgeoning field of historical research, Sunila S. Kale identifies another lacuna: 'Yet in much of this scholarship, what is arguably the twentieth century's most vital technology—electric current—is largely absent'.[65] The subject of electrification in the 'Global North' has fascinated historians and social scientists for long, but the history of power generation and supply industry and subsequent electrification in the 'Global South' is a less-frequented area of study.

Y. Srinivasa Rao emphasizes the initial urban-centric character of electricity in Madras. With the expansion of electricity, the boundaries of the urban area also expanded; the urban growth, in turn, created greater demand for electric power. Madras city, as colonial administrative and power centre, slowly began to acquire all the electric utilities (tramcars, lights, fans, electric lift, etc.), and began to blur the difference between the metropolitan and periphery. The electricity generation and distribution system, and various other technological systems powered by electricity came to India with colonial patronage. Initially, electricity faced oppositions in Madras, like several other western technologies, but gradually the relationship stabilized. '... lack of interest from the natives and lack of interest from the government had created some sort of regional inequality in electricity within the presidency'. This article dwells on how electricity as a motive power in urban Madras transformed from 'luxury' to 'necessity'.

The question of technology systems or technology projects cannot be adequately addressed unless engineering education is taken care of. Suvobrata Sarkar makes a plea for a serious engagement with 'academic engineering' by discussing how, in a colonial setup, the status of engineering searching for recognition in India, citing examples from the history of Bengal Engineering College, Sibpur. One might wonder why a college of engineering was opened at Roorkee (1847), and subsequently at Calcutta (1856), at a time when Britain itself did not provide academic training to engineers except for military purposes. The Bengal Engineering College produced several efficient engineers who excelled in their profession later. Several of them accepted government, semi-government, or private employment. But the combination of engineering and entrepreneurship was extremely rare in those days. The college produced a large cadre of civil engineers, but in the fields of mechanical and electrical sciences, its performance was meagre. Why this was so? Unlike the predominant historiography which emphasizes 'centre–periphery' relationship, the influence of the Public Works Department, military, and so on, Sarkar elucidates several positive points. Based on the archival sources, college materials, and other contemporary sources, the essay explores the struggle of the college to transcend the barrier imposed by colonialism, and appropriation of modern technical knowledge by the Indian elites.

Kapil Subramanian traces the story of how interwar India became a global pioneer in the use of tube-wells for irrigation and the crucial role these tube-wells played in the Green Revolution later that some have called it a 'tube-well revolution'. The Geological Survey of India played a significant role in providing information on groundwater for the purposes of irrigation, municipal and military. Knowledge about groundwater was, Subramanian opines, a 'heavily contested concern' in mid-twentieth-century India. Here comes another twist in the story! The transmission of techno-scientific knowledge between Europe and Asia has been the subject of several historical studies during the past decades. These studies explore the long nineteenth- and early twentieth century. However, by the mid-twentieth century, the rise of the USA as the 'Superpower', added multiplicity in the whole story. Subramanian explores the shift from British to American expertise in the development of water-divining science vis-à-vis role of the Geological Survey of India. He also traces Indian participation, initially as water-diviners, and, later as collaborators in the generation of engineering geology and groundwater knowledge in the late years of the Raj to the post-colonial India.

India by 1940 was among the eight most industrial countries in the world; it also had one of the largest scientific communities to be found anywhere outside Europe and North America. The early twentieth century saw far-reaching developments in technology. Among the most momentous was the advent of electric power. Like hydro-electricity, the rapid rise of the internal combustion engine, according to David Arnold, had several 'spin-off' effects for Indian science as well as industry. One was the fresh stimulus given to geology by the search for the country's own oil resources; another was the growth of transport and communications.[66] Oil's multiple uses made it a very profitable industry. The widespread use of petroleum was a calculative move, encouraged by the private industry as well as the state, once it entered in the households, it was not simply a 'colonial instrument'.

Ever since the 1980s, there has been great disappointment among renewable energy supporters that the shift from fossil-fuel dependence has been painfully slow.[67] Timothy Mitchell observes, 'Fossil fuels helped create both the possibility of modern democracy and its limits'.[68] A search for deeper understanding of the historical development of reliance on petroleum in India, and, society's present-day oil addiction, underline Sarandha Jain's contribution, 'From Battlefields to Homes'. The transformation of oil, from an agent of illuminant to lubricant, to fuel, contributed it an entirely 'new identity and meaning'. Jain rightly claims oil's chemical properties have the capacity to generate multiple substances, created versatile political and social possibilities. The two world wars demonstrated the contradiction of oil conservation and the growth of the oil industry *inter alia* national security. With independence, India found, 'Building a nation was about making a future with oil'. With the passage of time, being 'import-dependent' oil country, her dilemma increased endlessly: The 'oil shock' (the 1973 Arab–Israeli War) reveals oil was now a 'big enough issue', which was directly related to the political stability of the country. Citing examples from the National Archives of India extensively, Sarandha Jain concludes,

> The story of oil captures a double movement in which big and complex technology makes possible the everyday technical, political and social worlds, but simultaneously, the everyday use of that technology is what enables it to become big, and therefore useful for the state.

Environmental issues

In a densely populated country like India, environmental issues have both an ecological and a human dimension. Is the high Himalayan region the right place for large hydel projects? Mining schemes, if not controlled properly, may devastate hillsides, and pollute rivers. Thus, in India, ecological stewardship is not a luxury, but the very basis of human survival.[69] Indian social scientists and historians had previously ignored the environmental underpinnings of human life. There were several studies on the relationship between peasantry, zamindars, and the state; but few of them asked how agrarian life was conditioned by natural environs – forests, waters, minerals, etc. Richard Grove and others have explained environmental history as a historically documented part of the story of the life and death, not of human individuals, but of societies and of spices, both others and our own, in terms of their relationships with the world around them.[70]

South Asia has begun to develop its own distinctive contribution to environmental history since the early 1990s. Environmental historians have mostly been focused on the last two centuries, especially the period from 1858 when India came under the rule of the British Raj. Its intellectual origins as a 'self-conscious domain of enquiry' can probably be traced to the encounter of seventeenth- and eighteenth-century Western Europeans (naturalists, medical officers, army engineers, bureaucrats, etc.), with the entirely unfamiliar environments of the tropics, and with the damage done to these environments by them.[71] The scientists and medical men, employed under the East India Company, played a revolutionary role in the evolution of ecological consciousness, much before any such development in Europe. Both forests and water, according to David Arnold and Ramachandra Guha, have played a significant role in shaping South Asian history which seems unimaginable in the western eco-cultural systems. In the colonial era, these resources have increasingly come under state control. The colonial state powerfully influenced environmental changes by formulating legislation pertaining to and assuming control over, resources which were earlier under more informal and decentralized systems of management.[72] The focus upon the agency of the state is crucial to the discussion of the environmental history of South Asia.

The contributors, in this sub-section, dwell upon the ecological encounter between Britain and India – changes in the ownership and management systems of Indian flora, fauna, minerals, etc. Sahara Ahmed engages with the issue of mining in the Bengal Presidency and its socio-political, economic, technological, and environmental consequences both in colonial and post-colonial contexts. With the growth of technological system, here scientific mining, a supporting culture also grew – trained manpower to handle the machines, and institutional facilities to burgeoning them: '... importance of "scientific miners" received recognition with the creation of mining academies'. They, in turn, advocated conservation measures to preserve the seemingly unlimited resources. However, Ahmed emphasizes, exploitation (of the resources) and conservation were alien to each other from the beginning of the mining industry. 'The crux of the problem was extensive mining in the watershed basins'. Not only the health of the miners but also the general health of the people, residing near the coalmines, was deteriorated due to adverse environmental impact. Numerous committees were set up, but their recommendations were only meant to shelve – no question of implementation. Independent India learnt the art from their colonial masters and Indians have excelled it!

Is pastoralism inimical to the environment? Majoritarians believe that pastoralists' animals contest with wildlife inhabitants, that their feeding habits negatively affect the regenerative capacities of forested landscapes, and that their unpredictable movements make it impossible to put institutional arrangements in place that would regulate environmental use. With the growing presence and authority of the forest department in British India, according to Arun Agrawal and Vasant Saberwal, the rhetoric against heralding acquired greater urgency. This was in line with greater hostility towards a variety of land-use practices that interfered with the regeneration of Indian forests – including shifting cultivation and pastoralism.[73] Earlier, Neeladri Bhattacharya has asked, who owned the grazing runs, the uncultivated land, the open pastures, the forests? He underlined, the extension of cultivation was synonymous with progress – 'Uncultivated tracts where pastoralists grazed their cattle were outside the pale of culture' – they had to be 'claimed' for the people through cultivation.[74] However, the environment – pastoralism equation resolves itself in many different ways.

Himanshu Upadhyaya believes that a close attention to the policy documents on grazing lands and approach of the botanical scientists to grasslands helps to understand pastoralists in colonial and post-colonial phases. The colonial experts and botanists, in the early decades of the twentieth century, were not serious about the improvement of cultivated fodder in India. The grazing lands were seen through the prism of 'efficient management' – some of the remedies – relying on the appropriate indigenous species in the permanent grasslands, assigning value to the concept of 'rotational grazing', experimenting with foreign grass and cultivated fodder crops to address the animal nutrition issue, etc. Upadhyaya elaborates on how the preference for cultivated fodder crops hid the issue of grazing areas and grasslands. Thus, the colonial bureaucracy attempted to tackle the seasonal shortages of rainfall, with 'an agrarian and forest management' mindset, rather than 'pastoral' one. Following Bhattacharya, Upadhyaya shows that there were larger initiatives to convert grazing lands, into the cultivable lands, to increase food production, just after independence.

The history of Indian forestry is, according to Mahesh Rangarajan, to a great extent, an account of the systematic growth of intervention in the processes of natural regeneration to upgrade the value of forests for the Raj.[75] For Marlene Buchy, much of the literature on the management of natural resources has attempted to understand the degree to which colonialism has to blame for the destruction of the tropical environment.[76] Nirmal Kumar Mahato seeks to explore in what way the scientific forestry was established on the principles of Deitrich Brandis in Purulia, West Bengal. The institutionalization of forest service started with the appointment of Brandis (1824–1907) as an Inspector General of Forest of India. Here, one can find two contradictory views regarding forest conservation and utilization of forest resources: As Mahato emphasizes that the forest officials of Purulia were concerned to establish scientific forestry based on Brandis's principles while the district officials were interested to extract more revenues. Independent India has chosen the second path – generating profits from the forests – even at the expense of her lash green jungles.

Medical encounters

Since the late 1990s, health and medicine have emerged as major concerns in South Asian history. This is a dynamic and innovative field of research, covering many facets of health, from government policy to local therapeutics. The British became the dominant colonial power in India in the eighteenth century and, along with establishing their territorial power and market monopolies, they introduced their own medical institutions, practices, drugs, and marginalized indigenous ones. As Pratik Chakrabarti argues, European medical experiences in the tropics led to the integration of environmental, climatic, and epidemiological factors within modern medicine.[77]

In the nineteenth century, with the consolidation of the British Empire, colonial medicine was firmly established through colonial medical services, hospitals, dispensaries, educational institutions, vaccination campaigns, etc. Roy Porter had raised a question long ago (1986), and is now familiar to scholars interested in the history of medicine in modern South Asia: 'What is colonial about colonial medicine?'[78] This is an indication of the ways in which the history of medicine in India has been shaped by the experiences and legacies of colonial rule. Several studies have examined the extent to which modern medicine emerged from intercultural encounters in colonial contact zones such as South Asia.[79] Medicine has also been a 'tool of empire', which informed the ideological justifications as well as technologies of colonial control.[80] Colonial practices of managing diseases, health, populations, in turn, redefined general understandings about these categories as well. In the early twentieth century, as a response to these colonial medical interventions, indigenous medical practitioners and doctors reorganized and revived Indian medical traditions.[81]

Right from the late-nineteenth-century questions relating to public health engaged both the official and public minds in India and the debates gradually became more intense in the wake of major cholera and plague epidemics. Public health usually refers to organized efforts made under the direction of medical experts for preventing disease and improving the health of the people. Deepak Kumar asks a pertinent question, '...how "public" was public health?'[82] Later he emphasizes, it ranged from 'assertions of imperial altruism to allegations of colonial callousness'.[83] Recently, an anthology tried to 'locate the medical' in an already-established subfield of historical research: according to the editors, they explore some of the ways in which the 'medical' was put together as an 'object of knowledge, as a subject impacting others, and as an ethico-moral organizing concept in specific moments in modern South Asian history'.[84]

The first essay in this sub-section, by Jayanta Bhattacharya, dwells upon Leo Tolstoy's *The Death of Ivan Ilyich* (1886) and several problems related to medicine, health, body, and disease arising out of reading this classic. If Ivan Ilyich is eager to know of organ localization of his disease, the doctor appears to be omnipotent (and omniscient too) regarding medical decisions. Ilyich was solitary and alone. He seems to be sincerely in search of some metaphors which could fill in the vacuum of his excruciating pain and long drawn illness. Bhattacharya asks, did he also think of a few moral and ethical questions which could redress his suffering? In Ilyich's case, nay in the modern world too, the entire cosmos of everyday life seems to be completely filled with metaphors of fabricated 'health and youth' of the commodity world or objective scientific metaphors which have destroyed traditional morality and the normal range of predictable moral expectations derived from religion or interpersonal subjective network and bondage. Ivan's story, in its extension, poses before us multiple layers of questions regarding subjectivity, person, metaphors of life, and the body of a patient. The agony of Ivan Ilych might be felt from this position of medicine. Here, medicine is not a theory pertaining to 'curing machine'. Bhattacharya believes looking through the philosophical matrix of Ayurveda can benefit us providing some insight.

Epidemics stand at a juncture of medical knowledge, and the ever-changing relationship between health of the individual and the imperatives of larger colonial economic formations. This has been explored by Arabinda Samanta in the context of colonial Bengal.[85] But there were other major challenges like tuberculosis which has been dealt by Niels Brimes and Achintya Kumar Dutta.[86] Brimnes claims that the late colonial debates about tuberculosis control offer illustrative examples of how colonial authorities saw their obligations towards the Indian population, while post-colonial public health initiatives reveal how this relation was transformed by decolonization.[87] In this sub-section, B. Eswara Rao analyses the history of tuberculosis in the Madras Presidency during the twentieth century. It was a widespread assumption among the colonial health authorities in the first half of the twentieth century that tuberculosis was, if not an entirely new disease in India, became a serious public health concern. He focuses on the conflict between Western medicine and indigenous perceptions that led to the advent of fascinating results on the therapeutic practices. How the disease was perceived and how the colonial state had struggled to bring it under control? Rao thus settles, '...both preventive and curative methods were rationalized under a state-controlled Western medical system'.

Homeopathy is a controversial medical system widely practiced in the world – India is no exception. The public discussion surrounding homeopathy, then and now, has tried to reduce the problem to a simple question: whether homeopathy is effective or not. Despite homeopathy's ambiguous position within public health establishments and academic sectors, many licensed health practitioners, pharmacists, and patients endorse and publicize it. This wide presence suggests long and profound roots that remain to be thoroughly examined. Recently, Shinjini Das argued that the historiographic attention in South Asia has remained overwhelmingly divided between studying aspects of state medicine promoted by the British Government on the one hand, and that of indigenous healing traditions like Ayurveda, Unani, and Siddha on the other. These medical histories have largely ignored those mostly sectarian, dissenting medical ideologies that flourished in Europe since the late eighteenth century, whose scientific values were hotly debated in the western world throughout the nineteenth century – Homeopathy. Heterodox healing practices such as homeopathy were routinely curbed by the colonial state in India.[88] Thus, in the absence of any substantial state records on homeopathy's history in British India, Das identifies the systematic publication of biographies as a significant arena of 'assertion for a heterodox, family-based practice like homeopathy'.[89]

Western medicine did not only mean allopathy, but there were also several heterodox strands of medical practices that originated in Europe and acquired a new meaning in South Asia, including homoeopathy. Dhrub Kumar Singh's essay in this anthology examines the role of introducers, their credentials and interaction with the state professed medicine, professional practice, commercial activities, certification, and the public reception of their work. Local contexts framed the development of homeopathic institutions in India. From the homeopathic dispensary records, Singh shows several founders of these dispensaries wanted to elevate their dispensary-status to that of hospital-status. Emphasizing the issues concerning the introduction, adaptation, and acceptance of homeopathic practitioners and products in the context of colonial Bengal, Singh appeals for a serious engagement with the histories of homeopathy dispensaries and hospitals as these institutions have endured the test of times and have become an integral part of India's plural medical traditions.

Where are the women in science, technology, and medicine? Why don't we see them more often? Do techno-scientific and medicinal knowledge reinforce gender-based prejudices or liberate us from its bondage? Women's unequal position in various spheres of social life is an important area of research for social scientists. Though recent studies on women's participation in scientific professions show an increase in fields such as medicine, they are still underrepresented in the natural sciences, engineering, and mathematics. Neelam Kumar rightly notes, women in science resemble a pyramid – with many women at the bottom and a few at the top in India.[90] The relation between gender and health was a major theme of the colonial Indian history. Scholars have probed the questions related to childbirth and motherhood, and some have examined curative care, female sexuality, and debates surrounding birth control.[91] Another subject arising from recent research is the life and work of women who were practitioners either of traditional medicine[92] or what is normally referred to as western medicine.[93] The education of women has also suggested itself as a fruitful area of research, as has the treatment of women in hospitals and dispensaries.[94] Studies of the nursing profession in India are still very much in their infancy.[95] Recently, other broader issues like professionalization, status, and recruitment, etc., are also being taken up by scholars.[96]

However, most of the scholarship on health and gender over the last two decades has examined the work of women medical missionaries and its connections to the so-called 'civilizing mission' of the churches and the colonial administration.[97] In a society like India in the nineteenth century, missionary women had an advantage over their 'male brethren' and were able to access women of the local communities, to whom male missionaries had no access. Medical work for women usually began as auxiliary to zenana outreach, because missionary women found it a useful instrument of access to local women living within gender-segregated structures. The early medical work for women in non-Western societies, Maina Chawla Singh argues, was initially tentative and experimental in nature. It grew as an offshoot of missionaries organizing children's schools, holding sewing and literary classes for women, and 'zenana-visiting'.[98]

The last essay in this sub-section by Ch. Radha Gayathri explores the initiatives of the trained missionaries of Delhi Female Medical Mission of Society (DFMMS) and institutionalization of St. Stephen's Hospital, Delhi, along with other medical colleges and schools in and around the region. St. Stephen's Hospital was started as a 'veranda dispensary', and 'medical chest' treatment gradually grew into a hospital. The medical missionaries not only cured patients but started training programmes for dais, nurses, and dispensers. The growth of this institution reflects the initial struggles, efforts to reach out, the personal experiences of single female missionaries, the disappointments, success, etc. Based on the personal experiences of the doctors, Gayathri shows the immense struggle carried out by these missionary women. Undeniably, the female medical missionaries had worked selflessly for saving the lives of Indian women. Their lasting contribution was the unintentional creation of a 'feminist' consciousness in local women. The often-neglected area of Indian contribution in such enterprises is also taken up – at the grass-root level, it was the Indian Christians who worked as assistants and their contributions also need to be studied further, argues Gayathri.

Epilogue

Can studies in HISTEM help us get a better understanding of colonial modernity? Prakash Kumar recently argued, South Asia should not merely be a 'site' to which methods and analytics of history of science and technology may be extended, rather, South Asia should emerge as a 'site' for making new theories and methods for HISTEM. Prakash emphasizes writing of HISTEM and modernity in a South Asian context is to participate concurrently in the task of decolonizing ways of knowing the past.[99] Thus, there is a growing concern not to universalize western techno-scientific concepts to other areas and histories. Why should we be so concerned with the national or regional characteristics of our science and technology, so much so that we should construct our own case studies that are quite different from the western mainstream cases? This anthology showcases various new trends in the historiography of science, technology, environment, and medicine in South Asia. Taken together, the essays in this volume endorse the significance of the new knowledge in shaping the histories of colonial and post-colonial Indian subcontinent. Is focusing our scope of HISTEM inquiry on South Asia – a specific geographical and historical area – a reasonable and potentially fruitful strategy for doing research?

The South Asian histories of modern Islamic medicine (Unani Tibb and Tibb-ul-Nabi) were largely silent on regional variations within South Asia. Projit Mukherji has pointed out the gap between elite (*ashraf*) and popular (*atrap*) culture in Islam along with the absence of Bengal from histories of the institutionalization of Islamic medicine in South Asia.[100] Likewise, though scholars have discussed the changing attitudes of Hindus towards science and engineering, they mostly neglect to mention the attitudes of Muslims. Now that some influential people are talking about rewriting history, why stick to only political history? Why not politics of knowledge and politics of culture? Most recently, Ajantha Subramanian has provided interesting insights into the colonial history of engineering education and associated racialization of caste and the making of IITs in post-colonial India as a Brahmin-upper caste space. She explores the making of upper casteness and its inherent linkages with opposing reservations *inter alia* the making of IIT as a global brand. The institutional kinship of upper castes in the US reiterates that caste doesn't vanish amongst the IIT diaspora.[101]

Almost 125 years ago, the noted geologist Pramatha Nath Bose (referred earlier), had given a similar self-critical assessment of our engineering capabilities. He knew that

> Remains of temples, roads, bridges and reservoirs testify to the engineering skill of the Hindus in pre-British times. But, though some Sanskrit books on engineering subjects have come down to us, they had long before the establishment of British rule ceased to be taught in schools.

He also argued,

> The caste system had no doubt aided progress in the earlier stages of their civilization; it had also served to maintain some kind of order for centuries since the decay of their civilization. But caste did so at the sacrifice of progress; progress such as it is understood now in Europe and America. It was not to be expected that illiterate weavers, or illiterate dyers, or illiterate miners, would apply the scientific methods of modern industries to their professions. Not being able to do so, they have gone to the wall.[102]

In an era of (pseudo?) revivalism, Deepak Kumar emphasizes, HISTEM acquires special significance and makes insistence on scientific temper and definitive sources even more important. HISTEM research studies may provide a better understanding of the historical processes and forces. Quite exciting possibilities![103]

Notes

1 Deepak Kumar, 'Developing a History of Science and Technology in South Asia', *EPW*, Vol. 38, No. 23, 2003, pp. 2248.
2 Ibid, pp. 2249–2250.
3 Steven Shapin and Simon Schaffer, *Leviathan and the Air-Pump*, Princeton and Oxford: Princeton University Press, 2011, p. xvii.
4 The authors thus explain, 'There are three senses in which we want to say that history of science occupies the same terrain as the history of politics. First, scientific practitioners have created, selected, and maintained a polity within which they operate and make their intellectual product; second, the intellectual product made within the polity has become an element in political activity in the state; third, there is a conditional relationship between the nature of the polity occupied by scientific intellectuals and the nature of the wider polity'. Ibid, p. 332.
5 Zaheer Baber, *The Science of Empire*, Delhi: OUP, 1998, pp. 140–144.
6 Anna Winterbottom, 'Science', in William A. Pettigrew and David Veevers (eds.), *The Corporation as a Protagonist in Global History, c. 1550–1750*, Leiden and Boston: Brill, 2019, p. 233.
7 Rohan Deb Roy, 'Review: Science, Medicine and New Imperial Histories', *BJHS*, Vol. 45, No. 3, 2012, p. 444 (443–450).
8 This is not only the histories of colonial governance, medical knowledge, pharmaceutical commerce, and Indian response, Deb Roy argues, but also portrays the manner British India was linked simultaneously to events and processes in other colonial territories. Rohan Deb Roy, *Malarial Subjects*, Cambridge: CUP, 2017, pp. 1–16.
9 The Opium War of 1840 convinced China that if it remained closed to the outside world, it would quickly fall behind. Moreover, erudite Chinese at that time started believing that the reason of European supremacy was their technological advances. The twentieth-century Chinese scholarship recognized well 'science-technology' as the most important productive force. When science is mentioned in Chinese terminology, it is generally mentioned together with technology as 'Keji'. Guoyan Wang, 'Science as Technology: What does Science Mean for the Chinese', *Science as Culture*, Vol. 30, No. 2, 2021, pp. 315–319.
10 Lorraine Daston and Peter Galison, 'The Image of Objectivity', *Representations*, No. 40, Autumn 1992, p. 82 (81–128).

11 For more information consult Lorraine Daston and Peter Galison, *Objectivity*, New York: Zone Books, 2010.

12 Lorraine Daston, 'The History of Science and the History of Knowledge', *KNOW: A Journal on the Formation of Knowledge*, Vol. 1, No. 1, 2017, pp. 131–154.

13 Epstein has argued that the recent research to extend studies of science and technology 'outward' beyond scientific settings (laboratory for example) has created new possibilities for interchange with the sociology of culture. Steven Epstein, 'Culture and Science/Technology: Rethinking Knowledge, Power, Materiality, and Nature', *The Annals of the American Academy of Political and Social Science*, Vol. 619, 2008, pp. 165–182.

14 *The Calcutta Review*, Vol. I, No. II, 1844, p. 257.

15 Brajendranath Seal, 'Mechanical, Physical and Chemical Theories of the Ancient Hindus', in Prafulla Chandra Ray, *A History of Hindu Chemistry*, Vol. II, London: Williams and Norgate, 1909, p. E.

16 While delivering the Presidential Address to Modern Indian Section, Indian History Congress (Diamond Jubilee Session, Calicut University, 2000), Deepak Kumar remarked the following: 'What made these scientists look into distant past? Intense nationalism, quest for an identity or concern for the future? Probably all three. All this part of self-exploration and self-criticism was considered vital. They used the term "Hindu" but not in a religious sense (or in the way it is being used now) … Be it Bankim or Afghani, their aim was just to show that modern science was compatible with their respective culture and traditions. It was not retrogressive revivalism'. Later published as Deepak Kumar, 'Science and Society in Colonial India: Exploring an Agenda', *Social Scientist*, Vol. 28, Nos. 5/6, 2000, pp. 24–46.

17 Sankari Prasad Basu (ed.), *Swami Vivekananda in Contemporary Indian News*, Vol. II, Kolkata: The Ramakrishna Mission Institute of Culture, 2007, pp. 48–50.

18 Rabindranath Tagore, 'Sadhana', in S. K. Das (ed.), *English Writings of Rabindranath Tagore*, Vol. 2, New Delhi: Sahitya Akademi, 1996, p. 304.

19 Jahnavi Phalkey, 'Introduction', Focus: Science, History, and Modern India, *Isis*, Vol. 104, No. 2, 2013, pp. 330–336.

20 One of the main emphases of the literature has been to critique and overthrow George Basalla's tripartite model of the 'diffusion' of science from the core to the periphery. His model, informed by modernization theory, further viewed non-European societies as passive recipients of science, which had a formative influence on much historical scholarship through to the 1980s. George Basalla, 'The Spread of Western Science', *Science*, Vol. 156, No. 37, 1967, pp. 611–622.

Gradually, historians were beginning to find that a single model of development of science could not cover its wide-ranging trajectories in different parts of the empire. In this context, Roy Macleod's concept of 'moving metropolis' is worth a mention. In his scheme, local centres such as Sydney, Toronto, or Calcutta could achieve autonomy to a significant extent while remaining within the parameters of imperial control. Roy Macleod, 'On Visiting the Moving Metropolis: Reflections on the Architecture of Imperial Science', *Historical Records of Australian Science*, Vol. 5, No. 3, 1982, pp. 1–16.

The early Basalla-inspired studies (Daniel R. Headrick, Michael Adas, among others) granted limited agency to the colonies themselves in the spread of Western science and technology. As one critique emphasizes that 'these studies embody a non-interactive approach, for they merely view the non-West as laboratories for the performance of scientific experiments'. Dhruv Raina, *Images and Contexts*, New Delhi: OUP, 2003, p. 177.

21 Deepak Kumar, *Science and the Raj*, New Delhi: OUP, 2nd Edition, 2006.

22 David Arnold, *Colonizing the Body*, Berkeley: University of California Press, 1993.

23 Roy Macleod and Deepak Kumar (ed.), *Technology and the Raj*, New Delhi: Sage, 1995.
24 Warwick Anderson, 'Remembering the Spread of Western Science', *Historical Records of Australian Science*, Vol. 29, No. 2, 2018, p. 78 (73–81).
25 Paolo Palladino and Michael Worboys, 'Science and Imperialism', *Isis*, Vol. 84, No 1, 1993, pp. 98–99 (91–102).
26 Steven Shapin, 'Placing the View from Nowhere: Historical and Sociological Problems in the Location of Science', *Transactions of the Institute of British Geographers*, Vol. 23, No. 1, 1998, pp. 6–7 (5—12).
27 Bruno Latour, *The Pasteurization of France*, Alan Sheridan and John Law (trans), Cambridge, MA: Harvard University Press, 1988, p. 140.
28 Mary Louise Pratt, *Imperial Eyes*, London: Routledge, 1992, p. 6.
29 Sandra Harding, 'Is Science Multicultural? Challenges, Resources, Opportunities, Uncertainties', *Configurations*, Vol. 2, No. 2, 1994, pp. 301–330.
30 Sandra Harding, *Is Science Multicultural*, Bloomington: Indiana University Press, 1998.
31 Warwick Anderson, 'Postcolonial Technoscience', *Social Studies of Science*, Vol. 32, Nos. 5–6, 2002, pp. 643–658.
32 Warwick Anderson, 'From subjugated knowledge to conjugated subjects: science and globalisation, or postcolonial studies of science?', *Postcolonial Studies*, Vol. 12, No. 4, 2009, p. 395 (389–400).
33 Satpal Sangwan, *Science, Technology and Colonization*, Delhi: Anamika, 1991.
34 Dhruv Raina and S. Irfan Habib, *Domesticating Modern Science*, New Delhi: Tulika, 2004, pp. 83–181.
35 J. Lourdusamy, *Science and National Consciousness in Bengal*, New Delhi: Orient Longman, 2004.
36 Gyan Prakash, *Another Reason*, Princeton: Princeton University Press, 1999.
37 Pratik Chakrabarti, *Western Science in Modern India*, New Delhi: Permanent Black, 2004.
38 David Arnold, *The New Cambridge History of India*, Cambridge: CUP, 2000, pp. 169–170.
39 Deepak Kumar, 'Reconstructing India: Disunity in the Science and Technology for Development Discourse, 1900–1947', *Osiris*, Vol. 15, 2000, pp. 241–257.
40 S. Irfan Habib and Dhruv Raina (eds.), 'Introduction', *Social History of Science in Colonial India*, New Delhi: OUP, 2007, p. xxiii.
41 David Chambers and Richard Gillespie, 'Locality in the History of Science: Colonial Science, Techno-science, and Indigenous Knowledge', *Osiris*, Vol. 15, 2000, pp. 221–240.
42 Kapil Raj, *Relocating Modern Science*, Basingstoke: Palgrave Macmillan, 2007, p. 23.
43 Ibid, pp. 11—18. See also, Kapil Raj, 'Beyond Post-colonialism…and Post-positivism: Circulation and the Global History of Science', *Isis*, Vol. 104, No. 2, 2013, pp. 337–347.
44 Richard H. Grove, *Green Imperialism*, Cambridge: CUP, 1995; Richard Drayton, *Nature's Government*, New Haven: Yale University Press, 2000; Mark Harrison, 'Medicine and Orientalism: Perspectives on Europe's Encounter with Indian Medical Systems', in Biswamoy Pati and Mark Harrison (eds.), *Health, Medicine, and Empire*, Hyderabad: Orient Longman, 2001, pp. 37–87.
45 Zaheer Baber, *The Science of Empire*, op. cit., pp. 205–212.
46 Russel Dionne and Roy Macleod, 'Science and Policy in British India, 1858–1914: Perspectives on a Persisting Belief', Proceedings of the Sixth European Conference on Modern South Asian Studies, CNRS, Paris, 1979, pp. 55–68.
47 Sujit Sivasundaram, 'Sciences and the Global: On Methods, Questions and Theory', *Isis*, Vol. 101, No. 1, 2010, p. 155 (146–158).

48 Deepak Kumar, 'HISTEM and the Making of Modern India – Some Questions and Explanations', *IJHS*, Vol. 50, No. 4, 2015, p. 619 (616–628).

49 Pratik Chakrabarti, 'Science', in Gita Dharampal-Frick et al. (eds.), *Key Concepts in Modern Indian Studies*, New Delhi: OUP, 2015, pp. 247–50.

50 Fa-ti Fan and John Mathew, 'Negotiating natural history in transitional China and British India', BJHS Themes, Vol. 1, 2016, pp. 43–44.

51 Ibid, pp. 58–59.

52 Deepak Kumar, 'The Culture of Science and Colonial Culture, India 1820–1920', *BJHS*, Vol. 29, No. 2, 1996, pp. 195–209.

53 James A. Secord, 'Introduction', *BJHS*, Vol. 26, No. 4, 1993, pp. 387–389.

54 Shiju Sam Varughese, *Contested Knowledge*, New Delhi: OUP, 2017, pp. 1–7.

55 https://www.historyoftechnology.org/ (accessed on 13 May, 2019)

56 Melvin Kranzberg, 'Technology and History: "Kranzberg's Laws"', *Presidential Address*, SHOT, Henry Ford Museum in Dearborn, Michigan, October 19, 1985.

57 Wiebe E. Bijkar et al (eds.), *The Social Construction of Technological Systems*, Cambridge, Massachusetts, and London: The MIT Press, 1987.

58 Ruth Schwartz Cowan, *A Social History of American Technology*, New York: OUP, 1997, p. 3.

59 Daniel R. Headrick, *The Tools of Empire*, New York: OUP, 1981.

60 Michael Adas, *Machines as the Measure of Men*, Ithaca: Cornell University Press, 1989.

61 David Arnold and Erich Dewald, 'Everyday Technology in South and Southeast Asia: An Introduction', *MAS*, Vol. 46, No. 1, 2012, pp. 1–17.

62 David Arnold, *Everyday Technology*, Chicago and London, University of Chicago Press, 2013.

63 Ritika Prasad, *Tracks of Change*, Delhi: CUP, 2015.

64 Smritikumar Sarkar, *Technology and Rural Change in Eastern India*, New Delhi: OUP, 2014.

65 Sunila S. Kale, 'Structures of Power: Electrification in Colonial India', *Comparative Studies of South Asia, Africa and Middle East*, Vol. 34, No. 3, 2014, p. 455 (pp. 454–475).

66 David Arnold, *The New Cambridge History of India*, op. cit., pp. 205–206.

67 Stephen McGlinchey, Review–Crude Reality. https://www.e-ir.info/2013/07/11/understanding-iran-a-summary-of-recent-scholarship/ (accessed on 12 May, 2019)

68 Timothy Mitchell, *Carbon Democracy*, London and New York: Verso, 2011, p. 1.

69 Ramachandra Guha, *Environmentalism*, Gurgaon: Allen Lane, 2014, p. 3.

70 Richard H. Grove, Vinita Damodaran and Satpal Sangwan (eds.), 'Introduction', *Nature and the Orient*, New Delhi: OUP, 1998, p. 3.

71 Richard H. Grove, *Green Imperialism*, op. cit.

72 David Arnold and Ramachandra Guha (eds), 'Introduction', *Nature, Culture, Imperialism*, New Delhi: OUP, 1995, pp. 12–13.

73 Arun Agrawal and Vasant Saberwal, 'South Asian Pastoralism: The Environmental Question', in Mahesh Rangarajan (ed.), *Environmental Issues in India*, New Delhi: Pearson, 2007, pp. 288–289.

74 Neeladri Bhattacharya, 'Pastoralists in a Colonial World', in David Arnold and Ramachandra Guha (eds.), *Nature, Culture, Imperialism*, op. cit., pp. 67–72.

75 Mahesh Rangarajan, 'Production, Desiccation and Forest Management in the Central Provinces 1850–1930', in Richard H. Grove *et al* (eds.), *Nature and the Orient*, op. cit., pp. 575–576.

76 Marlene Buchy, 'British Colonial Forest Policy in South India: An Unscientific or Unadapted Policy?' in ibid, p. 636.

77 Pratik Chakrabarti, *Medicine and Empire*, Basingstoke: Palgrave Macmillan, 2014, p. 101.

78 Shula Marks, 'What is Colonial about Colonial Medicine? And What has Happened to Imperialism and Health?', *Social History of Medicine*, Vol. 10, No. 2, 1997, pp. 205–219.
79 Mark Harrison, *Medicine in an Age of Commerce and Empire*, New York: OUP, 2010; Pratik Chakrabarti, *Materials and Medicine*, Manchester: MUP, 2010.
80 Mark Harrison, *Public Health in British India*, Cambridge: CUP, 1994; David Arnold, *Colonizing the Body*, op. cit., Anil Kumar, *Medicine and the Raj*, New Delhi: Sage, 1998.
81 Projit Bihari Mukherji, *Nationalizing the Body*, London: Anthem Press, 2009.
82 Deepak Kumar, 'Perceptions of Public Health: A Study in British India', in Amiya Kumar Bagchi and Krishna Soman (eds.), *Maladies, Preventives and Curatives*, New Delhi: Tulika, 2005, p. 44.
83 Deepak Kumar, 'Probing History of Medicine and Public Health in India', *IHR*, Vol.37, No. 2, 2010, p. 267.
84 Rohan Deb Roy and Guy N. A. Attewell (eds.), 'Introduction', *Locating the Medical*, New Delhi: OUP, 2018, p. 9.
85 Arabinda Samanta, *Living with Epidemics in Colonial Bengal*, New Delhi: Manohar, 2017.
86 Niels Brimnes, *Languished Hopes*, New Delhi: Orient Blackswan, 2016. Also see, Achintya Kumar Dutta, *Trauma in Public Health*, Kolkata: K. P. Bagchi, 2018.
87 Niels Brimnes, ibid, pp. 2–3.
88 The expulsion of Mahendra Lal Sircar (1833–1904), a physician of the highest repute and the second MD of the Calcutta Medical College, from the medical faculty of the Calcutta University in 1878 due to his inclination towards homeopathy, established the rising government intolerance towards any unorthodox medical practices.
89 Shinjini Das, 'Biography and Homeopathy in Bengal: Colonial lives of a European heterodoxy', *MAS*, Vol. 49, No. 6, 2015, pp. 1732–1771.
90 Neelam Kumar (ed.), 'Introduction', *Women and Science in India*, New Delhi: OUP, 2009, p. xiv.
91 Geraldine Forbes, *Women in Colonial India*, New Delhi: Chronicle Books, 2005; Sara Hodges (ed.), *Reproductive Health in India*, New Delhi: Orient Longman, 2006.
92 Charu Gupta, 'Procreation and Pleasure: Writings of a Woman Ayurvedic Practitioner in Colonial North India', *SIH*, Vol. 21, No. 1, 2005, pp. 17–44.
93 Geraldine Forbes, 'Introduction to Memoirs', in Tapan Raychaudhuri, trans. and ed., *The Memoirs of Dr Haimavati Sen*, Delhi: Roli Books, 2000.
94 Sujata Mukherjee, *Gender, Medicine and Society in Colonial India*, New Delhi: OUP, 2016.
95 Rosemary Fitzgerald, 'Making and Moulding the Nursing of the Indian Empire', in Avril A. Powell and Siobhan Lambert-Hurley (eds.), *Rhetoric and Reality*, New Delhi: OUP, 2006, pp. 183–222.
96 Madelaine Healey, *Indian Sisters*, New Delhi: Routledge, 2013; Sneha Sanyal, 'Institutionalization of Nursing as Profession in the Early Twentieth Century Bengal', *IJHS*, Vol. 52, No. 3, 2017, pp. 297–315.
97 Kumari Jayawardena, *The White Woman's Other Burden*, New York and London: Routledge, 1995; Antoinette Burton, 'Contesting the Zenana: The Mission to Make "Lady Doctors for India", 1874–1885', *Journal of British Studies*, Vol. 35, No. 3, 1996, pp. 368–97; Narin Hassan, *Diagnosing Empire*, Surrey and Burlington: Ashgate, 2011; Jharna Gourlay, *Piety, Profession and Sisterhood*, Kolkata: K. P. Bagchi, 2017.
98 Maina Chawla Singh, *Gender, Religion, and "Heathen Lands"*, New York: Garland, 2000, p. 89.
99 Prakash Kumar, 'Introduction: New histories of technology in South Asia', *Technology and Culture*, Vol. 60, No. 4, 2019, pp. 933–952.

100 Projit Mukherji, 'Lokman, Chholeman and Manik Pir: Multiple Frames of Institutionalising Islamic Medicine in Modern Bengal', *Social History of Medicine*, Vol. 24, No. 3, 2011, pp. 720–738.
101 Ajantha Subramanian, *The Caste of Merit*, Cambridge, Massachusetts and London: Harvard University Press, 2019.
102 P. N. Bose, *A History of Hindu Civilization during British Rule*, Vol. I, Calcutta: W. Newman & Co, 1894, p. lxxx.
103 Deepak Kumar, 'Why HISTEM and how to do it?', Kuruvila Zachariah Memorial Lecture, A J C Bose Auditorium, Presidency University, Kolkata, 7 November, 2019.

Section I
Science and society

1 Medicine, natural history and the curious case of Soorjo Coomar Goodeve Chuckerbutty

John Mathew

The role of native contributors to the making of natural history, most significantly, the relatively new field of zoology in its taxonomic guise, in India, was conspicuous for its seeming absence through most of the 19th century. This was quite in contrast to strides that Indians were taking in the applied field of medicine, an interesting fact seen as so many of the Scottish medical men that had made their way to India had been at the vanguard of zoological and botanical research in their country of professional domicile. This surprising lacuna on the part of the indigenous community under Western tutelage did not pass unnoticed at the time. As late as 1890, Bengali intellectual Sarat Chundra Mitra in Calcutta bewailed the lack of input on the part of his countrymen on the subject, saying that while they had rendered themselves proudly conspicuous in literature and science generally, "Natural history pursuits, as intellectual recreations, have never been popular amongst the people of India, whether of past or modern times".[1] Irish entomologist Edwin Felix Thomas Atkinson (1840–1890) in his Presidential address to the Asiatic Society of Bengal in 1887 addressed the matter as well:

> "I should be glad to see our native members take more interest in Natural Science, and thus wipe away the reproach that, perhaps with the exception of the late Babu Harimohun Mukherji, and one gentleman in Bombay (K. P. Kirtikar), there is not a single native of India, known outside its limits, for proficiency in either botany or zoology."[2]

The Scots-Irish linguist and congregationalist cleric, Bishop Robert Caldwell (1814–1891), while addressing to Madras University a little over a decade earlier (1878), had similarly expressed regret that no Indian student had elected to study the rich flora and fauna of the country.[3]

At one level, this criticism seemed founded. Elsewhere, I have argued that natural history in India was largely undertaken at the behest of individuals that I have denominated 'translocates', typically people of European descent, whose presence in a colonised region was precisely on account of

DOI: 10.4324/9781003241980-3

such a turn of events, but who, as a consequence, were as much a part of the milieu of the region under thrall (oftentimes more so) than that of their ancestral roots.[4] Such translocates were significant in mediating the flow of information between and among colonial centres and hinterlands without losing sight of European metropoles as contested sites of reference. Arguably, such activities were not restricted to natural history alone and extensions could be made to other branches of science. For instance, Irish-born William Brooke O'Shaughnessy (1809–1889), later the pioneer of the telegraph system in India, undertook pioneering research in chemistry through his role as a Professor of Materia Medica at the Calcutta Medical College in seeking to understand the underlying properties of medicinal plants, through such seminal works as *A Manual of Chemistry* (1841) and *The Bengal Dispensatory and Pharmacopoeia* (1842). As a promising forensic chemist emerging from the University of Edinburgh, he was frustrated in his efforts to gain a chair in medical jurisprudence in Great Britain, and so accepted the opportunity to move to India where he soon was appointed to the Chair in Chemistry at the Calcutta Medical College, a position that enabled him to make significant interventions.[5]

Natural history itself was largely dependent upon translocates, such as Sir William Roxburgh (1751–1815), Francis Buchanan-Hamilton (1762–1829), Nathaniel Wallich (1786–1854), Robert Wight (1796–1872) and William Griffith (1810–1845). Some of these naturalists also had interests that extended to faunal studies. Buchanan-Hamilton was given charge of the menagerie at the gubernatorial residence of Richard Wellesley (1760–1842), Earl of Moira (later Marquess of Wellesley) and Governor-General of India, and undertook what came to be known as the Indian Natural History Project (if limited to animals), the first dedicated study of the kind in the country by any European power.[6] While much of zoological study (along in some cases, admixed with botanical research) in the subcontinent was undertaken, interestingly enough, by French workers, such as Pierre Sonnerat (1748-1814), Jean-Baptiste Louis Leschenault de la Tour (1773–1826), Jean-Jacques Dussimier (1792–1883), Pierre-Médard Diard (1794–1863), Alfred Duvaucel (1792–1825) and Victor Jacquemont (1801–1832), especially in the Bengal and Madras Presidencies, particular attention was paid to some faunal groups. If Buchanan-Hamilton would pen *An Account of the Fishes found in the River Ganges and Its Branches* based in Calcutta, Patrick Russell (1727-1805), would contribute a similar work on edible fishes in the Coromandel region while in Madras, although his major attention would be paid to snakes.[7] While French activity in zoology would wane as the nineteenth century lengthened, its place was largely assumed by more directed activity by the British, in the main, those bred and / or educated in Scotland, especially those trained in medicine, who had a mandatory paper in natural history. One of them, Thomas Caverhill Jerdon (1811–1872), a translocate who spent over 30 years in India, would be responsible for writing both *The Birds of India* and *The Mammals of India*, while Francis Day (1829–1889), another old India hand, would produce *The Fishes of India*. In the interim, one of the most significant of the surveys instituted by the British Government in the subcontinent, the Geological Survey of India, would produce a number of workers that contributed magnificently to the study of animals of the area, so much so that one of their number, William Thomas Blanford (1832–1905), would be named the first editor of what would become the most ambitious regional taxonomic study anywhere in the British Commonwealth at the time, *The Fauna of British India* (henceforth *The Fauna*), comprising a series of 81 ½ volumes.

The Fauna itself was a state project that was commissioned at the instance of a number of the most distinguished biologists in Great Britain, including the prime originator of the theory of evolution through natural selection, Charles Robert Darwin (1809–1882), his close friend and at the time Director of the Royal Botanical Garden at Kew, Sir Joseph Dalton Hooker (1817–1911), the celebrated comparative anatomist Thomas Henry Huxley (1825–1895), the Conservator of the Hunterian Museum and later Director of the British Museum of Natural History, Sir William Henry Flower (1831–1899), the noted banker with decided zoological inclinations, Sir John Lubbock(1834–1913), later Lord Avebury, and the Secretary of the Zoological Society of London and founder of the journal *The Ibis*, Philip Lutley Sclater (1829–1913). As one of the lone Indians (a botanist) working for the then newly established Bombay Natural History Society (1883), Kanhoba Ranchoddas Kirtikar (1849–1917), a surgeon in the covenanted Indian Medical Service was fulsome in his praise of the commencement of the effort, as he fawningly described in 1905:

> "This appeal of Mr. Charles Darwin and his Co-Memorialists was fully and promptly accepted. The day when this was done will ever remain a red letter day in the annals of Indian zoology. Indian naturalists, of all shades and capacities whatsoever, cannot be sufficiently grateful to the learned and disinterested British Memorialists for the mightily encouraging stimulus they have independently and unsolicitedly given to the further progress of Indian zoology. May the beaming torch they have lighted shine brighter and brighter in days to come, and show us the bright-beaming light we have hitherto wanted. The light comes from West to East."[8]

The process of writing *The Fauna* was admittedly aided by many researchers in the British Museum of Natural History in London, but there were also specialists in the Indian Museum in Calcutta, the former holdings of the Asiatic Society started by Wallich in 1814 and given over soon after the Rebellion of 1857–1858 to the Government, with the East India Company's fall and the handover of Indian possessions to the British Crown. Of the more than 30 authors of *The Fauna*, only two were Indian, and that itself was a postlude to an attempt to incorporate native workers by a Superintendent of the Indian Museum and the Founder-Director of the Zoological Survey of India (1916), another Scotsman and translocate, Thomas Nelson Annandale (1876–1924).[9] Otherwise, Indians had been largely absent in the forging of knowledge regarding natural history (particularly of the fauna), which the beginning of this article makes explicit. The irony resides in the fact that Indians found themselves more involved in natural history as a discipline at just the time that nationalist leaders were embarking upon an attempt to relieve the country of its colonial oversight.

And yet it seems that European-directed zoology might have been purveyed in India even earlier, as early as mid-19th century, if a single 'native' person had championed its cause. The candidate in question, who bore the somewhat unlikely name of Soorjo Coomar Goodeve Chuckerbutty (1826–1874), was the first Indian to become a Covenanted Member of the Indian Medical Service, and was also the protégé of Robert Grant (1793–1874), the first professor of Zoology and Comparative Anatomy anywhere in England. Chuckerbutty's unusual tale tells in capsule a larger story involving flows and circulation of personnel and knowledge between England and India which merits some recapitulation here.

An attritional battle between Orientalists, who sought to give native literature and learning its due without compromising on the centrality of Western learning, and Anglicists, who did not even wish to dignify the former with attention, support and time, had resulted in a semantic feud that would decisively swing in favour of the latter during the 1830s and beyond.[10] This was due in no small measure to a controversial 'Minute on Indian Education', presented in 1835 by Thomas Babington Macaulay (1800–1859), Law Member in the Council of the then Governor-General in India, Lord William Bentinck (1774–1839).[11] It was not surprising therefore, that the triumph that attended the first scientific dissection in 1836 of a human body by an Indian, Madhusudhan Gupta was heralded with a 21-gun salute.[12] The surgical event had been overseen by Professor Henry Goodeve (1807–1884), the only other teacher in the recently founded Calcutta Medical College (1835) apart from its Principal, Assistant Surgeon Mountford J. Bramley (1803–1837).

There had been some preliminary context to the event. Bramley and Goodeve had the previous year instituted some level of exposure to parts of human bodies, beginning with a course in osteology, where bones were handled by native students without any visible sign of repugnance, after which larger sections of cadavers were systematically introduced. The success that marked the admission of a native cadre emboldened the administration to keep the medical programme afloat, despite the untimely demise of Bramley at only 34 years of age.[13] Key subjects in the curriculum were Anatomy, Chemistry, taught by William Brooke O'Shaughnessy, and Botany (in the service of teaching Materia Medica), led by Nathaniel Wallich (1786–1854),[14] who at the time was the Superintendent of the Botanical Gardens in Calcutta, and the major spirit behind the museological collections of the Asiatic Society of Bengal. If, by the beginning of the 1840s, the vast majority of students admitted were Hindu, to their number were soon added Europeans, Ceylonese, Armenians, Burmese, Christians and Muslims.[15]

It was after nearly a decade of the founding of the college that Goodeve and O'Shaughnessy were impelled to consider sending some of their students to London to compete on equal terms with their English counterparts. The East India Company (henceforth EIC) was initially opposed to the idea but ultimately four students were permitted to leave with some grudging governmental support, but additionally with private sponsorship from Indian and English quarters.[16] The major Indian sponsor was Dwarkanath Tagore (grandfather of the only Indian Nobel Laureate in literature, Rabindranath Tagore), who offered to pay all expenses for two students following the example of Goodeve, who had undertaken the same commitment for one, while public donations accounted for the support of a fourth. Thus, it was that Bhola Nath Bose, Dwarka Nath Bose, Gopal Chunder Seal and Soorjo Coomar Chuckerbutty left for England on the 18th of March 1845, setting sail on the Bentinck, accompanied by Goodeve, with Dwarkanath Tagore also on board.[17] Reaching London, the students were admitted to the University College, and took up residence with Professor Goodeve.

The choice of the medical school at University College London (henceforth UCL) rather than another in the city or even at Oxford or Cambridge was significant. It was widely considered the most progressive and radical in Europe, with freethinking, often dissenting faculty, whose unremitting efforts were matched by a particularly gifted student body from all over the continent. The four Indians, despite the unfamiliarity of their surroundings in what was their first long departure from their homeland, adapted remarkably well, 'winning gold and silver medals and certificates of honour in various subjects from anatomy, botany, and chemistry to zoology'.[18] Bhola Nath Bose, Dwarka Nath Bose and Gopal Chunder Seal became Members of the Royal College of Surgeons (MRCS) on the 27th of July 1846, with Soorjo Coomar Chuckerbutty joining their ranks on the 12th of May 1848,[19] when he had attained to the admissible age.[20]

Chuckerbutty's case was of particular interest. Born around 1826 (the exact date is unknown) in the district of Dacca (now in Bangladesh), he lost his parents at the age of 6 and thereafter had to fend for himself. It was only at 13 that he first heard English and would come to learn the language in exchange for cooking meals for a well-to-do Indian gentleman, Golak Nath Sen, in Comilla.[21] Subsequently, he made his way to Calcutta, where he studied at Hare School. In 1843, he tried unsuccessfully to gain entrance to the Calcutta Medical College, despite an undertaking of financial assistance from a Bengal civilian, Mr. Alexander, but the fates were kinder the following year and he commenced his studies at the institution, in all likelihood on the encouragement of Goodeve, who recognised his clear potential.[22] Chuckerbutty relinquished the sacred thread of his Brahminism upon admission to the Medical College – four years later, he converted to Christianity, adopting as a middle name that of his great supporter and guiding light, Dr. Henry Goodeve, who felt constrained to emphasise that Chuckerbutty's change of religious heart was by no means coerced.[23] Notwithstanding such protestations of innocence on the part of his mentor, the newly minted doctor would henceforth bear the name Soorjo Coomar Goodeve Chuckerbutty.

But whence zoology and its promise for translocation back to India via a 'native' son? The moment, if at all, might have resided between the relinquishing of the Brahmin's thread and the assumption of the cross. For in those four years, Chuckerbutty fell under the tutelage of the Edinburgh-born and trained Francophile and transmutationist Robert Edmond Grant (1793–1874). The first professor of Zoology and Comparative Anatomy anywhere in Great Britain had already held the position at the University of London (later UCL) for 18 years by the time Chuckerbutty became his student, and was renowned for his meticulous lectures, the result of painstaking work in comparative anatomy both in Britain and continental Europe, particularly France, where he had come to know both Georges Cuvier (1769–1832) and Etienne-Geoffroy St. Hilaire (1772–1844). Grant would ultimately privilege St. Hilaire's unity of plan in nature across organisms, simplest to most complex, over Cuvier's four distinct and unrelated 'embranchements', the Vertebrata, Mollusca, Articulata and Radiata, where form was subordinated to function.[24] With specimens from his constituted 'Zootomical Museum', Grant sought to demonstrate changes from the simplest organisms to the then fancied telos of complexity that was humankind. Each lecture was divided into sections, which dealt in turn with different systems of the body such as digestion, respiration, circulation or reproduction, for instance, across different classes of the Animal Kingdom, beginning with the Protozoa and culminating with the Mammalia.[25] The Museum itself enshrined the same principles, organised as it was to illustrate "the whole continuous chain of beings, from the lowest corals up to the highest animal forms that exist".[26]

This was battling material. But Grant was unfazed. Dedicated as much to the ideals of the former Professor of 'Insects and Worms' at the Muséum national d'Histoire naturelle in Paris, Jean-Baptiste Lamarck (1744–1829), as much as to those of St. Hilaire, he returned with a strong sense of the unity of composition and plan connecting all living organisms to Edinburgh, where he soon became an acknowledged expert on small marine invertebrates, even coining the still extant name 'Porifera'[27] for the sponges. At the time, he also took under his wing a young diffident student, who was attending lectures by Robert Jameson (1774–1854) on natural history whilst training for a medical career that would ultimately be discarded, named Charles Robert Darwin (1809–1882). Grant and Darwin became walking companions in and around Edinburgh to the coast, the latter benefitting from his new mentor's pioneering work on sponges, and even presenting a paper on the hitherto unnoticed cilia of the genus *Flustra* to the Plinian Society on the 27th of March, 1826 at Grant's proud behest.[28] When Grant moved south in 1827 to assume the chair of zoology and comparative anatomy at the University of London, his portmanteau of ideas, generated and refined in the crucibles of Paris and Edinburgh, accompanied him, investing his lectures and underscoring his anathema to the religiously inflected, wealth-absorbed worlds of Oxford and Cambridge to which the radical realm of the University of London was so doctrinally opposed. In

Grant's world view (and Geoffroy St. Hilaire and from the perspective of influence, Lamarck's as well), all animals, from polyps to people, were possessed of similar organs, the differences only residing in the complexity thereof.[29]

Eighteen years later, Grant had new walking companions. This time, however, one of them was imbued with a darker hue and traced his footsteps on the European continent rather than along the periphery of the Firth of Forth.[30] Soorjo Coomar Chuckerbutty was one of a select group of students who had gained Grant's favour for their diligence and aptitude and would, with him, attend lectures and visit collections of eminent comparative anatomists from Europe.[31] Chuckerbutty responded positively to these influences, learning Latin and gaining fluency in spoken French and German.[32] In these, he followed to some extent the multilingual Grant, who had mastered Dutch, Danish, French, German and Italian, apart from gaining proficiency in ancient Greek.[33] The inspiration was not merely linguistic. En route to obtaining the M.B. and later the M.D. in 1849, Chuckerbutty would win the Gold Medal for Comparative Anatomy.[34] Yet on his return to Calcutta soon after, there was no seeming concomitant spurt of interest in matters zoological engendered at his instance nor for the next several decades.

It is tempting to speculate as to why. Chuckerbutty, after all, had been exposed to Grant's lectures, brimming with transmutationist fervour. He had examined at first hand the vast possibilities of the animal kingdom, even winning top honours in the process. He had been witness to evolutionary concepts snaking their way inexorably from France to Britain. Whence then his reticence?

Part of the reason might be to do with the constraints under which he was labouring as a doctor. On his return to India, his patrons in England, including the President of the Asiatic Society in Calcutta Sir Edward Ryan (1793–1827), enthusiastically recommended his appointment to the Covenanted Medical Service as well as to a professorship at the Medical College, Calcutta.[35] The authorities, however, were implacable in refusal and Chuckerbutty was instead admitted to the Uncovenanted Medical Service in the capacity of Assistant Physician to the Medical College Hospital on the 10th of May 1850. Undeterred, he worked indefatigably, publishing an account of a singular case of epilepsy[36] which caught the editorial eye of the *Medical Times and Gazette* in London in 1852, the same edition in which the article was published.

> It is with sincere gratification that we today place before our readers an ably-written communication of considerable interest, by a native of our Indian empire. Dr. Chuckerbutty who is, we believe, the first native of that clime who has contributed to the progress of the science of medicine; and his friends – and he left many in this country – will rejoice that the high promise that he held out when a student here bids fair to be well fulfilled.[37]

In 1854, Chuckerbutty was named the Officiating Professor of Materia Medica and Clinical Medicine and Second Physician to the Calcutta Medical Hospital. In the same year, the EIC opened the examination to the Covenanted Medical Service to public competition, which meant that Indians could participate, if faced with the arduous prospect of voyaging to London in order to appear for the examination. Chuckerbutty's decision to sit the examination was considered; his 'ambition had always been to become a member of the Covenanted Service of the East India Company, and thus remove from his race, the stigma of a proscription which denied them a career of honourable ambition in their own land'.[38] As he wrote:

> "If I fail, it will be a satisfaction to me that I have used my best efforts to the service of my country and that is only physical difficulties thrown in our way by the Legislature which have been the cause of my disappointment."[39]

He would emerge second in order of merit. On the 24th of January 1855, he was appointed the first Indian to the Covenanted Medical Service as Assistant Surgeon.

During the remainder of his short life (he died at the age of 48), Chuckerbutty would hold a number of positions in Calcutta, both in the hospital and outside it. He taught, lectured and wrote extensively on infectious diseases including typhus, cholera, dysentery and smallpox. At the same time, he was caught in the bind of being a child of two cultures, his own and that of his Western medical education. The outcome could be paradoxical; on the one hand, he could excoriate indigenous health practitioners for their lack of professional qualifications,[40] proclaim that a "single day in London was of more value than a month in Calcutta", and discount training in Sanskrit and Arabic as "Oriental mania",[41] sounding suspiciously similar to the rhetoric employed by Macaulay in his infamous Minute; while on the other, he could criticise the University of Calcutta for representing 'only European opinion and interests' while ignoring 'the national element'.[42] Interestingly, he drew a rare zoological parallel in an attack on racialism.

> "The inhabitants of the colder latitudes are white because the sun is less powerful in them...In like manner the proteus, which dwells in caves, when exposed to the sun, becomes coloured, losing its former translucency of surface. The pride of colour, therefore, is as foolish in man as it could be in that humble creature."[43]

The foregoing account suggests that Chuckerbutty had his plate full with the need to choose his own battles – an overextended doctor doubly encumbered by the symbolic burden of being the first of his people to occupy a particular role and in some senses paving the way for others in his train. It may have been too much to expect him to open new avenues for fields that were ancillary to his chosen vocation of medicine, no matter how adept he might have been at them during his education in England.

Yet, I would hazard that there might have been another reason as well. Prominent as Grant was as a lecturer, he was an outsider as well. If Chuckerbutty had to pass an examination in England in order to join the Covenanted Medical Service, Grant's medical degree from Edinburgh was not acceptable in London unless he took and cleared another exam in the latter city to prove his credentials, something he on principle refused to do, thereby relinquishing the right to practise, until the rule was changed years later, at which point he was too old at 68 to serve as a doctor. Furthermore, his opinions on transmutation in an admittedly non-religious setting as provided by the University of London and later UCL, coupled with his fulminations against seats of established power swiftly alienated him from more conservative colleagues, not least the powerful Richard Owen (1804–1892), keeper of the Hunterian Museum, formulator of the term 'dinosaur' (terrible lizard) and contender with Grant for the position of leading comparative anatomist in the city. Darwin for his part found himself increasingly distant from the harangues of Grant, and at odds with his version of transmutation, which Darwin had long since found outmoded. It must have been particularly difficult, when, as Secretary of the Geological Society, he watched Grant being systematically dismantled by the Reverend William Buckland (1784–1856) and Richard Owen on the subject of whether the platypus was a reptile (which fitted with Grant's Lamarckian view of an upward sweep from lower to higher forms) or a mammal.[44] Grant's position was publicly dismantled, a significant defeat among others over the years to come, such that by the time when Chuckerbutty became his student, his star had dimmed significantly. Perhaps all too aware of the controversies surrounding his professor, not least on the subject of transmutation, Chuckerbutty might have deemed it prudent to clear comparative anatomy as best as he could (which as it turned out was the best anyone could, given his gold medal in the subject) and then relinquish it, thereby letting sleeping dogs lie.

In the absence of correspondence and documents (it remains a mystery what happened to the copious notes pertaining to Grant subsequent to his demise[45]), it is difficult to know precisely how to reconstruct this part of the puzzle. However, Chuckerbutty was on furlough in England in 1874 at the time of Grant's demise, and even attended his funeral in unconsecrated ground on the East Side of Highgate Cemetery in North London.[46] In a matter of weeks thereafter, Chuckerbutty himself would be dead at 48, the consequence apparently of the change in season from summer to autumn on a body racked by severe asthma He was laid to rest in consecrated ground (given his conversion to Christianity), another of London's Magnificent Seven of cemeteries, Kensal Green, (Dwarkanath Tagore, one of his early supporters was also interred there, if in dissenters' earth), under a recumbent monument of pink marble (Figure 1.1), on which the epitaph engraved was as fitting as it was touching:

Honour was his motto, and he promoted the welfare of his country.

Figure 1.1 Grave of S. C. G. Chuckerbutty at Kensal Green Cemetery, London. Photograph by the author.

Notes

1 S. C. Mitra, *Calcutta Review*, No. 91, 1890, p. 159.
2 E. T. Atkinson, 'Presidential Address to the Asiatic Society of Bengal, (1887)', quoted in R. Desmond, *The European Discovery of the Indian Flora*, Oxford: OUP, 1992, p. 188.
3 Ibid.
4 J. Mathew. 'To Fashion a Fauna for British India', Ph.D. thesis submitted to the Department of the History of Science, Harvard University, Cambridge, 2011.
5 M. Gorman, 'Sir William Brooke O'Shaughnessy, F. R. S. (1809–1889), Anglo-Indian Forensic Chemist', *Notes and Records of the Royal Society*, Vol. 39, No. 1, 1984, pp. 51–64.
6 M. Vicziany, 'Imperialism, Botany and Statistics in Early Nineteenth Century India: The Surveys of Francis Buchanan (1762–1829)', *MAS*, Vol. 20, No. 4, 1986, pp. 625–660. Also see Mathew, *To Fashion a Fauna for British India*, op. cit.
7 See F. Buchanan-Hamilton, *An Account of the Fishes found in The River Ganges and its Branches*, Edinburgh: Archibald Constable and Co., 1822, and P. Russell, *An Account of Indian Serpents, Collected on the Coast of Coromandel Containing Descriptions and Drawings of Each Species, Together with Experiments and Remarks on their Several Poisons*, London: The Court of the Directors of the East India Company, 1796.
8 K. R. Kirtikar, 'Progress in Natural History during the last Century', *The Journal of the Bombay Branch of the Royal Asiatic Society*, Extra Number – The Centenary Memorial Volume, Part V. Science Section. 1 (1905), 353–381.
9 Mathew, *To Fashion a Fauna*, op. cit.

10 See S. Nurulla and J. P. Naik, *A History of Education in India*, Bombay: Macmillan, 1951, pp. 131–52, S. Mahmood, *A History of English Education in India*, Delhi: Idarah-i Abidiyat-i Delli, 1895, pp. 28–29 and 50–52, and B. K. Boman-Behram, *Educational Controversies in India*, Bombay: K. K. Taraporevalla Sons and Co., 1943.

11 Bureau of Education. *Selections from Educational Records*, Part I (1781–1839), edited by H. Sharp. (Calcutta: Superintendent, Government Printing, 1920). Reprinted (Delhi: National Archives of India, 1965), 107–117.

12 D. Arnold. *Colonising the Body*, California: University of California, 1993.

13 M. J. Bramley. *Indian Public Consultations*. IOR, P/186/72, 17 March 1835, quoted in M. Gorman. 'Introduction of Western Science into Colonial India: Role of the Calcutta Medical College', *Proceedings of the American Philosophical Society*, Vol. 132, No. 3, 1988, p. 284.

14 Gorman, op. cit., pp. 286–287.

15 Ibid, p. 288.

16 Ibid, p. 289

17 P. C. Sen Gupta. 'Soorjo Coomar Goodeve Chuckerbutty: The First Indian Contributor to Modern Medical Science', *Medical History*, Vol. 14, No. 2, 1970, pp. 183–191.

18 Gorman, op. cit., p. 290.

19 *The Lancet*, 1847, No. ii, p. 138.

20 S. K. Ghosh. *Calcutta Medical Journal*, Vol. 28, 1934, p. 514. (based on an article by M. L. Sarkar, M.D., D. L., published in 1897).

21 A. N. Sen. *Swargiya Dina Nath Sener Jibani o Tatkaler Purba Banga* [Biography of the late Dina Nath Sen and an account of contemporary East Bengal], Calcutta, published by the author. 1948, Vol. 1, p. 15, Vol. 2, p. 400. Quoted in Sen Gupta, op. cit., p. 191. Also see D. G. Crawford, *A History of the Indian Medical Service*, Vol. II, London: W. Thacker and Co., 1914, 441–442.

22 Sen Gupta, op. cit., p. 183.

23 Ibid, p. 184.

24 S. E. Parker. *Robert Edmond Grant (1793–1874) and his Museum of Zoology and Comparative Anatomy*, London: Grant Museum of Zoology, UCL, 2006, pp. 10–11. For more on the debate between Cuvier and Geoffroy St. Hilaire, see T. Appel, *The Cuvier-Geoffroy Debate: French Biology in the Decades before Darwin*, New York and Oxford: OUP, 1987, and Adrian Desmond, *The Politics of Evolution*, Chicago: The University of Chicago Press, 1989.

25 Parker. *Robert Edmond Grant*, op. cit., pp. 17–18.

26 R. Grant. Evidence before Committee – Report of the select committee appointed to inquire into the condition, management and affairs of the British Museum (Parliamentary papers, 14th July 1836) X: 19–31, 84–95, 124–137. Quoted in S. E. Parker, *Robert Edmond Grant*, op. cit., p. 18.

27 Parker, *Robert Edmond Grant*, op. cit., p. 13.

28 A. Desmond and J. Moore. *Darwin*, London: Michael Joseph, 1991, pp. 35–38.

29 Ibid, pp. 39–40.

30 Sen Gupta, op. cit., p. 184.

31 Parker, *Robert Edmond Grant*, op. cit., pp. 21–22.

32 Gorman, op. cit., p. 293.

33 Parker, *Robert Edmond Grant*, op. cit., p. 9.

34 University College London. 'Report of the Council and Financial Statements.' *Proceedings at the Annual General Meeting of the Members of the College, 23rd February 1848*, p.212.

35 Ghosh, op. cit., p. 514.

36 S. C. G. Chuckerbutty. *Medical Times and Gazette*, 1852, N. S. 5. p. 406.

37 Ibid. Quoted in Sen Gupta, op. cit., p. 183.

38 *Medical Times and Gazette*, Vol I, 1855, p. 173.

39 Ibid. Quoted in Sen Gupta, op. cit., p. 185.

40 S. C. G. Chuckerbutty, 'Lecture on the Present State of the Medical Profession in India', February 2nd, 1864, in *Popular Lectures on Subjects of Indian Interest*, Calcutta, 1870, p. 78,quoted in Deepak Kumar, 'Medical Encounters in British India, 1820–1920', *EPW*, Vol. 32 No. 4, 1997, pp. 166–170.

41 S. C. G. Chuckerbutty, 'Introductory Lecture at the Commencement of the Thirty-Sixth Session of the Medical College of Bengal', *Indian Annual of Medical Science*, Vol. 27, 1870, p. 58,quoted in P. C. Sen Gupta, op. cit., p. 189.

42 S. C. G. Chuckerbutty, 'Necessity of forming a Medical Association in Bengal', May 27th, 1863, in *Popular Lectures on Subjects of Indian Interest*, Calcutta, 1870, p. 135, quoted in Deepak Kumar, 'Medical Encounters in British India, 1820–1920', op. cit., p. 167.

43 S. C. G. Chuckerbutty, 'Lecture on a Defence of Native Education', July 8th, 1858, in *Popular Lectures on Subjects of Indian Interest*, Calcutta, 1870, p. 85, quoted in Deepak Kumar, 'Medical Encounters in British India, 1820–1920', op. cit., p. 167.

44 Desmond and Moore. *Darwin*, op. cit., pp. 274–275.

45 Parker, op. cit., pp. 47–50.

46 Anonymous 1874–1875. Obituary Notices of Fellows of the Royal Society. *Proceedings of the Royal Society* 23: vi–x.

2 Examining the foundations of science

An essay on Ramendra Sundar Trivedi's epistemological inquiries

Santanu Chacraverti

Part I: introduction

Ramendra Sundar Trivedi (1864–1919) was a college teacher of physics and chemistry in Kolkata. A noted nationalist intellectual and polymath-essayist, his writings spanned a wide range of themes—from the history of astronomy, non-Euclidean geometry, electromagnetic waves, and the laws of thermodynamics to the nature of scientific knowledge, the sense of beauty, Bengali grammar, Buddhist philosophy, Vedic sacrificial rituals, and education policy. The mathematical-physicist S.N. Bose, while critical of Trivedi's brand of traditionalist nationalism, was an admirer of his formidable erudition and the range and quality of his popular science essays. Ramendra himself attached great importance to writing as a vocation. He wrote superbly and with delightful humour.

Many of Trivedi's philosophical reflections concern epistemological issues. Some of these found appreciation among his contemporaries. However, there is little in-depth scrutiny of Trivedi's philosophical engagements.[1] The present essay aspires to improve the situation a little by providing a basic introduction to the context, nature, and structure of Trivedi's epistemological reflections.

A note on the approach

One could have begun with providing a historical backdrop—for example, mentioning the colonial setting, the character of Hindu Bhadralok Bengal, and the cultural–intellectual climate at the turn of the century. However, these are fairly well-trodden grounds. For the same reason, references to Orientalist/Indological themes and how these were utilized in a nationalistic setting would be avoided. Rather, we shall pay more attention to elements with which the reader could be less acquainted—to Trivedi's personal intellectual inclinations and the philosophical trends that influenced him.

DOI: 10.4324/9781003241980-4

The thinker

Trivedi was not an academic philosopher. Nor did he write formal essays meant for professional philosophers. Further, he almost never wrote in English. All these factors possibly contributed to the scholarly neglect of his philosophical discussions.

His purely philosophical writings, including comments, can be divided into three categories: first, there were popular essays on various philosophical questions; second, his only large philosophical work, *Bichitra Jagat*. The nine chapters of this work were serially published in the periodical *Bharatavarsha* between February 1915 and August 1917 and published posthumously as a book in 1920;[2] and third, occasional comments on philosophical questions strewn across his numerous writings.

Before moving on to Trivedi's epistemological observations, let us spare a few moments to get an idea of his emotional–intellectual tendencies.

Intellectual temperament and tendencies

First, there was a natural fondness for science and mathematics. Second, his temperament leaned strongly towards analytical and the empirical. This temperament provoked Trivedi to be critical of empirically unfounded or logically indefensible ideas in both modern science and philosophy and *sceptical of positions claiming 'scientific' basis for various brands of spiritualism and traditionalist beliefs*. Third, there was a deep love of and voracious appetite for learning across domains. Naturally, he leaned towards the idea of the unity of scientific method—the notion that science was not a particular subject or a group of subjects but a method for critical evaluation of empirical facts with universal application. Thus, for Trivedi, all genuine pursuit of knowledge, in whatever field, was *scientific* pursuit and, hence, the study of the nation's history, society, culture, language, poetry, philosophical and spiritual traditions, and so on, was an integral part of the nation's scientific enterprise.[3] His extensive studies in Indian philosophical literature allowed him to compare and combine elements of Western and Indian philosophical traditions with enviable ease. Fourth, and in apparent contrast with his empirical-analytical temperament, Trivedi had profound attachment to what he considered to be the country's values and traditions. This tended to make him particularly soft towards traditional beliefs, customs, and rituals. This attitude of his can be seen as a part of the so-called Hindu revivalism that characterized Bengali culture of the late nineteenth and early twentieth century.[4]

However, Trivedi's 'revivalism' was *not* a strident assertion of overall Hindu/Indian intellectual superiority—an attitude present in various absurd forms and which today has acquired serious political backing. Trivedi simply discovered in Brahmanical Hindu culture a moral stance that strongly favoured the *transcendental* over the utilitarian—a stance he approved. He had no trouble linking this love of the transcendental to his ardent love affair with modern science—as we shall see.

This essay is largely concerned with *elements* of Trivedi's inquiries into the nature of scientific knowledge. For our purposes, it will suffice to confine ourselves mostly to *Bichitra Jagat*,[5] referring only occasionally to his other writings.

Part II: the philosophical backdrop

Empiricism and experience

The increasingly dominant philosophy of knowledge in the second half of the nineteenth century was strongly empiricist (as in J.S. Mill, A. Bain, T.H. Huxley, G. Kirchhoff, Claude Bernard, Ernst Mach, W.K. Clifford, William James, and Karl Pearson)[6], and often described as positivist.[7] In connection with Trivedi, it would be useful to mention briefly the various strains characterizing the empiricist approach of the period.

Phenomenalism, pragmatism, and the biological approach to knowledge

Empiricism, which sees experience as the foundation of knowledge, gave rise to phenomenalism—which holds only phenomena experienced by humans as truly real for them. Scientist-philosophers like Kirchhoff and Mach were inclined towards this position. Yet, scientific theories are compelled to invoke entities that are not experienced—atoms, forces, fields, ether[8], and so on. How could one reconcile this with the supposed empiricist foundations of science? Here, pragmatism came to the rescue. In the pragmatic approach, theories involving non-perceivable entities were fine, so long as such entities were taken not as real but as only pragmatically useful. In the last analysis, the task of science, even the most abstract and theoretical, was providing brief descriptive formulations that could stand the test of observation and *critical experiment*. As William James described it, this approach does not see scientific theories as true, but merely as 'conceptual shorthand descriptions' of phenomena.[9]

This view of theories as 'conceptual shorthand' was, of course, inspired largely by Mach, who argued that there was a natural human drive to look for succinct accounts of phenomena—to come up with finite accounts of infinite number of facts. The source of the above mentioned psychological drive lay in the biological need for survival and evolutionary success—the more succinct and wider the generalization, the better it would be retained and used to deal with phenomena and solve practical problems.[10] Obviously, the emergence of this biological approach to human knowledge was largely inspired by the tremendous influence of the Darwinian Theory, which sought to understand various aspects of animal and human life in biological-evolutionary terms.

An early articulation of this approach may be found in the mathematician-philosopher W.K. Clifford. For example, Clifford proposes a pragmatic solution to the familiar problem of justifying inductive inference, which assumes uniformity of nature. He writes:

> ...if you and I had not habitually acted on the uniformity of nature from the time we could act at all, we should not have been here to discuss the question.
>
> Nature is selecting for survival these individuals and races who act as if she were uniform; and hence the gradual spread of the belief over the civilized world.[11]

Another aspect of the biological approach to knowledge

Another aspect of the biological approach to knowledge is to be found in the writings of the psychologist-ethologist-philosopher C. Lloyd Morgan. He made a distinction between the act of *sensation*, which he viewed as a somewhat rudimentary neurological process (through which the impression is registered) and the act of *perception*, through which the 'object' is *constructed*. He explains this construction of an 'object' as an elaborate process by which the mind or brain adds past memories to existing sensations to fashion an object rich in qualities. Therefore, says Lloyd Morgan, the 'object is a *construct*'.[12] The perceived world was an assemblage of such 'constructs'. However, *as the sensory constitution* differed across species, genus, order, etc., *the nature of 'constructs', and consequently that of the world, also differed correspondingly*. Moreover, the world varied even across human individuals, according to the variation in their biological-cognitive constitution—'...for no two human beings is the world we live in quite the same'.[13]

Lloyd George's biologically based cognitive relativism received approval in Karl Pearson's *Grammar of Science*, first published in 1892,[14] which used it to develop the theme of perceptual relativism within the human world. Pearson says that there are differences in perceptual and cognitive apparatus across humans, and what we call the 'world' is nothing but reality as perceived and conceived by 'normal' human beings. He writes,

> The universal validity of science depends upon the similarity of the perceptive and reasoning faculties in normal civilised men.[15]

The term 'civilised' is of consequence here. Pearson argues that in addition to possessing 'normal' faculties, a certain degree of intellectual training was necessary to be able to view things in a genuinely scientific manner, only available in a civilized setting (it is apparent from the context and tone of discussion that Pearson has in mind 'western civilization').

For individuals differently enabled in sensory terms or mental functions, the world is different. Not only that, even individuals with 'normal' faculties might not be in a 'normal' state at a certain moment—they could be in a state of intoxication *or emotional exaltation. At those times, their perceptions would not yield scientifically reliable evidence*.[16]

Pearson's book was published in 1892, with the second edition coming out in 1900. It was influential and is known to have featured prominently in the reading list of young Einstein and his friends.[17]

Part III: elements of Trivedi's epistemology

Physical science

Trivedi's analysis is largely focused on what he calls 'physical science', and what we would call 'physics'. Nevertheless, as one can make out, his observations are applicable in various degrees to other branches of knowledge.

Remarks on orthography and usage

In using words and terms from Bengali and Sanskrit we have normally used diacritical marks. However, *diacritical marks have been avoided in the case of proper nouns*, i.e. person names, book names, etc.

Trivedi's century-old prose is slightly different from that of today's Bengali. For example, Trivedi almost never uses the term *abhijñatā*—the usual Bengali term used today to denote *experience*. He usually uses *pratyakṣa* (direct or immediate perception) as the general term denoting sensory experience. For example, he uses *pratyakṣavādī* to denote 'empiricist'.[18] This is in line with the Bengali usage of his times. At one point, Trivedi uses the term 'experience' several times in its English original.[19] Indeed, he often mentions the usual English terms, which is helpful as he is largely discussing concepts drawn from contemporary Western philosophy.

The notion of normalized experience

The first step in understanding Trivedi's arguments is to understand his critique of *experience*, which leads to his notion of *normalized experience*. Let us see how Trivedi builds up that notion.[20]

a. Experience and reality: As an epistemological inquirer, I must begin from what I understand as my experience. My experience at any moment is undeniably real for me *at that moment*. For example, the dream is real at *the moments* of dreaming. It doesn't matter that *later* I might 'wake up' and decide that the earlier experience was a dream. For, the *later* is a different moment with a different experiential reality. While dreaming, the dream *is* the reality.[21]

b. Defining the real: Hence, *my reality* at a particular moment is *my actual experience at that moment*.[22] This nominalist / phenomenalist definition is cardinal to Trivedi's analysis. The *true* and *real*[23] is not what *exists* in any *extra-experiential ontological* sense but is what we/I perceptually *experience*. Anything other than perceptual experience is inference, hypothesis, or imagination. Trivedi insists that this attitude towards what is real is helpful in clarifying matters.[24]

c. <u>The nature of my experience</u>: Trivedi calls the individual's entire private world of experience the *prātibhāsik jagat* (borrowing a term from the Advaita Vedanta with similar but not identical connotations)[25]. This *prātibhāsik jagat* is a fascinating succession of states. My perceptions depend on my temper and the nature of my perceptive faculties, which change with time, mood, disposition, state of health, and age. Fleeting thoughts and daydreams interrupt aware presence of the surroundings, excited or tired senses assail observation and judgement, and dreams, illusions, biases, and beliefs fashion our opinion on all fronts. My world varies in shape and flavour with time, often without rhyme or reason. It is a world without point, purpose, or plot.[26]

d. <u>Others</u>: However, my world of experience appears to include *other* experiencing entities like me. Yet, for me, my experience is compellingly real while your experience is merely inference, based on your behaviour, or what you tell me.[27] Indeed, I do not even really know whether you genuinely have experience or whether you are merely an automaton (*kaler putul*), an entity merely appearing to experience.[28]

e. <u>In practice</u>: Yet, despite the absence of any theoretically compelling reason to accept you and others as experiencing entities, I must, *in practice*, do so. For, that is unavoidable in the daily business of living.[29]

f. <u>The schism</u>: The acceptance of other experiencing entities brings about a schism in my *prātibhāsik jagat*. There appears to be portion of experience I share with others. This is the portion that seems to come through what I call my 'five senses'. The very fact that it is available to others seems to give it an independent existence. It appears to be external to me.[30] However, there remains another portion that seems to be available only to me. It is the domain of 'internal' sensations—pain, pleasure, joy, thoughts, desires, daydreams, etc.[31]

g. <u>Experientially incomparable</u>: But, even when I agree that everyone has experience, there remains a problem. I have no experiential access to your internal experiential realm, nor do you have access to mine.[32] Moreover, there is the problem of incomparability even with respect to the so-called shared or external experiences.[33] For example, perhaps your actual experience of the colour is very different from mine. However, as long as you and I can distinguish, say red from blue and blue from green, etc., and agree to use the same names for the different colours, there is no way of discovering that we differ in the nature of our sensations.[34]

h. <u>Getting along</u>: In real life, however, the above incomparability does not seem to matter. So long as you and I seem to respond to shades of colour and pitches of sound, etc., in a similar fashion, we can be on the same page and get on with the practical business of living. For, life in society is about getting along.[35]

i. <u>How even 'normal' accounts differ</u>: Even in practical life, some persons evidently have very different sensory experiences and their experiential worlds tend to differ from the general run. Such persons, or some of their experiences, are considered to be outside the 'normal' pale. However, the 'normal' majority appear to have broadly similar experiences. Nevertheless, even the normal person, in certain physiological or psychological states or under the influence of narcotics, will come up with 'unusual' or 'aberrant' experiences and estimates.[36] Moreover, we find that the sensory experience of any two 'normal' persons even in their healthy and wakeful moments is seen to differ in the finer details. Whenever minute precision in measurement is called for, even careful normal observers in their alert states record different values—mostly minutely different, but different nonetheless.[37] This idiosyncrasy of personal sensation is called the *personal equation* (a notion popularized by William James[38]). Thus, estimates of things and accounts of events given by normal individuals in their alert moments often differ, particularly in small details. (The reader today might be reminded of the *Rashomon effect*.[39] However, the latter usually refers to variation in reporting of social events, and the variation is seen to derive largely, if not entirely, from the diversity of biases, perspectives, political or other interests. Trivedi, however, has something more fundamental in mind— perceptual differences resulting from differences in cognitive patterns and states even among 'normal' persons witnessing an event in their wakeful moments. This is of course related to Trivedi's primary concern with physical observation, because, primarily, his concern is with the epistemology of physical science.)

j.   From the idiosyncratic to the normal: Therefore, even in so far as our *prātibhāsik jagat*s appear to be comparable, they differ. *How do we then go about living our daily lives, which consist of being able to agree with the experiences of others?* The answer is simple, yet counter-intuitive. We do so by *normalizing* our experiences. When someone's experience differs sharply from that of the *majority*, we consider it an aberration—an error, illusion, or hallucination. Persons who tend to witness occurrences that differ widely from those witnessed by the majority are considered to have defective or *dissimilar* cognitive apparatuses. Usually, the experiences of 'normal' healthy individuals in their sober wakeful states are considered to be the stuff of reality.[40] Note two things: (i) *normalcy* here simply means '*similar to what the majority perceives*', and *majority* here stands for the majority of the species, the 'man on the street'.[41] However, belonging to the majority does not ensure that all my experiences will be accepted as 'real'. Experiences unique to the person are discounted. So are observations in an ailing or tired state or under the influence of narcotics. The process of normalization consists of screening out vital parts of our experience—which may have personal, poetic, or mystic reality or significance, but is of little value in constructing a fabric of socially acceptable reality.[42] Therefore, conceding to normalization is the subordination and sacrifice of the personal to the common.[43] (Although, Trivedi is not explicit here, it seems that in his scheme the screening out of 'non-normal' experience

in the process of daily interaction is not rigorous. The boundaries sep-
arating the commonplace normal from the non-normal are somewhat
fuzzy, unlike in the case of the scientific construction of reality, where
acceptable experience is defined with greater rigour—of which more
soon.)

k. Similar to Pearson, but not same: In a sense, Trivedi is following
Pearson. However, there are two crucial differences. First, Trivedi
is not concerned with only scientific observation and knowledge,
but with knowledge generation as a social process. Second, as mentioned earlier, even when it comes to scientific observation, Trivedi's
'normal' human is not the 'civilised European', but declaredly 'just
the common person'—a representative member of the human species, so to say.

l. The pragmatic world (*byabahārik jagat*): Trivedi calls this socially
accepted physical reality the *byabahārik jagat* (in Bengali).[44] Once
again, this is harking back to a Sanskrit term, the *vyavahārika satta*
of the Advaita Vedanta. The term *vyavahāra* means practical use and
transaction. Trivedi says that the term *byabahārik* is best translated
into English as 'pragmatic'. He links it explicitly to Western pragmatism, à la William James.[45] However, in Trivedi, the 'pragmatic' has a
richer content. It means not just practically efficacious or convenient
but denotes an epistemic realm born of social interaction.

m. The *byabahārik* world is a constructed realm: But the *byabahārik jagat*
is *not* the real world of authentic experience. Authentic experience
belongs only to the *prātibhāsik jagat*, which denotes the realm of *all*
experience. But much of this is deemed inadmissible into the *byabahārik
jagat*. This consists only of experience regarding which there is rough
agreement among normal individuals. Therefore, the agreed-upon reality is a constructed and contrived entity. Thus, the world we all suppose
to be common to all our experiences, or at least, to the experiences of
the 'normal' majority in their wakeful moments, is common precisely
because it is no one's actual experience. It would differ, at least in the
details, from any single person's reality.[46]

n. This constructed realm is the domain that science investigates: Not only
does the common person accept the constructed realm of the *byabahārik
jagat* as reality, it is essentially this constructed domain rather than the
authentic experience of the individual that science accepts as its point
of departure. However, science, and more emphatically physical science, is not satisfied with the roughly defined *byabahārik jagat*. Particular care is taken in observing, and instruments are employed to improve
observational accuracy.[47] Yet, measurements taken by different 'normal'
observers, or even repeated measurements by the same observer, provide different, even if marginally different, values. But science aspires
towards exact values. Its realm of observation must be undisturbed
by the personal equations of observers. Therefore, from the general
range of observed values recorded by normal observers, an average is

taken, which is accepted as the most accurate measure. It would be the measure observed or estimated by the *ideal average* human being. Trivedi calls this ideal 'normal average' observer the 'mean man', who, it goes without saying, is again a conceptual construct.[48] And the world this 'mean man' observes is a contrived realm, not exactly observed by any actual human observer. (Thus, Trivedi not only avoids Pearson's equating 'normal' with 'normal civilised men', he strongly insists that scientific observations can have validity only if they can be observed by *any* 'normal' human observer—he literally writes, 'the man from the street'.[49]) It is this understanding of normalized experience as the domain that science investigates is what *distinguishes Trivedi's position*. (Science's process of normalization involves caution and rigour. Therefore, science's realm of normalized experience *is a more rigorously defined byabahārik jagat*. However, Trivedi does not give any new name for this world.)

The first incongruity

We have arrived at the first incongruity. Science, which swears by empirical evidence, refuses to take the *empirically authentic* perceptual world of the individual person as its domain. Consequently, the empirical evidence of science is not exactly what you or I might see, hear, or feel. It is in its essence an abstraction. And it is this abstraction rather than the concrete experience that proves to be scientifically useful.[50]

A clarificatory digression

Trivedi was aware that his exposition was susceptible to misunderstanding.[51] Emphasizing on the primacy of the normal and ordinary might appear to fly in the face of facts. For, obviously, science has progressed by listening to minority views—to unusual and even apparently preposterous theories. But as Trivedi explains, when highlighting the role of the *majority*, he is speaking only of empirical observation and not of hypotheses. Indeed, science is an abundantly creative enterprise, which makes huge advances only through bold and imaginative theorization. However, even the most brilliant hypothesis must be subjected to the discipline of rigorous empirical testing and this testing is based on observation—to which the rules of normalcy apply. *No scientist, whatever his merits as a creative and critical thinker, can claim any privilege when it is a question of evidence.* His or her evidences must stand up to the criterion of normal observation.[52]

The reasons for the usefulness of the byabahārik jagat

Why is the constructed *byabahārik jagat* of daily life *and its more scientifically refined version* more useful and effective than the individual's authentic experience?

The argument here has two parts. First, it shows why the perceptions of the normal majority tend to be practically advantageous. Second, it shows why it is desirable to focus on what are deemed the normal ranges of perception and choose their central values.

Due to the natural selection of features best adapted to the organism's surroundings, sensory faculties and ranges of sensation that provide most effective inputs come to *dominate* numerically in human populations. This sets the standard for the normal assortment of senses and the normal range of variability of sensation in a species. Further, as may be expected, the proportion of those who differ significantly in their sensory and cognitive faculties from the general run is rather low. (This, of course, applies both to those who are much better endowed and those who are grossly deficient. Thus, in any standard human population, the pattern of distribution of faculties would be a typical normal distribution.)[53] No wonder, perceptions and estimates that follow from these 'normal' or *common* sensory-cognitive structures tend to be close to each other. Further, in this evolutionary explanation of the nature of our faculties, it makes sense to accept the normal as 'true', as it is both biologically successful and close to the experiences of the majority, though it must differ from the perceptions of a small proportion of individuals who are 'not normal'. For the latter, the minority, the normal account of the world is wrong—as it does not agree with *their* perceptions. Therefore, says Trivedi, it would not to do say that the 'normal' view of the world is *truer* than that the view of the person who thinks he is in the minority. Rather, the 'normal' is more common and leads more often to practical success and is therefore accepted as *true*.[54]

Second, 'normal' represents not a single perception but a range of slightly differing perceptions. Hence, based on the reasoning outlined above, the midpoints would tend to denote the most successful. Therefore, it is desirable to identify roughly average values (broadly or grossly in daily life and more carefully in scientific observation) and use them as the bricks for building a reality that is acceptable to the majority if not to all.

Two related points

In the above connection, it is important to touch upon two questions that Trivedi clearly addresses.

> First: The only criterion that any observation must justify to be treated as scientific evidence is that it must be 'observable' for the normal majority of humans. An experience that fails on this count will not be acceptable as scientific evidence, even if it is strongly supported by the testimony of a group of scientific elites.[55] On the other hand, any experience, however strange, even 'miraculous', will begin to be admitted if it satisfies the above criterion. Trivedi tests this criterion by taking it to its logical extreme. What if, he says, someone argues that there was a time when the normal majority saw and

heard things that the public no longer seems to experience today—for example, the so-called supernatural or miraculous? Would it have been all right then to have admitted these as 'evidence'? Trivedi does not even flinch. He says that if there was a time when such was the case, then it would have been legitimate for the science of *that time* to have admitted such observations as evidence; but, until such observations become common today, they cannot be admitted as evidence by today's science'.[56] This is not meant as sophistry. Trivedi simply sees this as a logical corollary to his notion of scientific method.

Second: What would have happened if the conditions of life had been different and hence the 'normal' had been different, i.e. a different range of perceptions had been taken as 'normal'? This answer is equally straightforward. In that case, the contours of the *byabahārik jagat* would have been different. Had our species evolved in an atmosphere with a somewhat different proportion of gases—abilities and sensory ranges contributing to success and, therefore, the spectrum of normalcy would perhaps have been different. Consequently, the apparent contours of the world and perhaps even the scientific description of how nature behaves would have been different. Who knows what the normal sense of the world and the formulation of the laws of nature are for creatures in another planet with different environmental conditions?[57]

The law of nature—another digression

In a much earlier article (1899), *Niyamer Rajatva*, Trivedi had discussed problems inherent in the notion of natural laws. In *Bichitra Jagat*, he introduces a new angle with respect to the supposed uniformity in nature (the logical basis for assuming the existence of natural laws). He argues that since the phenomena that science takes into cognizance belong not to the *byabahārik jagat* but to the *prātibhāsik jagat* of personal experience, the uniformities we observe belong to the former and not to the latter. And since, of course, the former is essentially a contrived world, so are the uniformities observed in it.[58]

The second incongruity

As we have seen, the so-called scientific laws, theories, or hypotheses, are descriptions of phenomena not in *the perceived world of actual individuals* but *in the socially acceptable construct*—the *byabahārik jagat* and, more specifically, in its scientifically refined version. However, the creation of this refined version is a bit tricky.

The *byabahārik jagat* is created by working on and comparing the experiences of numerous 'normal' individuals. But subjective *qualia*—the private sensing of qualities—the basic stuff of experience, are not really comparable.[59] Therefore, science has to convert observations into numbers in order to compare multiple observations. If the numbers vary a little even after repeated measurements, one can take an average.

However, how can one convert human sensations into measurable entities? Trivedi refers to Jevons's comment that science measures all quantities by transforming them into length.[60] He also reminds us of Bergson's observation that it is only space that is susceptible to quantitative measurement.[61] Here, Trivedi finds the clue to solving the above problem. He points out that physics measures temperature not by sensation but by the height attained by the thermometric column. Similarly, pressure is measured by the height of the barometric column. Colour is specified by the degree of bending that coloured light suffers while passing through a prism. Pitch is approached in terms of the size of the string or the length of the column of air in the flute.[62] The weight of two objects is considered equivalent when they balance each other while being equidistant from the fulcrum of a balance, and even time is measured by the quantity of shift in the shadow or the distance traversed by the hands of a clock or, if we are using the sun as a clock, a shadow.[63] Thus, physics, the most complete of sciences, does not speak of sensations at all but of movements and their measurements, even in its discussion of such topics as Heat, Light, and Sound.[64]

Indeed, the degree of completeness of a science can be measured by the extent to which it has transformed many-splendored reality into austere arithmetic. The incompleteness of chemistry as a science is testified by its inability to purge itself entirely of reference to sensations.[65] Hence, science, which prides itself on its factuality, is compelled to move assiduously away from the true stuff of experience into measurable units of abstract space. It is only by doing so that science (in its most advanced form) can manage its empirical data. This is the second incongruity.

Space and its forms

Having shown that all measurement must involve space, Trivedi proceeds to examine the concept of space. Below, we provide a drastically summarized version of his ideas.

Trivedi follows Mach in distinguishing the real space of experience from the conceptual space of geometry.[66] In analytical detail, he shows how the conceptual space of geometry is an *idealization* from the perceptual space of different human subjects. Physical science since Newton had taken the Euclidean version of conceptual space as its framework, had distributed its material particles all over this conceptual space, and studied their movements.[67] In the Newtonian system, material objects had been conceived as elements of discontinuity, heterogeneity, and resistance in this conceptual Euclidean space. However, argues Trivedi, the entire approach could be reversed. Matter itself could be conceived as nothing more than local instances of undulation or bending in space, which accounts for the sensation of resistance that a subject experiences when passing through such locations in space.[68] And indeed, space as a whole need not be conceived as infinite, but as finite and curved on itself.[69] These ideas of Trivedi can be traced back to an early essay, *Clifforder Keet*, influenced by his early reading of W.K. Clifford.[70] Further, he points out that conceptual space might easily be conceived as non-Euclidean and n-dimensional and this might provide us with a more valuable model the world.[71]

Trivedi notes further that a new approach to space and consequently to time, motion, and mass seemed to be very much on the cards, as new developments in physics seemed to suggest light has absolute velocity and there was contraction of length in the direction of motion noticeable in fast moving objects, increase of inertia (or mass) with velocity, and time dilation.[72] All this showed the occasional need in science of altering its entire conceptual structure and foundations, in order to successfully tackle new findings in the realm of phenomena. Had Newton been born after Lobachevsky and Riemann and had proceeded to build the architecture of science in terms of non-Euclidean space, perhaps he would have produced a vastly different conceptual structure and the history of science would have taken a different course.[73]

The third incongruity

Human language functions by classifying reality—creating classes or categories based on our natures, interests and concerns, and reclassifying and creating sub classes when we will. Further, we can name our classes with equal freedom. These arbitrary class names do not indicate *actual perceived objects* but simply denote the classes we have created for our convenience. Further, not only do we play with reality by classifying experienced objects at our will, we also do so by employing imagined abstractions that are beyond experience to fashion the fabric we call 'knowledge'.[74]

What applies to human knowledge, applies to science in equal measure. Science classifies reality at will and conjures class names in hordes. Further, in order to present new theoretical structures, science must create new concepts and give them *new labels* or use old labels with *new meanings*. While constructing a theory, it thinks up new abstract entities with new properties and names them at will. It does all this with playful abandon. (As an example, Trivedi describes the amazing career of the *ether*, a conceptual construct whose properties are modified in accordance with the need to fit phenomena. Here, he refers to Fresnel's experiments and conceptualizations).[75] Science is much bolder than everyday knowledge, much more creative in designing hypotheses to account for phenomena. The reality of science is far removed from its portrait as a nose-to-the-ground pursuit that empiricism apparently suggests.

Science goes on to become hugely productive only when it removes itself far from experience to build a world of *pure concepts*. Consider the *simple pendulum*. This, explains Trivedi, is an abstraction of the purest water. The thread from which the ball is suspended is conceived as weightless, not susceptible to friction, erosion, or the vicissitudes of temperature. Let alone construct such a pendulum, none can visualize one, for even a visualization would generate definite form in the mind's eye—which would belie the ethereal abstractness of the notion.[76] The simple pendulum can only be viewed by the faculty that is able to 'view' concepts. This is the faculty of *divyachakṣu* (illumined sight)—which is essentially the eye of Reason (*prajñā*), capable of viewing the *formless form* of concepts.[77] As is most evident in the case of physical science, the astonishingly abstract nature of the world it fashions does not prevent it from being practically useful. In fact, it is their sheer abstractness that allows the theories to be 'universally' applicable. The more sweeping and more abstract the concept and formula, the greater its power and practical reach. Trivedi repeatedly refers to the Machian view of the economic nature of theories—their ability to summarize a huge amount of descriptive information in brief formulaic sequences.[78]

The world of physical science is the world of knowledge built out of names and concepts. Trivedi calls this the *Bāṅmay Jagat*. The Bengali term *bāṅmay* is derived from *vāṅmaya*—that which is made from *vāk*. What is *vāk*? The full significance of Trivedi's use of this term will become apparent only if we visit his book, *Yajna Katha*.[79] Here, in explaining the meaning and significance of certain Vedic sacrificial rituals, Trivedi explains that the closest analogue of *Vāk* is the Greek 'logos', which means not only *word* but also *idea* or *concept*. *Vāk*, in Trivedi's analysis, was central to ancient Vedic beliefs just as the *logos* was to the Greco-Christian philosophical culture. Both these cultures accepted the supreme world-generative power of concepts.[80] And concepts are generative of entire worlds, as is most dramatically evident in the case of science. For the world that science creates is not merely a conceptual structure, it is the entire framework of *our knowing the world*. And we use it not to know the world of our personal authentic experiences but *the abstract world of normalized experience*. Therefore, science's *jaḍa jagat* (inert or material world) is actually a *made up tale of an unreal world*.[81] Yet, it is a tale that yields great practical success and benefit. This is the third incongruity.

Creation and joy

Now we begin to get an idea of what Trivedi means when he sees science as a grand game of creation—involving all the playfulness, adventure, ingenuity, intensity, and commitment that a player can bring to the game. For the genuine sportsman, the reward of game is the game itself, with the occasional bonus of winning and, perhaps, tangible rewards. Similarly, science may result in utilitarian advantage, in improvements in practical application. However, for Trivedi, the true spirit of scientific inquiry and creation is finding the reward in the activity itself—in the transcendental joy of discovery and creation.

Why was Trivedi not interested in the practical implications of scientific inquiry? He answers this question in his essay *Mayapuri*.[82] He wasn't interested because, for him, this aspect of science was linked to power and worldly advancement and would not necessarily lead to the reduction of violence, rivalry, and suffering in human society. On the contrary, this was likely to lead to heightened rivalry for power and wealth. On the other hand, the game of science for science's sake, by bringing joy, could contribute to consoling a spirit troubled by the ceaseless competition and violence of modern society. Through its ability to discern order in the apparently untameable chaos of natural phenomena, science provides its practitioner, connoisseur, and student with a joy that approaches the ecstasy of the mystic experiencing the absolute.[83]

Here, Trivedi found the link between modern science and Vedic-Brahmanical wisdom, which he saw as consciously reaching out for the transcendental. No wonder, his attitude was seen as leaning towards the retrogressive.[84] However, he stuck to his guns to the very end—trying to combine his insights into different areas into a unifying mesh of philosophical–sociological understanding. But that is a story for another day.

Part IV: in lieu of an epilogue

It is easy to detect the similarity of elements in Trivedi's approach with elements in Kuhn, Feyerabend, and some so-called postmodern thinkers. Some readers might detect similarity with Tagore's epistemology. The similarities are real and merit scrutiny, which, unfortunately, could not be attempted here.

Thoughtful readers might wonder how far Trivedi's ideas about scientific observation are significant in an era where almost all observations are heavily based on automatic instruments that reach far beyond human senses.[85] Moreover, instrument-based observational mode raises *other* epistemological issues that would seem to support a realist account of knowledge. The present author believes that a realist critique of Trivedi's epistemological position is possible and desirable, yielding a richer theory of *reality* and *truth*. However, all that is best attempted on another occasion.

We would like to end the present article with just the following brief comment.

Trivedi thinks he has proved the utterly unreal nature of all scientific 'truth'. Actually, what he seems to have succeeded in demonstrating are the following:

1) The source of empirical data used in science must be trans-personal and intersubjective
2) In becoming intersubjective, science sacrifices *personal reality* to the altar of public sanction and pragmatic efficacy
3) Precise quantification is an important step towards the intersubjective description of observation
4) Science employs great freedom in conjuring up concepts and conceptual entities with no direct perceptual correlates, but which, nevertheless, prove greatly useful in dealing with phenomena

5) Therefore, the ability to make up *stories* (possible scenarios involving imaginary entities) is as useful in the quest for knowledge as is the concern for being faithful to phenomena

6) Further, such stories, whatever their status as 'truth', because they fit the phenomena superbly, prove to be of immense practical value. (today, a popular example would be quantum mechanics, but others could be mentioned)

With that, we must call it a day.

Notes

1 Two early essays, though not exactly exceptions, are worthy of mention. They are Sisir Kumar Maitra, 'Baigyanik o Darshanik Ramendra Sundar', in Srikumar Bandyopadhyaya et al. (eds.), *Acharya Ramendra Sundar Shatabarshiki Smarakgrantha*, Kolkata: Acharya Ramendra Sundar Shatabarshiki Smaran Samiti, 1966, pp. 30–42 and Girijapati Bhattacharya, 'Mahamanasvi Ramendra Sundar', in Srikumar Bandyopadhyay et al. (eds.), op. cit., pp. 95–107. However, they are too brief to do justice to the range and depth of Trivedi's ideas. Other writers have also commented on Trivedi's philosophy. But, these, at best, are stray references or vague descriptions. A more recent article is Ashish Lahiri, 'Ramendra Sundar, Bigyan Darshan o Jiban Darshan', in Ratankumar Nandi and Ashok Upadhyaya (eds.), *Ramendrasundar Smaran Grantha, 150 Bachharer Sraddhanjali*, Kolkata: Bangiya Sahitya Parishat, 2019, pp. 240–250. This is an interesting read and attends to some of Trivedi's ideas regarding science and scientific method. However, even this article does not try to attend to the roots, character, and scope of Trivedi's epistemological analyses. It also appears to be unaware of Trivedi's discussions on relativistic physics of the pre-Einstein genre (à la Lorenz, Poincare et al.). I am grateful to Prof. Tapas Mitra of Gurudas College, Kolkata for sending me a copy of Lahiri's article. Prof. Mitra himself mentions the great significance of Trivedi's ideas on the philosophy of science and quotes in detail from Trivedi in his article. See Tapas Mitra, 'Punjibad, Adhunikata o Padarthobidya', in *Eshana*, Vol. 5, 2007, pp. 133–149. However, his article, though influenced by Trivedi's ideas, is about broader issues. A relatively detailed expository and analytical treatment of Trivedi's philosophical ideas occurs in Santanu Chacraverti, *Ramendra Sundar Trivedi and Bengal's Response to Modern Science*, Ph.D. Thesis, Jadavpur University, 1996. However, the work has remained unpublished.

2 Ramendra Sundar Trivedi, 'Bichitra Jagat', in Brajendranath Bandyopadhyaya and Sajanikanta Das (eds.), *Ramendra Rachanabali*, Volume III, Kolkata: Bangiya Sahitya Parishat, Phalgun 1356 BS (Feb-March 1950), pp. 205–482.

3 See Ramendra Sundar Trivedi, 'Bangiya Sahitya Parishat', in Sajanikanta Das (ed.), *Ramendra Rachanabali*, Vol. VI, Kolkata: Bangiya Sahitya Parishat, 1957, p. 135.

4 Sureshchandra Samajpati described Ramendra as the embodiment of reaction to the Derozian epoch. See Brajendranath Bandyopadhyaya, 'Ramendra Charit', in Sunitikumar Chattopadhyay and Anilkumar Kanjilal (eds.), *Ramendra Rachana Sangraha*, Kolkata: Bangiya Sahitya Parishat, 1965, p. xxxvii.

5 Ramendra Sundar Trivedi, *Ramendra Rachanabali*, Volume III, op. cit., pp. 205–482.

6 Except for Mill and Bain, all the others mentioned here were professional scientists (even William James, who was a professional psychologist with a doctorate in medicine.)

7 The terms 'positivism' and 'positivist' have significant issues of interpretation. Hence, we avoid using them in this essay, except sparingly.

8 We are talking of the late nineteenth century. Ether was only conclusively rejected in the twentieth century, beginning with gradual acceptance of Einstein's paper of 1905 on the so-called Special Relativity. Later, the ether seems to have made a different kind of a comeback, proving, once again, that the scientist will use whatever conceptual tool necessary to account for things. That, however, is another story.

9 William James, *Collected Essays and Reviews*, London: Longmans, Green and Co., 1920, p. 449.

10 Ernst Mach, 'On the Economical Nature of Physical Inquiry', in *Popular Scientific Lectures*, English translation, London: The Open Court Publishing House, 1898, pp. 186–213.

11 W.K. Clifford, *Lectures and Essays*, Vol. I, London: MacMillan and Co., 1879, pp. 293–294.

12 My italics, SC.

13 C. Lloyd Morgan, *Animal Life and Intelligence*, Boston: Ginn and Company, 1891, pp. 335–336. My italics, SC.

14 Karl Pearson, *The Grammar of Science*, London: Walter Scott, 1892, p. 50, pp. 101–2n.

15 Ibid., p. 47.

16 Ibid., p. 66.

17 See, for example, John Stachel, *Einstein from 'B' to 'Z'*, Boston: Springer, 2001, p. 147.

18 *Ramendra Rachanabali*, Vol. III, op. cit., p. 215, p. 222.

19 Ibid., pp. 234–235.

20 In what follows, we shall not be following Trivedi's own narrative sequence exactly, but pursue the essential logical structure.

21 *Ramendra Rachanabali*, Vol. III, op. cit., pp. 239–240.

22 Ibid., pp. 238–240, 258, 264.

23 Trivedi uses the terms (*satya*, i.e. 'true' and *jā aché* i.e. 'real') synonymously.

24 Ibid., pp. 238–239, 256–263.

25 In Advaita Vedanta literature, '*prātibhāsika satta*' stands for what the individual experiences in delusion, illusion, or hallucination. Ramendra, however, uses the term to denote the *entire* world of purely personal experience of what appears at the moment.

26 Ibid., pp. 241–245.

27 Ibid., pp. 279–284.

28 Ibid., p. 250. Had he been writing today, Trivedi might have used the English term 'zombie' or coined a clever translation, instead of using '*kaler putul*' (automaton).

29 Ibid., pp. 249–250.

30 Ibid., p. 247.

31 Ibid., p. 209.

32 Ibid., p. 209, 279.

33 Ibid., p. 284.

34 Ibid., pp. 284.

35 Ibid., pp. 284–286.

36 Ibid., pp. 211–213.

37 Ibid., pp. 219–220, 283–84.

38 Ibid., pp. 219–220. Trivedi uses the expression 'personal equation' in English, but does not provide any references. The *concept* owes its discovery to astronomers, going back to the late eighteenth century. The *term* seems to have been coined by practising astronomers in the 1830s and widely employed by them, before psychologists and philosophers noticed it. See, for example, E. C. Sanford,

'Personal Equation', *The American Journal of Psychology*, November, 1888, Vol. 2, No. 1, pp. 3–38, Raynor L. Duncombe, 'Personal Equation in Astronomy', in *Popular Astronomy*, 1945, Volume 53, pp. 2–13 and pp. 63–76, http://articles. adsabs.harvard.edu//full/1945PA.....53....2D/0000002.000.html; last accessed 27 December 2020; and Christoph Hoffmann, 'Constant Differences: Friedrich Wilhelm Bessel, the Concept of the Observer in Early Nineteenth-Century Practical Astronomy and the History of the Personal Equation', in *The British Journal for the History of Science*, Vol. 40, no. 3, 2007, pp. 333–365; last accessed 27 December, 2020, http://www.jstor.org/stable/4500747. The term and the concept were popularised by William James—William James, *The Principles of Psychology*, Vol. I, London: Macmillan and Co. Ltd., 1890, p. 194, 413n. However, from the nature of the prose, it appears that Trivedi's discussion of the 'personal equation' in the *Bichitra Jagat* was influenced by the discussion in E. Walter Maunder, *The Royal Observatory, Greenwich: A Glance at its History and Work*, London: The Religious Tract Society, 1900, pp. 176–177.

39 The first person to use this term in the above sense appears to have been James Scott, who called it the 'Roshomon Effect' (the spelling 'Roshomon' obviously being a mistake.) See James C. Scott, *Weapons of the Weak, Everyday forms of Peasant Resistance*, New Haven: Yale University Press, 1985, p. xviii. For examples of subsequent use, see Christian Davenport, 'The Rashomon Effect, Observation, and Data Generation', in *Media Bias, Perspective, and State Repression: The Black Panther Party*, Cambridge: Cambridge University Press, 2010, pp. 52–73.

40 Ramendra Rachanabali, Vol. III, op. cit., pp. 218–220, 225–227.

41 Ibid., p. 225.

42 Ibid., pp. 219–220, 225–227.

43 Ibid., pp. 233–234.

44 Ibid., pp. 246.

45 Ibid., pp. 230–231.

46 Ibid., pp. 233–238.

47 Ibid., pp. 215.

48 Ibid., pp. 222.

49 Ibid., pp. 18. The statement here might seem a bit odd, as untrained observers are usually not brought in from the streets for taking scientific readings. Trivedi, of course, was well aware of that. But as the language makes clear, Trivedi does not expect to be taken literally and is only emphasizing the idea that the process of normalization of observations works best when it comes closest to the species average. We later indicate the possible reasoning behind this assertion. Incidentally, Trivedi's notion that the taking of readings becomes most reliable when it is rendered relatively commonplace may be seen to be in line with the practice introduced by British Astronomer Royal George Biddell Airy, who revolutionized the organizing of observation and compilation at the Greenwich observatory. Airy felt that the actual task of observation and the keeping of astronomical records were best left to ordinary people who had been given a little basic training. See, for example, Simon Newcomb, *The Reminiscences of an Astronomer*, 1903, Project Gutenberg EPUB eBook, Chapter X.

50 *Ramendra Rachanabali*, Vol. III, op. cit., pp. 222–223.

51 The misunderstanding mentioned in this passage actually occurred when Trivedi's first essay was published. *Ramendra Rachanabali*, Vol. III, op. cit., p. 252.

52 *Ramendra Rachanabali*, Vol. III, op. cit., pp. 252–253.

53 *Ramendra Rachanabali*, Vol. III, op. cit., pp. 221–222.

54 *Ramendra Rachanabali*, Vol. III, op. cit., pp. 227–229.

55 *Ramendra Rachanabali*, Vol. III, op. cit., p. 213.

56 *Ramendra Rachanabali*, Vol. III, op. cit., p. 215.

57 Ibid. pp. 232. Trivedi is obviously thinking in terms of sapient creatures, i.e.

creatures that not only perceive the world but can describe their perceptions of the world in analytical detail, as humans can. For, otherwise, there was no need to have taken the discussion beyond the planet earth, where natural selection has resulted in very different cognitive abilities in different organisms, resulting obviously in widely divergent experiences of reality.

58 Ramendra Rachanabali, Vol. III, op. cit., pp. 242–243.

59 Of course, the terms 'quale' and 'qualia' were still in the future.

60 Note the discussion refers to *measuring* (as of continuous variables) and not merely *counting* (as of discrete variables).

61 Ramendra Rachanabali, Vol. III, op. cit., pp. 288–289. Trivedi mentions Jevons's *The Principles of Science* (the first edition of which appeared in 1874 and saw several editions; it is unclear to which edition Trivedi is referring). As regards Bergson, the source seems to be Henri Bergson, *Creative Evolution*, English translation, New York: Henry Holt and Company, 1911, p. 216.

62 Notice that Trivedi does not mention wavelengths of light and sound as examples of reducing phenomena to measurable units of space. The reason seems to be that Trivedi is talking of directly observable phenomena and, for him, waves and wavelengths are conceptual entities, valuable in producing beautiful explanations of certain phenomena, but, not being directly observable by humans, they are not *phenomena*.

63 Ramendra Rachanabali, Vol. III, op. cit., pp. 288–289. A digital clock might appear to fly in the face of Trivedi's (or Bergson's) position. However, Trivedi was speaking about humans doing the measuring and not about instruments that automatically measure. A digital clock simply records the number of oscillations of something—for example, a quartz crystal oscillator—and displays the number on a screen. We just read the number. Counting our pulses would appear to be more germane as a counter example. Unfortunately, human or animal pulses cannot stand as reliable measures of time as the periodicity easily varies. More stable measures are atomic clocks or the pulsations of a pulsar. Unfortunately, such pulsations aren't available to human senses directly.

64 *Ramendra Rachanabali*, Vol. III, op. cit., p. 282.

65 Ibid., pp. 302–303.

66 Ibid., p. 338.

67 Ibid., pp. 291–303.

68 Ibid., pp. 331–332.

69 Ibid., pp. 454–455.

70 For W.K. Clifford's remarkable ideas in this regard, see Robert Tucker (ed.), *The Mathematical Papers by William Kingdon Clifford*, London: Macmillan and Co., 1882, p. 21 and W.K. Clifford, *The Common Sense of the Exact Sciences*, London: Kegan Paul, Trench & Co., 1885, pp. 214–226 (A copy exists in the Trivedi collection at the Bangiya Sahitya Parishat, Kolkata).

71 *Ramendra Rachanabali*, Vol. Ill, op. cit., pp. 454–455.

72 Written in 1917, this portion was perhaps the earliest substantial reference by an Indian to relativistic physics. An earlier reference, though a quick one, is to be found in a 1910 essay by *Trivedi himself*: Ramendra Sundar Trivedi, *Bigyane Ptutulpuja*, in Brajendranath Bandyopadhyay and Sajanikanta Das (eds.), *Ramendra Rachanabali*, Vol. I, Kolkata: Bangiya Sahitya Parishat, 1949, pp. 455–479. It is interesting that although Trivedi refers to the notions of relativistic physics and to its foundations having been laid down by Lorentz, there is no reference at all to Einstein or the revolutionary significance of his 1905 paper, let alone to the General Theory. This is because, as textual comparison shows, here Trivedi is drawing almost exclusively from Henri Poincaré, *Science and Method*, English translation, London: Thomas Nelson and Sons, 1914 (although he makes no explicit reference to this work). Poincaré's book, which was originally written in 1908, does not mention Einstein's paper or its revolutionary

significance. One should also recall the information by Girijapati Bhattacharya, that books on relativistic physics were not available in India prior to 1920 (Girijapati Bhattacharya, 'Mahamanasvi Ramendra Sundar', op. cit., p. 105). It was also only in 1920 that M.N. Saha and S.N. Bose's English translation of Einstein's papers were also published. See A. Einstein and H. Minkowski, *The Principle of Relativity: Original Papers by A. Einstein and H. Minkowski*, English translation, Calcutta: University of Calcutta, 1920.

73 *Ramendra Rachanabali*, Vol. Ill, op. cit., pp. 454–455.

74 Ibid., pp. 264–268.

75 Ibid., pp. 282. A relatively recent portrayal of the wonderfully chequered career of the ether would seem to be in line with Trivedi's understanding. See Michel Janssen and John Stachel, *The Optics and Electrodynamics of Moving Bodies*, Max-Planck-Institut Für Wissenschaftsgeschichte, Preprint, 2004, pp. 1–2.

76 *Ramendra Rachanabali*, Vol. III, pp. 274.

77 Ibid., p. 252, p. 272, p. 438.

78 Ibid., pp. 267–268.

79 Ibid., pp. 485–635.

80 Ibid., pp. 589–613.

81 Ibid. pp. 277–282.

82 Ramendra Sundar Trivedi, *Mayapuri* in Suniti Kumar Chattopadhyay and Anilkumar Kanjilal (eds.), *Ramendra Rachana Sangraha*, op. cit., pp. 67–88.

83 Ibid. pp. 87–88

84 Satyendranath Bose, 'Acharya Ramendra Sundar', Ratanmohan Khan (ed.), *Satyendranath Basu Rachana Sankalan*, Kolkata: Bangiya Bigyan Parishad, 1993, pp. 176–189.

85 I am grateful to Prof. Mihir Chakraborty (formerly of the University of Calcutta) for reminding me to deal with this aspect, which I hope to do later.

3 Professor Balaji Prabhakar Modak

A forgotten science propagator

Abhidha Dhumatkar

Introduction

Science education began in Western India during the tenure of the first Governor Mount Stuart Elphinstone (1819–1829). In 1820, the native school-book and school committee set up by Mount Stuart Elphinstone awarded rewards of Rs. 1000/- to Rs. 5000/- for the best Marathi and Gujarati book, which contributed greatly to the publication of Marathi science literature on natural sciences. In 1823, the Engineering Institute was established by George Resto Jervis to train surveyors and builders for public works department in Marathi and Gujarati languages.[1] In 1845, the Grant Medical College was developed out of the medical school started in 1825.[2] The Marathi periodical of the Students Literary and Scientific Society contributed towards the foundation of original Marathi science literature.[3]

Balshastri Jambhekar, a scholar of mathematics, languages and modern sciences and a member of the Geographical Society wrote Marathi textbooks on mathematics, astronomy, history and geography. He published regularly, a glossary of scientific terms in his monthly *Digdarshan*. Hari Keshavji Pathare, a member of the Native Education Society and the Vice President of the Marathi wing of students Literary and Scientific Society, wrote Marathi books on physics and chemistry in 1833 and 1855, respectively, in the form of couplets. His Sanskrit-based Marathi science terminology was adopted by Marathi science writers subsequently.[4] Bhau alias Govind Mahajan acquainted Marathi readers with modern science, psychology and European history through his periodical *Jnan Darshan* from July 1854 to April 1856.[5]

Kero alias Vinayak Lakshman Chhatre (1824–1884), a second assistant at the Government Observatory at Bombay in 1840 and later Assistant Professor of Mathematics wrote Marathi textbooks on physics and mathematics and read papers on science to the students' Literary and Scientific Society. He observed eclipses with the help of telescope and delivered public lectures on astronomy and water supply. With the help of rainfall records of observations, he devised formulae to calculate the years of excessive rainfall and famine. His book, *Graha Sadhananchi Koshtake* (1860) contained

DOI: 10.4324/9781003241980-5

astronomical formulae, studied with the help of ancient Indian and modern western methods.[6]

Dr. Bhau Daji Lad (1822–1874), an eminent allopathic doctor and Indologist, was associated with the foundation of the Victoria and Albert Museum and the Victoria Gardens in Bombay. Out of his pharmacological research in his private botanical garden, with the assistance of his brother Dr. Narayan Daji, Dr. Bhau Daji devised a leprosy cure from a plant *Hydnocarpus Wightiana* known as *Tugarak* in *Sushruta Samhita*. Dr. Narayan Daji Lad (1828–1875), a professor of chemistry and pharmacology, Grant Medical College, was also a leading botanical researcher from Bombay and wrote Marathi books on chemistry in 1862 and pharmacology in 1865.[7]

Sakharam Arjun Raut (1839–1885), a professor of botany and surgery at Grant Medical College, the first Indian assistant surgeon at the J. J. Hospital and vaccination superintendent for Bombay Presidency, was a leading Marathi writer on gynaecology, child-care, child psychology and public health between 1865 and 1880. He wrote Marathi textbooks on principles and practices of medicine and obstetrics for the Marathi classes of Grant Medical College. He distributed free of cost, a Marathi pamphlet on prevention and cure of smallpox. He published a detailed account of the drugs in the Bombay Bazar to identify fake drugs. He discovered a plant called Dressina Furgussaina and was one of the founders of the Bombay Natural History Society. He lectured on science to the women wing of the Bombay Prarthana Samaj.[8] Dr. Kanhoba R. Kirtikar, a specialist of mental diseases and a botanist wrote Marathi books on botany and was the only Indian, contributing research papers to the journal of the Bombay Natural History Society in the nineteenth century. Atmaram S. Jayakar translated into English a medical lexicon from Arabic. Dr. Sir Bhalchandra Krishna Bhatavadekar (1852–1922) an allopath, coming from family of ayurvedic practitioners believed in the synthesis of Ayurveda with allopathy.[9] Women's medical education started in Maharashtra in the late nineteenth century with Dr. Anandibai Joshi, the first lady doctor of India.[10]

Grey, Directorate of Public Instruction (DPI), Bombay, proposed a chair for science in Bombay University. Poona Civil Engineering College (opened in 1868) was renamed as Poona Science College in 1879 by adding to it, the classes in agriculture, zoology and physics during the Governorship of Sir Richard Temple. B.Sc. Degree, the first of its kind all over India was introduced at the Bombay University in 1880. The Bombay Veterinary College and the Victoria Jubilee Technical Institute were started in 1888 and 1889, respectively.[11]

Marginal Marathi literature on geology agriculture, home industry and natural products was published in the last half of the nineteenth century. In 1880, there was emergence of specialized Marathi journals on medicine, agriculture and pure science in Bombay, Poona and Sholapur. The work of coining Marathi terminology commenced by Balshastri Jambhekar and Hari Keshaoji Pathare gathered momentum after 1880, culminating into the dictionary of Marathi Science Terminology by Marathi Sahitya Parishad in

1901.[12] The period also witnessed rapid expansion in transport and communications, mainly in the efficient postal and telegraphic service, railway network and steam engine, slowly challenging the old order in big cities like Bombay and Poona.[13] However, small towns like Kolhapur and Miraj remained feudal where Professor Balaji Prabhakar Modak spread emerging scientific culture from Bombay and Poona.

Professor Modak's life and work

Professor Balaji Prabhakar Modak was born on March 22, 1847, in a Chitpavan Brahmin family, in the village Achre, Taluka Malvan, Ratnagiri district of Bombay Presidency. He was distantly related to the well-known Prarthana Samajist Waman Abhaji Modak.[14] After his father's early death, he completed his primary education at Sangli where Dada Chhapkhane set a fertile ground for his scientific career through his instructions in arithmetic, geometry and trigonometry.[15] A scholarship from the ruler of Sangli enabled Modak to complete his matriculation from Sardar High School, Belgaum and pursue higher education in the Deccan College Poona. Unable to appear for B.A. examination due to illness, Modak joined Rajaram High School, Kolhapur, as a First assistant teacher in July 1869. He was a tutor to prince Shivaji of Kolhapur and superintendent of the school at Kolhapur founded exclusively for Maratha nobility. Abasahed Kagalkar and Balasaheb Mirajkar, two students from this school helped B. P. Modak in his work of spreading science.

A turning point in Modak's life was his training in theoretical and practical chemistry under Dr. Samuel Cooke (The Principal of the Poona Science College), under a Government project distributing science apparatus to the schools of Bombay Presidency. The practical experience in laboratory gained under Dr. Samuel Cooke's guidance generated in Modak a passion for science. His appointment as a professor of Natural Philosophy (combination of physics and chemistry) in 1880 in the newly opened Rajaram College, Kolhapur, paved the way for his science exhibitions resulting out of his public lectures on science. To keep himself abreast of latest developments in science, Modak visited Bombay and Poona regularly.[16]

Modak wrote chronological table with corresponding dates of the *Shaka* era, the *Raja Shaka* (Shivaji Calendar), *Hijri*, *Fasli* (Akbar's Calendar) and Gregorian calendars from 1728 AD to 1894 AD. In 1892, Modak translated into Marathi 24th Volume of the *Bombay Gazette* on the Kolhapur state and its principalities. Modak wrote the first history of Kolhapur and Bahamani rulers in Marathi. Modak, the Director of Industrial Museum (started in 1873 by Kolhapur Government on Modak's suggestion) held an industrial exhibition in 1891 at Kolhapur. Prompted by this exhibition, the Dewan of Kolhapur, Meherjibhai Kuwar asked Modak to conduct an industrial survey of Kolhapur.

On the orders of the Kolhapur Government, Modak conducted research on the problem of limited employment opportunities for teachers in the

Kolhapur state, with Sethna (the temporary Principal of the Rajaram College) and Kirtikar (Deputy Education Inspector, Kolhapur) in 1894–1895. He contributed to the development of the Native General Library of Kolhapur, presented Karvir Nagar Vachanmandir, as its secretary and president. He wrote a book on banking when he was entrusted with reorganization of Southern Maratha Bank by the Kolhapur government. Although not a member of any political organization, he was associated with the popular Maharashtrian leaders of late nineteenth century such as Bal Gangadhar Tilak, Justice M. G. Ranade, Vishnushastri Chiplunkar and M. V. Namjoshi and strongly condemned British colonial practices in the fields of science and education. Modak worked as temporary Vice-Principal of Rajaram College between 1894 and 1895 and retired as its Vice-Principal in 1900. He was offered the membership of Royal Chemical Society and Principalship of Kalabhavan, Baroda.

Modak's three decades of experience, as a lecturer of science had convinced him about the need of certain reforms in the education system. In 1876, he stressed the need to put natural sciences on par with social sciences, in the curriculum. In 1887, he advocated education through modern Indian languages. He recommended compulsory primary education and technical education in the industrial survey of Kolhapur, 1895. After researching curricular problems in 1895, he recommended closure of the Anglo-vernacular schools and proposed a plan for middle level education through modern Indian languages in 1897. He spoke for the first time about starting university teaching through Indian languages during his presidential speech at Mumbai Marathi Granth Sangrahalaya in 1904. He proposed a plan for national education in 1906. There were successful attempts to appoint him as the chairman of Kolhapur Representative Council in 1906. Modak spent his retired life studying the Bhagavad Gita and writing science books till his death on December 2, 1906.[17]

Marathi books on science

In order to disseminate scientific information among laymen and to facilitate Marathi science textbooks for future Marathi schools and colleges, Modak wrote and translated science books in Marathi. Out of his 38 published books (original writings and translations), 24 were on various branches of science such as physics, chemistry, mechanics, health and hygiene and zoology. His son and biographer, Ganesh B. Modak, credited him with an unpublished book on agriculture. Modak's first science book, *Rasayan Shastra Purvardha* was on inorganic chemistry and was dedicated to Dr. Samuel Cooke. The book was well received by readers, the Marathi Press and the Department of Public Instruction of Bombay Presidency. This detailed Marathi book on chemistry was a milestone, as the information on chemistry in Kero Lakshman Chhatre's book *Siddha Padarth Vijnan* was meagre and Dr. Narayan Daji's book *Rasayan Shastra* was used essentially by the students of Grant Medical College.[18]

Modak's book was reviewed in *Nibandhamala* of Vishnushastri Chiplunkar (a champion of Marathi language in the late nineteenth century) and in *Vividha Jnan Vistar* (a leading Marathi journal of the period). Vishnushastri Chiplunkar credited Modak with pioneering the movement to liberate Marathi from English domination, by producing original scientific literature in Marathi. Chiplunkar appreciated Modak's simple style of scientific writing, numerous illustrations as experiments and scientific explanations of natural phenomena from daily life quoted in the book.[19] The editor of the *Vividha Jnan Vistar*, Bhausaheb Gupte, compared Modak's task of introducing Western sciences in Indian languages, without leaving any trace of its foreign origin to that of Varahmihira in ancient India. In his critical appraisal of the book, Bhausaheb Gupte disapproved Modak's definition of atom and molecule to explain chemical combination and his adulterated Marathi terms like 'guru carbonated hydrogen', which Vishnushastri Chiplunkar had pardoned. He strongly objected to Modak's claims about Hindus not being credited with any discovery in natural sciences.[20]

Modak translated English books prescribed by DPI to English schools of Bombay Presidency either by orders of DPI or by his own will. Modak was also requested by T.K. Gajjar to translate Ganot's book on physics for Gajjar's series *Shri Sayaji Dnyanmanjusha*, which proved to be the first Marathi book on sound and light. While translating English books, the needs of Indian society were Modak's prime concern. For example, in order to acquaint Indian readers with health consciousness of an average European, Modak retained the information on European ideas about housing, clothing and food, in the translation of his books on health. Modak's science books containing information a little more than present high school textbooks might appear elementary to the present generation. However, his voluminous writing on almost all branches and sub-branches of science was a step ahead of scientific articles of Balashastri Jambhekar, Bhau Mahajan and generalized science books of Hari Keshavaji Pathare. It was indeed a Herculean task in view of low scientific literacy in India, absence of scientific terminology and apathy of an average educated Indian to science.

In his work, Modak readily accepted suggestions from experts like Dr. Bhalchandra Bhatvadekar, an eminent medical practitioner in Bombay and V. B. Sohoni, Principal, Poona Training College as well as criticism even from a matriculate like Bhaisaheb Gupte of *Vividha Jnan Vistar*. Modak was helped and encouraged by Chatfield, DPI, Bombay, who instructed all libraries through a circular to buy Modak's translation of the popular Natural Philosophy, *Vol. I Electricity and Magnetism* by Ganot. Maodak's permission was sought to translate his books in Guajarati and Kannada.

Modak's books with illustrations and guidelines for scientific experiments spread up-to-date scientific information through Marathi on gases, water, electricity, magnetism, gravitation, laws of buoyancy, sound, light, wind, clouds, atmospheric pressure, metal and metal compounds, Dalton's atomic theory, hydrogen, oxygen, alkalis, crystals, public and individual health, chemical products, medical properties of trees, animals, photosynthesis, pollination, the principles underlying the working of scientific instruments, like calorimeter, barometer, kaleidoscope, spectroscope, prisms, concave and convex mirrors, camera and magic lantern.[21] No other Marathi writer had written earlier in detail on such a variety of subjects.

Owing to Modak's path-breaking contribution, G. N. Sahastrabuddhe and G.D. Khanolkar called Modak the 'Standard Bearer' of Marathi science literature in the nineteenth century. According to them, Modak was convinced about the dire necessity to spread science, in view of the numerous problems encountered due to low level of scientific knowledge among students and the public at the time when G. V. Joshi had launched the movement for industrialization of Maharashtra. Through his prodigious Marathi books on various branches of science, Modak stove to shift the emphasis of nineteenth century, Maharashtrian Renaissance from social and political movements to spread of science. In fact, he was the major exponent of the translation movement of science literature in Maharashtra in the nineteenth century. Comparing his work of stemming the tide of Anglicization of Maharashtra in the field of science to the similar work undertaken by Vishnu Shastri Chiplunkarin the field of literature, G. N. Sahastrabuddhe attributed Modak's failure against Vishnu Shastri's success, to the numerous difficulties destined in the path of championing an unpopular cause and observed that his failure in this effort not only frustrated him but also discouraged all those who tried to emulate his example.[22]

Views on Indian language terminology and science literature

Modak believed that science literature in Indian languages ought to spread the most modern scientific knowledge with its practical application. For faster diffusion of scientific knowledge, Modak recommended Indian scientific terminology. According to him, science terminology coined in Indian languages should be simple, lucid, indicative and should be uniform to all Indian languages.

It should be derived from Sanskrit roots (having the system of prefixes and suffixes) with regional variations, like modern European languages adopting Greek and Latin oriented terminology with national variations. Modak did not hesitate to even use Pali and Persian words and retained English terms whenever he could not coin Marathi equivalents. He used the word 'Shrishti Shastra' for physics instead of the popular word 'Siddha Padartha Vidnyan'. Although an advocate of Sanskrit-based Indian terminology, he highly disapproved complicated, lengthy, incomprehensive terms. He welcomed Ranade's moves to incorporate scientific terminology in his English Marathi General Dictionary.

Modak was in favour of gradual introduction of Indian terminology. He argued that centuries-old learning of Ayurveda in India had simplified the task of coining medical terminology in modern Indian languages. However, a lot of time and a great amount of efforts would be required to coin Indian terminology for physics and chemistry (especially for chemical compounds like bromide, sulphate) which had developed mainly in Europe. Secondly, the scientific world required a universal language and terminology for faster exchange of scientific information. Thirdly, since chemical industries in India were in the hands of British, English chemical terms were in vogue in the market. Modak favoured provisional co-existence of English terms in literature, till science education was imparted in Indian languages and Indian industries were run by Indians themselves. Modak refrained from using the newly coined, simple and indicative terminology of Deshmukh and Gajjar. Therefore, he refrained from coining Marathi words for chemical elements including hydrocarbon compounds and formulae, a practice still followed in Marathi textbooks.

Some of the terminology coined by Professor Modak has been incorporated in the Marathi-science textbooks even today. In order to evoke public discussion to finalize accurate Marathi terminology, at the end of his books, Modak listed the terminology in Marathi coined by him for physics, mechanics, health and biological sciences, and appealed to other science writers to follow the same practice.[23] In his Marathi biography of Raghuveer, Shridhar Dattatraya Limaye, an eminent chemist, in post-independent Maharashtra credited Modak for his unparalleled, voluminous Marathi science literature as well as pioneering thinking on coining Marathi science terminology.[24]

Science exhibitions – a herald of science in the southern Maratha country

Encouraged by Abasaheb Kagalkar and Krishnaji Bhikaji Gokhale in the early stages and financed by Kolhapur Government throughout, Modak held annual science exhibitions in the Christmas vacation in Rajaram College from 1883 to 1896, in order to acquaint the public with the latest scientific inventions. With growing public response, their scope was widened by adding new equipment from time to time. In organizing the science exhibition, Modak was helped by his colleagues. His exhibitions were widely advertised, and police force was stationed to maintain law and order. Free but mandatory daily entry tickets were distributed to visitors. The first day of exhibition was reserved for honourable guests, nobles and organizers; the second and third day for students and rest of the days for public. The number of visitors of the exhibition of 1887 ranged between 10,000 and 12,000 per day. The smart students of Rajaram College, including N. C. Kelkar (later associate of Bal Gangadhar Tilak) and Govind S. Tembe (later famous harmonium player) with their experienced teachers, demonstrated scientific experiments to visitors with the help of instruments used in diverse branches of science placed in separate rooms on elevated platforms. According to N.C. Kelkar, a special room was allotted to physics and the experiments connected with electricity enjoyed precedence over those with chemistry.

In these exhibitions, Modak displayed the latest machinery like x-ray machines, cameras, surgical instruments, telescopes, microscopes and coils illustrating Faraday's law, Morse's telegraphs, Edison's phonogram, water mills, automobiles and fire engines. Volunteers disseminated scientific information with practical demonstration on electricity, sound, light, manufacture of dyes from trees, flying of guns using hydraulic pressure, magnetization and demagnetization, law of gravitation (by dropping a coin and a paper simultaneously from same distance).Various types of snakes, crocodiles, birds stuffed with straw and coins from diverse periods of history were brought from the Kolhapur Industrial Museum especially for science exhibition. An anonymous visitor was delighted to see through microscope enlarged portraits of the English Royal family, Bengal grams looking like pomegranates, eggs of caterpillars and the system of blood circulation in frogs. The science exhibition convinced him about potentiality of science to master the five elements of nature. Through his science exhibitions, Modak introduced electric lamps in Southern Maratha Country.[25]

N. C. Kelkar noted in his memories that spreading of Dynamo wires in verandah and lighting of colourful lamps chandeliers and hanging lamps of clay with the help of electricity at night reminded the students of the Rajaram College of a marriage ceremony or a Diwali celebration. They were terrified at the strong burning power of the arc light in the experiments connected with electricity. They were delighted to see the artificial rainbow appearing in the vacuum tubes. Girls standing in the circle shrieked and gesticulated when they received mild electric shocks.[26] Inspired by Modak's exhibitions, his semi-literate students too spread science in other villages by demonstrating experiments.

Modak's activities in spreading science were encouraged and financed by the princes of Kolhapur and Miraj. In fact, without the sponsorship of Kolhapur Government, Modak's science exhibitions would not have materialized, Abasaheb Kagalkar, the natural father of Shanu Maharaj, the Regent of Kolhapur and a student of Modak in the Rajkumar High School Kolhapur, solely sponsored Modak's Science exhibitions till 1884. Influenced by Modak, Abasaheb conducted science experiments. He furnished the common laboratory of Rajaram High School and Rajaram College with scientific instruments purchased during his annual Bombay visits with Modak, and a binocular worth Rs. 1000/- from England. Modak attributed continuing his scientific pursuits to the financial assistance and encouragement of Abasaheb Kagalakar. The princes of Kolhapur entrusted to Modak, the common laboratory of Rajaram High School and Rajaram College during his teaching career from 1870 to 1900 and permitted him to work there after retirement.

Gangadhar Patwardhan alias Balasaheb Mirajkar, the prince of Miraj, another student of Modak in Rajkumar High School Kolhapur was his sole assistant in the initial science exhibitions. On Modak's inspiration, he wrote two books, *Mutra Pariksha* and *Rasayan Shastra* and delivered public lectures on science at Miraj. In the preface to 'Rasayan Shastra' dedicated to Modak, Balasaheb Mirajkar noted that he had full faith in pleasant and encouraging nature and empirical knowledge of his great teacher Modak, others could hardly imagine the extent to which he was benefited by Modak's perpetual love and selflessness. He could master chemistry only because he treads upon the path shown by his great teacher Modak. Balasaheb delivered public lectures on science and opened a well-furnished laboratory at Miraj named '*Ganesh Kala Griha*'.

He sponsored printing of 200 stencilled figures in Modak's translation of Ganot's volumes on Electricity and Magnetism, when Baroda Government discontinued '*Shri Sayaji Jnan Manjusha*'. Balasaheb conducted experiments in the '*Ganesh Kala Griha*' on Modak's articles on dyeing and calico printing published in '*Shilpa Kala Vijnan*'. However, a book based on these articles could not be completed due to Modak's sudden death. Thus, with the help of his associates, Modak created scientific ambience in Kolhapur and its vicinity, so much so that in Kolhapur his name became synonymous with science.[27]

A champion of science education and a missionary

Mass education in natural sciences remained neglected till the 1870s. Science was excluded from the curriculum of vernacular primary schools. Scientific instruction in High Schools was on par with European Primary Schools. Neglect of scientific education in training colleges led to a low level of scientific instruction in the vernacular schools, generating low demand for science books which affected growth of original science literature in Marathi. Science being an optional subject, attracted few students in colleges. Bombay Presidency fell short of specialized science schools, colleges and science professors. Little was done by the British Government and private institutions to promote scientific research. Peile, DPI, Bombay, introduced physics and physical geography in class VI of vernacular primary schools and in training colleges but failed to include geology due to absence of Marathi books on the subject. Modak strongly advocated inclusion of natural sciences in the curriculum on equal if not on higher footing with social sciences right from the primary level, in order to inculcate scientific temper and to accelerate Indian industrialization. He argued that the study of physical science was imperatively needed in India as physical arts were extremely backward. Indian public, devoid of scientific knowledge and untrained in scientific experiments ascribed everything to supernatural causes. Religion seconded by superstition had divided Indians into several castes and produced intellectual stagnation by discouraging the spirit of inquiry due to unbounded reverence for old institutions. Therefore, science based on empirical knowledge could alone dispel the darkness of ignorance by proving that the nature is governed by some fixed and invariable laws.

Thus, science would further the aim of education, i.e., intellectual advancement of the regenerated Indian mind and the development of physical arts. No branch of knowledge other than natural science had exerted so much decisive and beneficial influence on the progress of civilization in a very short period. If the claim of science was readily allowed in the

'Civilized European Continent', it must be allowed with still greater force in a country like India, which could not claim the credit of having made a single discovery in physical sciences and yet had a lot to learn from the western nations in that field.[28] It must be pointed out that Modak's above advocacy of the cause of science to promote scientific and rational spirit was as early as 1876, the very year when the well-known Bengali promoter of science for rationalism and national regeneration, Dr. Mahendralal Sircar established his Indian Association for the Cultivation of Science.

Dissemination of scientific information among Indian masses being main thrust of his writing, Modak sold his books at subsidized price, despite large expenditure incurred in printing stencilled figures. While serving his mission, Modak had to struggle often with discouragement from a section of educated people, who doubted utility, readership and circulation of scientific literature in Indian languages. However, Modak continued to publish science books with a missionary zeal, despite waning support from the DPI, Bombay and reduced chances of financial help from the princes of Kolhapur and Miraj during famine.

As Modak considered diffusion of knowledge, a solemn duty of educated people, he tried to persuade his scholarly friends to write on subjects of their interest in Marathi. When they harped on the financial problems involved in the work, Modak sighted examples of the medieval poets like Moropant, Waman Pandit, Mukteshwar and Tukaram, who composed poetry out of pure personal interest, and as a religious service. Modak was convinced that his books, owing to their social utility, like those of medieval Marathi poets, whose poetry survived even in the absence of printing technology would stand the test of time.[29]

Modak adopted spread of science as is life's mission with the conviction that India would lag behind the rest of the world without her progress in science. He urged the government, voluntary organizations and the Indian princes to lend active patronage to science education and scientific research. In 1904, he advised DPI, Bombay and the Mumbai Marathi Grantha Sangrahalaya to reward the best Marathi science book on the model of DPI, Bengal. In order to produce scientific literature in Indian languages, Modak appealed to Indians to emulate Russia and Japan by inviting foreign professors to teach science through foreign languages for the first five years and to teach and write books on science in Indian languages in the next five years. He urged to send Indian artists abroad for learning modern techniques of Industrial crafts. He requested Indian princes to follow the footsteps of Abasaheb Kagalkar, Balasaheb Mirajkar and Sayajirao Gaikwad in patronizing the movement for spread of science, as they alone had resources to finance expensive science equipment and research ventures yielding no immediate financial returns.[30] Modak awakened the princes of Southern Maratha Country to play their due part in the movement for spread of science. He revolutionized the hitherto theoretical, urban and elitist movement for spread of science by extending it to the princes and masses of the Southern Maratha Country through his innovative science exhibitions carrying demonstrative effects and added a new dimension to the movement for spread of science in Maharashtra. In this sense, he was the first active science propagator from Maharashtra spreading science at grass-root level.

Science and Swadeshi movement

Modak was actively associated with the Swadeshi shop at Kolhapur and criticized British rulers for letting dwindle many labour-intensive industries in India. For the revival of once-flourishing dyestuff manufacture and calico printing industry, with the help of modern techniques and ample Indian raw material, Modak wrote a series of articles entitled 'Ran Dene va Chhite Chhapane' in *Shilpkala Vijnan*, discussing in detail manufactures of blue, yellow and red dyestuffs from plants and bleaching techniques for cotton, jute, wool and silk, with guidelines for improving existing Indian techniques.[31]

In the nineteenth century, when the superior British finished goods, a product of European scientific inventions, resulting out of widespread science education in Modern European languages began to capture Indian markets, educated Indians, devoid of science education, precipitated the ruin of Indian industries. Preponderance of history, economics and logic in education produced a plethora of lawyers and politicians capable of only ventilating political and economic grievances under colonial rule through political associations. However, even after half a century of establishment of the Bombay University, the Bombay Presidency lacked botanists, zoologists, mechanics and metallurgists, electricians, researching in Indian flora and fauna, functioning and production of imported machines, application of electricity in daily life, respectively. Dearth of Indian metallurgists enabled the British to exploit Indian metal deposits.

Consistent attempts of Swadeshi Movement to regenerate Indian industries and create employment for ruined artisans by training them in the manufacture of match sticks, soaps, paper and holding exhibitions of indigenous products failed drastically. Without general diffusion of scientific knowledge through modern Indian languages, a precondition to original research in industrial techniques, Swadeshi Movement could not and would not achieve success in its major objective of reviving Indian industries – 'we regard the cloth manufactured in Indian mills as Indian cloth, however all the machines used for its production are imported'. Regeneration of Indian industries being Modak's main concern, he regarded the instruction in physics and chemistry, more urgent than that of biology, agriculture, botany and zoology.

According to Modak, education imparted in Indian languages could alone be termed as national education. He called the nineteenth-century-Indian universities as foreign, because they were a few in numbers and taught through foreign language, English. He pleaded for treating education in Indian languages as an issue of paramount importance on the agenda of Swadeshi Movement. He was critical of the first generation of English educated Indians, the intellectual leaders, for tolerating in India the education system suited to white colonies like South Africa. He also cudgelled the Indian National congress for not raising the demand for higher education in Indian languages along with its political demands. He urged to open one school per village, one high school per taluka and one college per district to persuade British Government to remove foreign character of the Government educational institutions. He pointed out that Indian backwardness in education arrested growth of science education and scientific research.

As the British were reluctant to diffuse higher level of science education in India, refrained from establishing research institutions and controlled universities through legal hurdles. Modak called for Indian initiative in national and science education. Modak taught chemistry voluntarily after retirement in Samarth Vidyalaya, a nationalist school at Kolhapur. He appealed to the Committee of National Education at Calcutta, to open National Primary School, a National High School and a National University in each Presidency teaching in modern Indian languages, assigning equal positions for science and mathematics in curriculum. According to Modak, National High Schools, with a curriculum of 4–5 years should teach elements of all branches of natural science along with *Bhagvad Gita*. Hindi should be one of the subjects rather than a medium of instruction in the non-Hindi-speaking regions of India, where medium of instruction ought to be a local language. In higher scientific instructions, students must be provided with science equipment and familiarized with science-based industries and English scientific terminology.[32]

The industrial survey of Kolhapur

Modak conducted an industrial survey of Kolhapur from 1892 to 1895. Inspired by the industrial and mineral wealth of the state and its potential for development, Modak presented a report on industrial conditions in Kolhapur and his scheme of fresh survey at the Poona Industrial Conference of 1892. For the survey, he collected information about the existing industries in the territory through village and district agencies by distributing a simple questionnaire and got the data checked through educational inspectors, headmasters and school masters. He travelled extensively for the survey on an average of 13 miles per day visiting even localities with natural mineral deposits and forests as well. He motivated even the illiterate people to cooperate by explaining the purpose of the survey.

After analysing the problems of Industries like cotton and blanket-weaving, tanning, parched rice making, oil processing, sugar refining, dyeing, lacquer work and bangle making, Modak proposed to grant loans to artisans, capital at a low interest rate of 6% and incentives to promote settlement of Shahapur weavers in Kolhapur, in order to train Kolhapur weavers. For marketing of the finished goods, he put forth the idea of central shops financed partly by Government and partly by people (in the form of shares) and managed by Mamlatdars, at large centres where weavers and artisans could buy their raw material and sell finished goods. Profits earned by the state in this venture could be utilized wholly or partly, for training of weavers and purchase of expensive looms. In order to improve the lot of sugar industry (ruined by imposition of British tax on Indian sugar exports and the discovery of beet sugar in Europe) he suggested provision of better seeds and establishment of a large sugar factory in Kolhapur based on the English or the Mauritian model.

His other suggestions were employment of trained potters from Calcutta and Jaipur, improvement in cattle breeding, cultivation of bark for colouring leather and that of bamboo, mulberry trees (for silk), coffee, tea, pepper and cardamom on the Mysore model. He proposed agro-based industries to generate rural employment, use of better agricultural implements, better irrigation and transport facilities and training of artisans. He advised the Kolhapur Government to export groundnut to Bombay, to use small hand presses for extracting linseed oil and castor oil, and to establish oil mills in Kolhapur. Citing success stories of other countries, he strongly pleaded for adopting modern agricultural and industrial machines like manually operated vacuum pan and small centrifugal machines for easier crystallization of sugar. He supported Mahadeo Govind Ranade's proposal of a bank on European model for development of agriculture and industry but did not wish to displace traditional creditors or Marwaris altogether. Modak opposed monopolies and favoured government support for development of industries. Modak was the first Indian, both to plan and execute an industrial survey.[33] Bal Gangadhar Tilak, in his review of the Industrial Survey Report of Kolhapur in Kesari, congratulated the Kolhapur Dewan for the industrial survey, a venture not attempted hitherto even by the British Government in India and appealed to him to honour Modak for his valuable contribution, in this regard.[34]

Modak, the historian

Modak developed interest in history when he was asked to teach the subject at Rajkumar High School, Kolhapur. His pioneering books on the history of Kolhapur (from first century AD to the first decade of the twentieth century), served to revise interest in Shivaji's descendants ruling in Kolhapur but forgotten after Peshwa's hegemony in the eighteenth century. Modak's pioneering attempt is compiling the history of Kolhapur was based on the historical sources such as *Bakhars*, *Vakas* (the personal daily accounts of Maratha nobles) and the material from Kolhapur archives, English books, articles and report in order to write history of Kolhapur. Modak toured south Maharashtra extensively for 8–10 years collecting source material and old legends. He appealed to the princes to furnish him with their family chronicles, documents and historical legends about the great men of their dynasty to correct inaccuracies. As the source material on the Patwardhans, who ruled over the Southern Maratha Country or the principalities of Kolhapur state was easily available to Professor Modak, he published the 4th volume on Kolhapur on Patwardhan history before the 2nd and the 3rd volume consisting of the history of the Brahamani and Bhosale rulers in Kolhapur.[35]

Modak wrote first detailed book in Marathi on Bahmani rulers. Consequently, justice M.G. Ranade requested Modak to contribute a chapter on Brahamani history in his book *The Rise Maratha Power*. Modak wrote the ancient history of Kolhapur with the help of inscriptions and English books and Bahmani history from English translations of Farishta's accounts and books of Elphinstone and Elliot. His main source for the history of the Adil Shahi was the Marathi translation of *Busatin-i-Saltim*. According to Modak, the knowledge of the Brahamani history was essential to trace the causes for the rise of Marathas and the socio-cultural transformations witnessed in the Deccan during the medieval period. He covered the history of Delhi sultanate and the Mughals in his textbook, *Muslim Rule in India*. Modak was perhaps the only Maharashtrian who wrote extensively on the Brahamani rulers and Mughals in Marathi.[36]

While compiling the Kolhapur history, Modak felt the need of chronological table in order to synchronize dates from various calendars. Improving upon the chronological tables, he found in the Kolhapur archives, Modak compiled a chronological table giving corresponding dates of Shaka, Rajashaka, Hizri, Sheehur, Fasli and Gregorian calendars, from 1723 AD to 1855 AD, in 1889, Modak's son Vishwanath published a revised version of Modak's chronological table after adding subsequent dates. His table was extremely useful to historians and lawyers; however, Balvant D. Apte later corrected a few errors in Modak's table regarding the Hizri and intercalary months.[37]

Far from a dynastic record, history was pursued by Modak as a mirror of changing social, economic, political and cultural life of mankind. In his first volume of Kolhapur history, he argued that the complete history of a period cannot be compiled without the information about trade, defence system, average expenditure of a family, extent of foreign contacts, means of transport and communication, exports, imports, amount of rainfall and nature of crops, extent of popular support for the king, status of various classes in politics and conflict among various religious groups. Modak's advocacy of a benevolent government is evident from the criticism of the Brahamani rulers for not helping citizens during the 12 years of famine. As a nationalist from justice M. G. Ranade's generation, Modak viewed history as an instrument of inculcating among the youth during the colonial period. However, his appreciations of Akbar and Muhammad Govan testify that he did not allow nationalism to degenerate into communalism. As a historian, Modak regretted the absence of historical sense among Indians, their indifference to European research on Indian history and the death of historical journals in Indian languages. He appealed to Indian historians to set up societies and publish history periodicals in Indian languages. He advised editors of various journals in Indian languages to compile books from the articles on history published in their journals. He appealed to Indians to set up historical societies, launch historical journals and publish books out of historical articles from journals. He advised Shastris to study

Sanskrit literature with modern analytical methods, modern Indian scholars to collect Persian books in order to study the medieval history and British Government to fund scholars' pursuits instead of just spending on defence and administration. He appealed to Muslims to save Persians from decline resulting from the loss of the state patronage.[38]

Place in the history of modern Maharashtra and India

Modak enjoys a significant place in the history of modern Maharashtra for being an 'avantgarde' in many respects. He expounded technical and scientific education in Indian languages, when primary education was not free and compulsory even in British India. He envisaged Marathi University (half a century before the establishment of Poona and Shivaji University, Kolhapur) when the sharpest Marathi intellectuals like Justice M.D. Ranade were agitating to introduce Marathi language as only as an optional subject in university curriculum. Using holistic approach, Modak considered socio-economic factors while interpreting history, more than 75 years before Marxian interpretation of history gained currency in India. He was the first Indian to conduct an industrial survey. By his preference for manufacture of capital goods with modern technology over the popular advocacy of production of consumer goods, Modak raised the fundamental issues in the process of Indian industrialization in the nineteenth century. By his penetrating grasp of an issue and application of scientific method, Modak broke new ground even in history and economics, the subjects not of his regular pursuit. However, out of his diverse activities, Modak's most remarkable contribution to mass education was his work of spreading science, undertaken as a life mission, with the conviction that India would lag behind other countries in the struggle for survival, in the absence of her progress in physical science.

Himself, a product of the nineteenth-century Maharashtrian Renaissance, Modak tried to shift its exclusive emphasis from social and political movements to spread of science. In fact, he was the major exponent of the translation movement of science literature in Maharashtra in the nineteenth century. In this sense, he extended Vishnushastri Chiplunkar's movement of de-anglicization to science literature. Devoid of research accomplishments in a particular branch of natural science unlike Bhau Daji, Sakharam Arjun and K. L. Chhatre, Modak, a hard-core science propagator, greatly enriched Marathi literature by his translations and original writings on almost all branches of natural and social sciences then taught in schools and colleges. His public lectures on physics and chemistry in Kolhapur in the 1870s coincided with K. L. Chhatre's astronomy lectures in Poona and Sakharam Arjun's lectures on gynaecology and general science in Bombay.

However, Modak's attempts to revolutionize the hitherto theoretical, urban and elitist movement for spread of science by extending it to the princes and masses of the southern Maratha country. His innovative science exhibitions carried demonstrative effects and added a new dimension to the movement for spread of science in Maharashtra. In this sense, he was the first active science propagator from Maharashtra (spreading science at grass-root level) and the predecessor to Marathi Vijnian Parisad and like-minded organizations. He awakened the princes of south Maharashtra to play their due part in the movement for spread of science. He was one of the few Maharashtrian intellectuals who emphasized upon the scientific context of the Swadeshi Movement. Although a committed nationalist, Modak had a universal view of the world of science as testified by his views on scientific terminology. Although no evidence is available so far, to trace Modak's direct contacts with non-Maharashtrian science propagators except T. K. Gajjar of Baroda, the striking similarities between his ideas and work with his counterparts in Bengal and elsewhere present him as an active Maharashtrian representative of the wider all-India scientific community of the period. He resembled Mahendralal Sircar in championing the cause of science for self-reliance by exposing British reluctance to promote science education and research in India and stressing the need for Indian initiative in this respect. However, unlike Sircar, who tried to bring together scientific community of Bengal through his Indian Association for Cultivation of Science (IACS), Modak's individual efforts in Kolhapur and the vicinity evoked hardly any repercussions in the scientific circles of Bombay and Poona. He was a loner, without followers to perpetuate his movement.

Modak's well-organized annual science exhibitions attracting more than 10,000 visitors per day had no parallels anywhere in India in the nineteenth century. Just as Syed Ahmed Khan got financial support from the landlords of North Western Provinces, Modak drew patronage from the princely states of Kolhapur and Miraj. In his advocacy of Indian languages as medium of instruction, the Baroda School of T. K. Gajjar directly inspired him. Modak was however, opposed to the Banaras School, which favoured Hindi as medium of instruction throughout the country. While championing the cause of science education, Modak's shift of emphasis from promotion of scientific attitude in 1876 to consolidation Swadeshi Movement in 1906, reflected the changing politico-intellectual climate in India. Like Pramath Nath Bose, a geologist (who wrote *A History of Hindu Civilization*), and Prafulla Chandra Ray, a chemist (who wrote *History of Hindu Chemistry*), Modak, a science teacher, compiled history of Kolhapur to educate young generation about past Maratha glory. Like P. N. Bose, Modak desired marriage of modern science with Indian cultural values. Like the Bengali supporters of Swadeshi Movement, Modak looked upon Japan as a role model for Indian modernization and advocated higher scientific training of Indians abroad.

Despite his penetrating vision and outstanding contribution to the Renaissance in Maharashtra, Modak is unknown to the present scholars of modern Maharashtra for several reasons. He played no active part in the social and political movement, the major concern of the historical research on modern Maharashtra. The main field of his activities was Kolhapur, the capital of a princely state, away from Bombay and Poona, the centres of modernization in the Marathi-speaking regions in the nineteenth century. More of a science propagator than a research scientist, Modak was forgotten even by the scientific community. Bereft of any scientific inventions to his credit and a group of dedicated followers, Modak failed to leave a long-lasting movement after him. The movement for science education through Indian languages, the mission of Modak's life received a blow, with the failure of the translation movement and its sole purpose in equipping Indian languages with the most modern scientific knowledge.[39]

Notes

1 Pramod Oak, *Elphinstone*, Pune: Rajhans, 1990, pp. 190–209; also see J. V. Naik 'Marathi Madhyamacha Puraskarta Captain George Resto Jervis', *Lokrang Loksatta*, April 27 (1997) pp. 3–7.
2 Mrudula Ramanna, *Western Medicine and Public Health*, New Delhi: Orient Longman, 2002, pp. 16, 17, 20, 21, 24, 25, 36, 37, 45, 46, 137, 140, 150, 167, 150, 190, 195, 208, 212.
3 G. N. Sahasrabudhe, 'Marathi, Shastriy Vangmayacha Itihas', *Vividha Jnan Visitar*, Vol. 39, No.4 (1908) p.p. 143–146.
4 Niranjan Ghate, 'Marathi Science Recorded from 1818 to 1997', *Times of India*, Pune edition.
5 J. V. Naik, 'Bhau Mahajan and his Prabhakar, Dhumketu and Dnyaandarshan Maharashtrian response to British rule', *IHR*, Vol. 13 Nos. 1 and 2,1986 and Jan 1987, pp. 142–152.
6 K.L. Chhatre, *Parjanya Vishyak Vyakhyana*, Pune: Jnan Prakash Press, 1878, pp. 14–21.
7 Dr. Bhau, Daji Lad, *Vyakti Kal Va Kartutva*, Bombay: Mumbai Marathi Sahitya Sangh, First Edition, 1971, pp. 28–36, 273–284, 315–318.
8 Mohini Varde, *Dr. Rakhmabai, Ek Arta*, Bombay: Popular Prakashan, 1984, pp. 7–14.
9 Bhalchandra Krishna Bhatavadekar, *Arya Vaidyak ani Paschmatya Vaidyak*, Bombay: Nirnay Sagar Press, 1888, pp. 6–43. This was originally Bhatavadekar's lecture delivered at the Hindu Union Club for Hemant Vyakhyanmala, year 3, lecture 7.
10 Anjali Kirtane, *Dr. Anandibai Joshi Kal Ani Kartrutva*, Mumbai: Majestic Prakashan, 1997, pp. 222–346.
11 Deepak Kumar, *Science and the Raj*, New Delhi: OUP, 1995, pp. 114–115, 143–150, 192–194, 196–197, 218-, 228–229, 229–230, 236–238.
12 Niranjana Ghate, op. cit., p.
13 A. K. Priyolkar (ed.), *Personal Diary of K. L. Chhatre*, published in *Marathi Sanshodhan Patrika*, not dated, pp. 1–3.
14 Moro Hari Khare, *Modak Kulvrittant*, Poona: Adarsh Publication, 1946, pp. 6–19, 356–358.
15 G.D. Khanolkar, *Arvaachin Marathi Vanmay Sevak*, Vol VI, Poona: Venus Publication, 1963, p 28.
16 Ganesh Balwant Modak, *Prof. Balaji Prabhakar Modak*, Kolhapur: Dnyaan Sagar Press, 1931, pp. 2, 4–5, 32–35.
17 G. B. Modak, *Prof. Balaji Prabhakar Modak*, op. cit. p. 35.
18 G. D. Khanolkar, *Aravacin Marathi Vanmay Sevak*, op. cit., pp. 28–29.
19 Vishnu Shastri Chiplunkar, *Nibhandhamala*, Part I, pp. 466–470.
20 Bhausaheb Gupte, 'Rasayan Shastra, Pustak Pariksha', *Vividh Jnan Vistar*, August – October, 1899, pp. 184–191.

21 B. P. Modak, *Yantra Shastra Purvardha*, op. cit., pp. 6–8; also see B. P. Modak, *Padarth Varnan*, Part II, (*Udbhijja V Pranija Padartha*), op. cit., p. 2.

22 G. N. Sahasrabuddhe, 'Marathi Shastra Vangmayacha Itihas', op. cit., p. 149; also see G. D. Khanolkar, *Arvachin Marathi Vangmay Sevak*, Vol. VI, Pune: Venus Prakashan, 1967, pp. 34–35; also see B. P. Modak, *Sendriya Rasayan Shastra*, Vol. 1 Part I, Preface pp. 1–2, 6–9.

23 B. P. Modak, *Arvachin Rasayan Shastra*, op. cit., Preface p. 1.

24 Narayan Kanolkar, *Raghunath Panditacha Adhunik Avatar*, Pune: Arya Sanskriti Mudranalaya, 1949, pp. 4–7.

25 Govind S. Tembe, *Maza Sangitacha Vyasang*, Kirloskar, Kirloskar Wadi, pp. 36–38; also see G. B. Modak, *Balaji Prabhakar Modak*, op. cit., pp. 14–17, 21–23–31; also see 'Kolhapur Yeti Rajaram Collegatil Sastriya Yantranche Chote Pradarshan', *Shilpkala Vijnan*, Vol. 4, op. cit., pp. 1–15. Also see 'Kolhapur Yetil Sastriya Pradarshan', *Karmarnuk* (January 1897), Vol. VII, pp. 2, 66–67, An Interview with Purushottam Modak.

26 N. C. Kelkar 'Gatgoshti', *Rajaraman*, Rajaram College, Kolhapur, Maharashtra, 1956.

27 G. B. Modak, *Prof. Balaji Prabhakar Modak*, op. cit. pp. 28–33.

28 B. P. Modak, *Rasayan Shastra Purvardha*, Poona: Jnan Prakash 1876, Preface to the 1st Edition Marathi, pp. 1–2.

29 B.P Modak, *Marathi Bhashechya Abhivrydhi Karita Marathi Bhashechyadvare Shikshan Denyachi Avashakyata*, op. cit., p.17; also see B.P. Modak, *Rasayan Shastra Uttarardha*, op. cit., Preface, p.1; also see B.P. Modak, *Sendriya Rasayan Shastra*, Vol.1 Part I, Preface, pp. 1–2, 6–9; also see K.D. Naniwadekar, 'Pochva Abhipray', *Rajaramian*, Vol.19, Issue 3, 1932, p.96.

30 B. P. Modak, '*Marathi Bhashechyaa Abhivriddhikarita Marathi Bhashechyadvare Uchh Shikshan Denysachi Avashyakata*', an address to Mumbai Marathi Grantha Sangrahalaya, September 25, 1994, Bombay: Jhansagar press, 1905, pp. 22.

31 B. P. Modak, 'Kapade Va Sut Svach Karne', 'Rang Dene Va Chhite Chapane', *Shilpkala Vijnan*, Vol. 1, No. 9, 1888.

32 B. P. Modak, *Sendriya Rasayan Shastra*, Vol. I., Part II, Mumbai: Nirnay Sagar, 1906, Preface, op. cit., pp. 9–14.

33 B. P. Modak, *Report on the Industrial Survey of the Kolhapur Territory*, Byculla: Education Society Press, pp. 5–8, 10, 11–27, 33–34, 38, 47–50, 50, 52, 62, 71–101, 103–112, 116, 121, 126–130.

34 'Kolhapur Ilakhyachya Audyogik Pahanicha Report', *Kesari* (25th September 1895, 11.10.1895, p. col 3, 1st October 1895 on Microfilm)

35 B. P. Modak, *Kolhapur va Karnatak Prantatil Rajye va Sanstane Yancha Itihas*, Vol II, Part II, Belguam: Belgaum Samachar Press, 1880, Preface, pp. 2–3, p 132; also see S. M. Garge, *Karvir, Riyasat, Karvir Chatra pati Gharanyacha Itihas*, Poona: Shree Shahaji Chhatrapati Museum, 1968, p. 8.

36 B. P. Modak, *Kolhapur va Karnatak Prantatil Rajye va Sansthane Yancha Itihas*, Vol I, Part I, Belgaum: Belgaum Samachar Press, 1877, Preface, pp. 1–4, 111, 173, 194, 214–218, 102–131–135–149, 200–256.

37 B. P. Modak, *Shaka va San Vanchi Tithi va Tarikhvar Jantri*, Poona: Chitrashala Prakashan, 1889, Preface, pp. 1–3.

38 B. P. Modak, *Dakshinetil Musalmani Rajyancha Itihas*, Part I, Poona: Chitrashala Press, 1891, Preface, p. 2; also see B.P. Modak, *Hindusthanatil Musalmani Rajyancha Sansskhipta Itihas*, Bombay: Nirnay Sagar, 1818, p 10; also see B. P. Modak, *Dakshinetil Musalmani Rajyancha Itihas*, Part III, Poona: Chitrashala Press, 1891, op. cit., Preface, pp. 24.

39 Abhidha S. Dhumatkar, 'Balaji Prabhakar Modak and Spread of Science in Maharashtra in the Nineteenth Century', unpublished M.Phil. Dissertation, University of Mumbai, 1998, pp. 207–213; also see Abhidha S. Dhumatkar, 'Vijnaan Prasaarak Balaji Prabhakar Modak-Ek Ashtapailu Vyaktimatva', *Lokarang Loksatta*, 6 December, 1998, p. 5.

4 Cultural politics of engagement

Kerala Sastra Sahitya Parishad and the shaping of a scientific-citizen public in Kerala

Shiju Sam Varughese

Introduction

In this chapter, my attempt is to retell the story of Kerala Sastra Sahitya Parishad (KSSP) from a new vantage point. The social movement, which is often praised as one of the largest and foremost People's Science Movements (PSMs) in Asia, is discussed in scholarly literature as a unique experience of developing a people-oriented science that has the potential to catalyse social revolution. KSSP in Kerala was, according to the literature, transforming the society into a progressive and scientifically tempered one with its development into a mass movement in the late 1970s from an earlier phase of being a science writers' forum. This growth of the movement, it was pointed out, happened when the educated middle classes joined it during the successful campaign against the construction of a dam project in Silent Valley, one of the dense tropical forests in northern Kerala with great biodiversity. KSSP is seen largely as catalysed the unique developmental achievements of Kerala and epitomising the region's characteristic leftist political culture. The movement presented by most of these accounts emerges from the unique features of the modernity of the region, which has its roots in the 'Kerala Renaissance' and the anti-feudalism struggles and the reformist movements of the nineteenth and early twentieth century.[1] Often in these historical narratives, the uniqueness of the 'Kerala model of development' is tightly coupled with the KSSP's successful existence.

A quick glance at the scholarly literature on KSSP and its umbilical connection with the developmental modernity of Kerala[2] indicates that there is a wide appreciation of the accomplishments of the movement in promoting scientific temper and catalysing a people-oriented science in a post-colony. In the present chapter, my attempt is to critically assess this claim by situating KSSP's emergence in the 1960s and the 1970s within the history of the debate on social relations of science in the twentieth century and its influences on Indian science after independence. The ideological orientations of KSSP are examined against this background by decoupling it from the hagiographical accounts to suggest that the movement and PSMs in general were contributing to the shaping of a specific category of citizens who

DOI: 10.4324/9781003241980-6

were appreciative of the politico-epistemological contract between the state and technoscience in the country. The *scientific-citizen public* formed in this way through the PSMs, the chapter argues, should be contrasted with the wider population of India who formed a *quasi-publics* that belonged to the political society existing at the margins of the civil society.[3] I will argue in this chapter that the emergence of KSSP as a successful PSM in Kerala was closely linked to the constitution of a specific kind of scientific-citizen publics and the transformations in their engagement with the question of social function of science.

Emergence of science activism in Europe

The transformations of science around World Wars I and II from an earlier 'little science' to a 'big science' created an atmosphere of great enthusiasm in and appreciation of science as the vanguard of progress and development. This euphoria about the capabilities of science to solve the problems of the world emanated as part of a particular ideological formation, where science became part of government. Science since the 1920s became increasingly seen as with the potential to solve societal problems and it became established that national democracies could benefit from it. This has led to the development of science eventually as closely coupled with the developmental agenda of the nation-states all over the world in the post-World-War period.

This consensus between nation states and scientific communities about harnessing the epistemological potentials of science for national development was but achieved through extensive deliberation on the role of science in society. This debate was initiated by the forums of scientists in the early twentieth century, particularly in Britain. In the 1930s, British scientists were facing political discrimination; they had been excluded from participation in the political process as experts for the politicians and policy makers did not show any interest in the social impacts of scientific research.[4] Therefore, it became a serious concern for the British scientific community to raise their social status by assuring their social role as experts, and the 'Social Relations of Science' (SRS) movement became the medium for expressing these concerns. The movement was active among the British scientists between 1932 and 1945.[5] However, this does not mean that the SRS movement was monolithic or cohesive. Ideologically there were two prominent groups within, which worked through different organisations—the reformers and the radicals.[6]

The radicals believed that 'only a society transformed along socialist lines would be prepared to make the fullest and most humane use of scientists and their discoveries'.[7] The radicals were inclined to Marxism as a political ideology. They felt that Marxism offers a methodology to connect science with society, a position that found its forceful expression in John Desmond Bernal's book, *The Social Function of Science* (1939). The political background of the threat of War, fascism and economic depression deeply influenced the radicals' views and they looked towards the Soviet Union as the socialist hope of the world. Based on the experiences of the Soviet society, the radicals argued for the alliance of scientific community with those political forces who were most committed to the advancement of science for the benefit of the world. They exemplified the Russian scientific community and contrasted it with the scientists in Nazi Germany.[8] J.D. Bernal, J.B.S. Haldane and others

> repeatedly emphasized the superiority of Soviet scientific organisation, the scientific ethos of Russia's leaders, and the comparatively high status accorded to scientists in Russian society. Above all, they stressed the way in which scientific resources were devoted to the solution of important economic and social problems.[9]

The Social Function of Science that epitomised the radicals' standpoint argued that science is for everybody and it has a social function to perform, which gradually became the foundation of science policy frameworks in many countries including India.[10] Bernal argued for the need of a 'science of science', which can analyse how political economy influences the orientation of science. If used in a planned way, science could improve the life of the people. And the insights for this planning, he suggested, can be offered by Marxism as a methodology. According to Maurice Goldsmith,

> Bernal was the embodiment of the socially responsible scientist, a fine product of the immediate after-years of the Russian Revolution, when the new Soviet Marxism captured Europe's intellectuals. In his life, he sought to show how atheist ethics and socialist morality could be combined in a libertarian rational humanism. The dominant idea that inspired him was a belief in the possibility to achieve human perfection with reason. He believed that science should only be neutral ethically, but that scientists themselves should be committed ethically.[11]

The radicals' position was about the need for dissociating science from capitalism, to make it a vehicle for socialist transformation. Bernal suggested that only in a socialist system that science can actualise its full potential, and the Soviet society was the best example for the same.

The reformists differed. They feared that any linkage of science with politics and government would finally throttle their own intellectual freedom and professional autonomy. They were sceptical about the systematic social control of their profession, and pointed out that there was a total integration of scientists into the political systems of both Nazi Germany and Stalinist Soviet Union, which ultimately became detrimental to the growth of science.[12] The reformists defended a more open, internationalist system of science, which is not limited by nation-states. They believed that 'the fervent of nationalism which informed scientists' attitude in the two countries was opposed to the values of an international scientific community'.[13] While the radicals underscored the experts' role in public policy making (science for policy), the reformist asserted that scientific community cannot afford such a call, as science is purely a cognitive activity which has nothing to do with politics.

As mentioned earlier, these groups were working through separate organisations before 1938. Reformists actively participated in the British Association for the Advancement of Science and the British Science Guild. On the other hand, the radicals involved not only in the Association of Scientific Workers and the Cambridge Scientists' Anti-War Group,[14] but also in Labour and Communist Parties. Nevertheless amidst such intense political disagreements both the groups came together in 1938 to form the Division for the Social and International Relations of Science in the British Association in order to study 'the effects of the advances in science on the well-being of the community, and, reciprocally, the effects of social conditions upon advances in science'.[15] Beyond their ideological differences, they continued working together until 1945. What unified them was the nationalist spirit that hiked during the Second World War—their shared concern was the effective utilisation of scientific expertise for the victory of Britain in the Second World War.[16]

There were also other groups. In 1941, John R. Baker and Michael Polanyi launched the Society for Freedom in Science, among others, to provide a 'liberal' alternative to the radicals' standpoint.[17] They were more inclined to the ideas of the reformists, though occasionally attacked them as well. The radicals were strongly criticised for their uncritical admiration of Soviet Communism and the Stalinist Regime.[18] After the demise of the Division for the Social and International Relations of Science (where the radicals and reformists worked together), in the 1950s, there was a burgeoning of several scientific organisations, in which the World Federation of Scientific Workers, Science for Peace and the Campaign for Nuclear Disarmament were the prominent ones.[19] The Pugwash Movement had been formed in the wake of a declaration by Albert Einstein, Bertrand Russell and eight other scientists in July 1955 condemning the development of hydrogen bomb and the nuclear fallout from the Bikini Test in March 1954.[20] The main objective of the movement was to foster friendship and understanding among international scientific community in order to facilitate peace and disarmament.[21]

All these groups had different standpoints on the science–society interface but agreed on a single point—science for them was a cognitive process untouched by social impurities. They perceived ideologies and economic processes of the times as affecting science only externally, which broadly shaped the orientation of science. Their contestations were around the degree of this social influence that made the external shell in which the internal cognitive kernel of science is embedded. None of these groups believed in the radical possibility of understanding science epistemologically as purely a social practice, a position that started emerging in the 1970s with the advent of new scholarship and new social movements that challenged the epistemological legitimacy of modern science.

The optimism on the potentials of modern science however started collapsing in the 1960s in the West due to the new advancements in the chemical, biological and nuclear warfare, ecological crisis and the 'unforeseen social and economic consequences of pursuing industrial automation for its own sake'.[22] Among many groups that came to fore in response to the changed image of Science and Technology (S&T), the British Society for Social Responsibility in Science (BSSRS) was the prominent one.[23] The New Radical Science Movement of the 1968 broke away with the old movement on ideological grounds. There were different standpoints on science held by various groups within the new movement. A moderate position held by Jerome Ravetz, among others, criticised the old movement for its attempt to extend the scientific method to all spheres of life, for overlooking the multiple effects of S&T to highlight only the positive aspects and for their ecological insensitivity. Propagation of an idealist picture of science by the radicals also was critiqued. The second group was more radical, and their position was expressed by Hilary Rose and Steven Rose. They pointed out that the question of the abuse of science is deeply rooted in the contradictions within science itself and that science is a non-neutral, ideology-laden activity.[24] They anticipated a self-managed, autonomous science, which is a collective enterprise, a 'science for the people'. On the contrary, those like Bob Young and David Dickson, the ultra-leftists, proposed that science is not part of the economic base of society, but that of the ideological superstructure, and challenged the earlier argument of scientific knowledge as truth and technology as a tool.[25] Following these contentions, the movement split apart into so many issue-based movements in the 1970s (ibid). The emergence of second-wave feminism, Afro-American thought and third-world standpoints along with different strands of ecological thought had very strong influence on the science movement's rethinking on the epistemology and social relations of science.

Cultural politics of public engagement in India

The debate on science and society in India has close linkages with the European deliberations that we have discussed above. The Rationalist Movement as well as the PSMs emerged in response to the wider discursive transformations in the post-War period, although there had been intellectual organisations in colonial India, which propagated science and its values.[26] While the colonial organisations worked towards cultural reception of modern science by translating the idiom of science into vernacular languages, the late-twentieth-century science movements were different in their ideological inclinations. They were influenced more by the Bernalist–Nehruvian ideology of science,[27] which was a culturally specific nationalist perspective that developed in the post-independent India.

The nascent nation-state of India under the leadership of Jawaharlal Nehru as the prime minister had a great optimism on the powers of modern S&T to help the nation to progress in the path of modernisation. The foundation of this belief was laid by the intellectual collaboration between British radicals like J.D. Bernal. There was already a strong debate on social relations of science available in the first half of the twentieth century in India that found strong contestations about the role of western science in national reconstruction. The critique of western science and its values, which was identified of late as the 'Gandhian critique',[28] gradually eclipsed under the industrialisation paradigm proposed by the 'Nehruvians'.[29] The Nehruvian point of view was radically shaped by the British intellectuals' perspectives on social relations of science and many of them were frequent visitors to India and had strong friendship with Nehru and his team of scientists and policy makers.

There were two major influences on the development of the period that created a national S&T system in India. These were A.V. Hill[30] and J.D. Bernal who offered scientific advice to the nascent nation-state. While the former's influence was more in the capacity of an official commission to assess the state of scientific research in India in 1944, the latter's influence was long-lasting as a policy advisor and theorist of science–society interface.[31] Jawaharlal Nehru had read *The Social Function of Science* while he was in jail. Between 1947 and 1955, Bernal visited India several times.[32] He was present at the inauguration of National Chemical Laboratory (Pune) and National Physical Laboratory (New Delhi),[33] and had addressed the scientific community of India at the annual meeting of the Indian Academy of Science in December 1949.[34] In 1954, he was again invited by S.S. Bhatnagar, the then minster of science to review the country's scientific institution building process in detail.[35] His book was well accepted in India as providing guidelines to developing an Indian S&T system and the formulation of the Scientific Policy Resolution of 1958.

By the beginning of the nation state, and especially with the second five-year plan known as the 'Nehru–Mahalanobis plan', the Nehruvian paradigm of development was established, based on the twin imperatives of industrialisation and national security. The Bernalist view of social function of science was its corner stone. According to Dhruv Raina, the Nehruvian paradigm consisted of three elements; 'centrality was accorded to a particular kind of knowledge, a well-defined strategy for its deployment in social transformation, and a path of action: these being modern science, planning and industrialization, respectively'.[36] As many scholars have pointed out, science became a reason of the state at this juncture; it provided epistemological support to the nationalist agenda of development by assisting the creation of large scale technoscientific projects as well as the national security programme. It provided epistemological legitimacy to the nation-state. In return, science was insulated from public exposure and social audit, and the nation-state assigned a special status to science.[37] In this sense, there was a contract that came into being between science and the state, from which both the partners benefitted. Indian nation-state was hence formed as a *state-technoscientific duo* on the basis of the contract, an important aspect of the national sovereign which is less studied in the Indian context.

The governmental operations of the duo were through the *technoscientific complex* that was built to ensure national security and to shore up the developmentalist agenda:

> The industrial and nuclear imperatives of establishing the S&T system within the frame of big science in the 1950s and 1960s thus discriminated against university-based academic science in favour of the state-centric technoscientific complex. Economic development and national security were the dual objectives of that complex. 'Self-reliance' was used as a rhetorical claim to garner public support for large technoscientific projects (both civilian and military) developed by the state with the help of the technoscientific complex. The coming of age of the technoscientific complex in the 1960s further marginalised the ill-developed academic research system and its potential to develop human resources through research-based education. As mentioned earlier, an elite group of scientists, engineers, technocrats and policy makers represented the technoscientific complex, frequently disguising technoscience as academic science. They worked in close association with the country's political elite to actualise the state's developmentalist dreams. Hi-tech projects were promoted as peak achievements on India's path of rapid modernization.[38]

The politico-epistemological contract between science and the state found a large number of admirers, who wanted to participate in the nation building process by employing their technical expertise as scientists and engineers. These admirers also included educated citizens who were appreciative of the prowess of science to solve the societal problems and the ability of the state-technoscience duo to deliver it through the operations of the technoscientific complex. This group of elite experts and citizens formed the scientific-citizen public of postcolonial India. The state-technoscience duo wanted the elite citizenry who was appreciative of the statist ideology of progressivism to transform the wider population that was scientifically illiterate and uneducated—the 'ignorant masses' who had to be governed through welfare mechanisms and developmentalist programmes. However, since its inception, the duo's governmental operations were made possible by this very split between the elite scientific-citizen publics and the quasi-publics constituted by the population groups. These population groups had to be educated to be scientifically tempered, so that the nation can progress. The responsibility to educate and deliver the state's welfare programmes to the quasi-publics was attributed to the scientific-citizen publics.[39]

The Nehruvian–Bernalist ideology provided an overarching frame to the scientific-citizen publics in the 1960s and the 1970s to perform their duties towards the society at large as responsible citizens who believed in the transformative potential of modern science. As in the context of Britain, here too, the socialist experiments of the Soviet Union (USSR) under the communist regime became a constant reference point. Nehru himself was very much excited about the progress of the USSR than the capitalist system of the USA. The origin of KSSP in the 1960s and the mushrooming of PSMs thereafter were unique manifestations within this discursive field of politics of science in India.

KSSP and the constitution of the scientific-citizen publics

The origin of the movement was in the form of three independent initiatives. The first was a science writers' organisation called *Sastra Sahitya Samithi* (Science Literary Forum) in 1957 at Ottappalam in Palakkadu district, following the organisation of a special section of the annual meeting of the *Sahithya Samiti* (Literary Forum) to discuss science literature.[40] Their emphasis was on publishing science literature in Malayalam for the general public.[41] They planned a tri-monthly publication called *Adhunika Sastraam* (modern science) modelled on the *Penguin Science News*, and decided to translate *The Origin of Species* into Malayalam. However, they could not sustain their activities for long.[42] In 1962, another group assembled in Kozhikode and organised a forum for science writers, named Kerala Sastra Sahitya Parishad. Some of the persons involved in the first initiative were also part of this new venture. While the first initiative was under the leadership of a group of left-inclined writers and activists, the second one was largely initiated by practicing scientists under the leadership of Dr. K.G. Adiyodi, a renowned zoologist. Since the late 1950s, Dr. Adiyodi had shared his concerns about the need for such an organisation to propagate science.[43] This was largely envisioned as a popular science writer's forum in the wake of a meeting organised at Kozhikode by him on April 8, 1962.[44] Initially, there were around 30 science writers as members. The organisation was officially begun with its inauguration at Kozhikode on September 10, 1962. Its objectives were publication of popular science in Malayalam and organisation of public lectures, seminars and film shows towards the cultivation of enthusiasm and awareness of science among the masses.[45]

The third initiative was by a group of scientists in Bombay who were actively interested in writing popular science in their respective mother tongues. Many of them were scientists trained in the Moscow Power Institute of the USSR. As Dr. M.P. Parameswaran, who was one among them,[46] pointed out, discussion on how to contribute to science writing in their own mother tongues had already been started among these scientists during their Russian days itself. After coming back to India in January 1966 as a scientist at the Bhabha Atomic Research Centre (BARC), Parameswaran took the initiative to start organisations for the propagation of science in different Indian languages: Sastra Sahitya Parishad (Malayalam), Bombay, was one among them.[47] A federation of all these organisations called the Federation of Indian Language Science Associations (FILSA) was also launched. Before his initiatives, Parameswaran had contacted the organisers of KSSP in Kerala such as Dr. Adiyodi and P.T. Bhaskara Panikkar and since its inception the Bombay unit was in constant touch with KSSP.

KSSP's origin(s) and development as an organisation to enhance science writing in Malayalam language however reflected the ideological differences among them. There was a common concern about the scientists' responsibility to connect science with society, which got transpired as a commitment to develop popular science literature in Malayalam. This concern was expressed in the words of Dr. Adiyodi:

> We are living in an age of science... The general laws of science and the discoveries of science, that exert such great influence on the development of mankind, should remain as the family property of certain experts. Either these experts themselves or some other people have to shoulder the responsibility of explaining them to the common man in a language that he can understand. Because of the step motherly attitude to the 'native languages' during the period of British rule, there was hardly any progress of science literature in various Indian languages... On the other hand, the conservative local 'sastries' insisted that we have a science created by the great sages of the past that cannot contain any mistakes. Consequently, the awareness of the people of our country lagged behind that of others. Only regional languages can enter into a dialogue with the heart of the common man. The task before the science writers in Malayalam as well as in other languages is to convey the message of the new knowledge to the hearts of the people in a style they can easily understand.[48]

The main inspiration to begin with science writing activities in Malayalam hence was the spreading of scientific values and perspectives, although the word 'scientific temper' was yet to be used by the organisation as the term to explain it.[49] The argument was about the social responsibility of scientists (experts) to explain their esoteric knowledge to common masses in a language and idiom they can easily comprehend. While this remained as the shared view, there were serious conflicts regarding the vocation of scientists with reference to the social relations of science. As mentioned earlier, this debate was largely set within the discursive frame created by the British discourse on science–society interface and the Nehruvian nationalist–socialist concerns. It is also important to note that at least three different groups of intellectuals were struck by the idea of propagating the method and values of science to the masses in Kerala by the late 1950s. It seems that there were several such attempts, although we cannot state it conclusively due to dearth of evidence.

Since the inception of the organisation, a large number of KSSP's members were of socialist leanings, and the interventions of the Bombay group accentuated the development of a perspective of social function of science as more than popular science writing. The ideological differences had manifested in the inaugural meeting of KSSP on September 10, 1962 at Kozhikode itself—there was an opposition to the decision of some organisers to invite Dr.Humayun Kabir, who was a renowned educationist close to Jawaharlal Nehru. It was suggested that Dr. J.B.S. Haldane, who was active in the SRS movement in Britain in the 1930s, should be a more apt person to inaugurate the organisation.[50] It seems that Haldane himself was involved in the controversy when he wrote to Dr. Adiyodi about the decision to call upon Kabir to the inaugural function.[51] This clearly indicates that there were reformers and radicals within the organisation, although these divisions might not be strictly in line with the divisions in the British movement.[52] However, the gradual development of the debate in the organisation, especially after the involvement of the Bombay group, was predominantly between those who opted for a reformist orientation of propagating science to the wider population and a radical group that wanted to combine this objective with the Bernalist position of using science as a weapon for social transformation by entering into other domains of social activism. The second group eventually became dominant in the organisation, which led to the transformation of the organisation into a vibrant social movement in the late 1970s.

The transformation into a social movement with specific ideological orientation was a gradual one.[53] Since the third annual meeting of KSSP in Olavakkodu in 1966, the possibilities of employing modern science for the well-being of the society was pondered over in a more systematic way. The first editorial of *Sasthragathi* (October 1966), a popular science tri-monthly magazine started by the KSSP followed by a decision in the third annual meeting reflects the debate on science–society interface. The editorial identified the gap between the common people and scientists in India as there is neither a sound tradition of science nor a 'scientific mode of thinking'. Therefore, the editorial declared 'science to common masses for a scientific revolution' as the organisation's objective.[54]

In the next few years, the organisation continued with its moderate agenda of propagation of science to the wider population. It has started magazines for schoolchildren,[55] and participated in the wider debate on creating technical words in Malayalam, with an aim to transform the language into one with the potential to hold scientific ideas. This inspired the publication of scientific and technical terms in Malayalam, the result of a creative collaboration between KSSP and the State Language Institute.[56] Several seminars and public lectures also were organised to propagate science. In 1971, Parishad launched its own publishing wing called STEPS (Scientific, Technical and Educational Publishing Co-operative Society).[57] With its children's magazines, KSSP also actively focused on schools as a site for public awareness of science campaigns and the major turning point on this regard was the campaign in 1973 to form science clubs in schools.[58] Public campaigns for science popularisation also were organised in the early 1970s.

These activities suggest that the organisation was concentrating on propagation of science among the public at large in the early years of the 1970s. The internal debate between different positions on the social relations of science seems to have continued during this phase, although there was no upper hand for the radical position regarding harnessing science for social transformation. At the same time, KSSP turned to be the embodiment of the

Nehruvian nationalist spirit of the times, believing in linking science with nation building, and the unique role the scientists, engineers and educated elite to play to support the nationalist programme. The activities organised by the KSSP in the 1960s and the early 1970s were oriented towards these goals. The members of the movement during the period were mostly scientists, science writers and school and college teachers. The membership strength of the organisation was minimal; in 1969, it consisted of approximately 500 members.[59]

It may appear ironical that KSSP accepted the Nehruvian–Bernalist standpoint as its official ideology by the mid-1970s, which was a period that indicated an eroding faith in the Nehruvian developmentalism. It was in December 1973 at the annual meeting in Thiruvananthapuram that KSSP accepted 'science for social revolution' as its official standpoint. It was also in the same meeting that the organisation reshaped its understanding of the general public more in tune with Marxist perspective—the public was redefined as lower economic classes who were engaged in the productive activities in society, especially the rural population and the factory workers. This was in sharp contrast to the yester years' view of the wider public as scientifically ignorant masses. The science in the everyday activities of the Indian working class had to be acknowledged and linked with the formal activity of modern science, KSSP realised.[60] This has culminated in a new form of organisational activity; Grama Sastra Samithys (Village Science Forums) were formed since 1973 for linking the knowledge potential of rural India with formal science, to transform society. In 1975, the first activists' camp at Peechi decided to accelerate the creation of Grama Sastra Samithys and by 1978, approximately 600 of them were active in Kerala.[61] The formulation of the slogan, 'science for social revolution' was also under the influence of a group of scientists with ultra-leftist leanings, Dr. Parameswaran points out.[62] In the Peechi camp, the development of Kerala became a serious topic for contemplation; within the emergent framework of harnessing science for social revolution, development of Kerala could not have been elapsed the attention of the organisation in the background of the emergence of development discourse in Kerala. As an aftermath of the deliberations in the camp, Dr. K.G. Adiyodi resigned from the movement, arguing that scientists must keep a distance from 'society' to safeguard science from politics.[63] This was also the year Dr. Parameswaran left his job and became a full-time activist of the organisation. The resignation of Adiyodi was the end of the reformist line of thought in the organisation and the dominance of radicalism represented by Marxist intellectuals like Parameswaran.

The new ideological emphasis of the organisation became stronger in the consecutive years with a series of programmes. KSSP held a seminar on industrialisation as part of its 13th annual meeting in Kannur in 1976. A second activists' meeting was organised in Idukki in the same year, where it was decided to collect data regarding the natural resources of Kerala.[64] KSSP also started a School for Technicians and Artisans (START) in 1976 to equip the artisans and technicians who lack formal education[65] and a series of lectures on 'Nature, Science and Society' and 'the Wealth of Kerala' were organised. In 1977, it organised a seminar on natural resources and the science caravan (*Sastra Kala Jatha*, 2 October–7 November) of the year raised the slogan, 'industrialise or perish'. The developmental emphasis of the organisation got a fillip when in 1978 they conducted research on the developmental crisis of Kuttanadu, the heart of paddy cultivation in Kerala.[66] The debates of these years were around two major axes. Firstly, KSSP participated in the debate on development initiated by the Nehruvian paradigm. Industrialisation was hence proposed as a catalyst of the region's development, and the natural resources available in Kerala were mapped to assess the region's industrialisation potential. Harnessing the knowledge and skill potential of the productive classes for development became a second emphasis that culminated in public awareness campaigns and starting training institutes and educational programmes. Although the emergent debate on environmental crisis had been discussed in the organisation, it did not become a serious concern until its involvement in the Silent Valley struggle of 1978.[67]

The year 1978 was a milestone in the history of the movement. The idea of 'people's science movement' (*Janakeeya Sasthra Prasthanam*) took shape during the first all India convention of the like-minded organisations under the auspices of the KSSP from November 10 to 12, 1978. This initiative helped various voluntary groups all over India to share their perspectives and it fostered the emergence of a national perspective.[68] The role and experience of these groups in broader socioeconomic and political context of the country was discussed in the meeting.[69] The deliberations were around four general themes such as formal and non-formal education, people's health movements, scientific research and technology and the harnessing of science for social revolution.[70] The deliberations in the meeting successfully consolidated different groups and organisations under the banner of PSM and the Nehruvian–Bernalist ideology of science became their broader framework for social action. This move was further consolidated by a second All India Convention organised on February 9–11, 1983 at Thiruvananthapuram, which focused on defining the term PSM from a national perspective.[71] Krishnakumar lists out the objectives of PSMs collected from bulletins and announcements of similar groups.[72] They are (a) popularisation of science and the creation of a scientific outlook among the masses; (b) challenging the forces of supernaturalism, obscurantism and superstitions, (c) equipping the poor with knowledge and skills to analyse and articulate their demands and rights; (d) reassessment of modern 'western' S&T which has grown mainly within the historical and economic context of colonialism; (e) re-evaluation of indigenous S&T systems; (f) development and propagation of appropriate technology; (g) motivating professional scientists to work on poor people's problems; (h) involving researchers, teachers and scientists in mobilising the people for social change; (i) creating pressure on the state to ensure the use of local resources for evolving problems related to health, education, housing and industry; and (k) the development of critical awareness of the pedagogy of the present education system and the creation of an alternative method of education, especially science education.

The emergence of KSSP as a PSM in terms of ideological orientations and broadening of membership hence occurred only in the 1980s, a decade that saw a growing public scepticism about modern science and its linkages with the idea of developmentalism and progress.[73] And this is a paradox as far as the Indian context of public engagement with science is concerned. As I have pointed out elsewhere,[74] by the second half of the 1970s itself, the contradictions of the operations of the technoscientific complex had started manifesting in the country, leading to the development of a critical engagement with science among the scientific-citizen publics. The expressions of this critical engagement were much diverse, emanating from Gandhian to Socialist and Marxist perspectives.[75] Among these groups of scientific-citizen publics who sought for better science, KSSP's standpoint was hardly critical of the state-technoscience duo and its operations. Shiv Viswanathan expresses this point acerbically:

> The KSSP, like the Hoshangabad experiment [in Madhya Pradesh], was generally leftist and modernist…. It was a populist drama of science with Desmond Bernal, Albert Einstein and C.V. Raman as heroes…. Its attitude to traditional knowledge verged on the illiterate and its theory of science was desperately positivist. The DSF [Delhi Science Forum] and the BGVS [Bharat Gyan Vigyan Samiti] were lesser clones of this same imagination and worked at the diffusion end of the map. As a result, they often became extension counters of the regime.[76]

KSSP's labour at its best was to create a science that is people-oriented and their fight was against a hegemonic view of science as a cognitive practice of special status safeguarded by an elite scientific community. In this endeavour, the movement developed new perspectives and social practices, blending Nehruvian socialism with a Marxist theory of science (Bernalism) which attracted the left-inclined middle-class intellectuals of Kerala to the movement in the late 1970s, immediately after the Rule of Emergency (1975–1977) declared by Indira Gandhi, the then Prime minister of India.

The Silent Valley controversy was the crucible for the epistemological squabbles over modern science in Kerala. The controversy raised serious questions about the developmentalist agenda of the state endorsed by KSSP in the 1970s. KSSP as an organisation became active in the struggle a little late, although some of its activists were at the forefront of the struggle in their individual capacities. A large number of its members were strong supporters of the Dam project and the movement itself had a statist developmentalist approach. There were several environmentalist groups and nature clubs actively led the Silent Valley campaign against the construction of the dam in the dense tropical forests in the Palakkad district of Kerala.[77] This has created a strong crisis for KSSP, which was broadening its base as a leftist mass movement around the time. It had to think seriously about shaping an ecological outlook that did not contradict with the foundations of its ideological framework.[78] KSSP's gradual participation in the Silent Valley Movement helped the movement readjust its ideological framework to incorporate the problem of ecology, while vociferously opposing other ecological groups with a more radical vision by denigrating their views as 'romantic environmentalism' which has to be discarded for a more 'scientific' understanding of environment.[79]

Conclusion

In this chapter, my attempt was to decouple KSSP from the construction of its history as a heroic struggle in the socialist path of liberating science from the clutches of capitalism and harnessing it for social revolution. The KSSP started constructing this valiant narrative in the 1980s, when the movement became established as constitutive of a scientific-citizen public in Kerala who had a largely uncritical appreciation of science and its values. KSSP thus became the vehicle of a particular kind of public engagement with science in Kerala that represented the progressive liberal left among the growing middle classes of the 1980s and the 1990s. The Nehruvian–Bernalist perspective adopted by the movement in the 1970s was actually the official ideology of the nation state that claimed its legitimacy through the operations of the technoscientific complex. KSSP and PSMs in general embodied this vision in the field of public engagement with science, and their practices were constitutive of a scientific-citizen public. Even in the late 1970s when there was a mushrooming of multiple publics and epistemological fault-lines appearing in the political field of public engagement with science, PSMs in general and KSSP in particular continued to harbour a largely uncritical scientific-citizen public. In other words, Parishad's endeavour was to save science from the attack of the quasi-publics by offering a theory of science that had scientism at its core while developing a sophisticated external shell that accommodated everything from feminism to environmentalism and economic development to deepening democracy. This has made the organisation increasingly conservative in recent times, and it struggles to be politically relevant in a shifting field of politics that opened up new sites of public engagement with science and new political imaginations of democratising science in Kerala.[80]

Notes

1 See, for example, T.M. Thomas Isaac and B. Ekbal, *Science for Social Revolution: The Experience of Kerala Sasthra Sahitya Parishat*, Trichur: KSSP, 1988.
2 A representative sample of this large corpus of work includes K.P. Kannan, "Science for Social Revolution", *EPW*, Vol. 11, No. 26., 1976, pp. 943–944; K.K. Krishna Kumar, "'Science for Social Change': The Kerala Sastra Sahitya Parishad", *Social Scientist*, Vol. 6, No.2, 1977, pp. 64–68; Krishna Kumar, "'People's Science' and Development Theory", *EPW*, Vol. 19, No. 28, 1984, pp. 1082–1084; Mathew Zachariah and Sooryamoorthy, *Science for Social Revolution?* New

Delhi: Vistaar Publications, 1994; T.M. Thomas Isaac and B. Ekbal op. cit.; T. M. Thomas Isaac, Richard W. Franke and M.P. Parameswaran., 'From Anti-Feudalism to Sustainable Development: The Kerala People's Science Movement', *Bulletin of Concerned Asian Scholars*, Vol. 29, No. 3, , 1997, pp. 34–44.

3 Shiju Sam Varughese, 'Where are the Missing Masses? The Quasi-publics and Non-publics of Technoscience', *Minerva*, Vol. 50, No. 2, 2012, pp. 239–254; for the theorisation of political society, see Partha Chatterjee, *The Politics of the Governed*, Ranikhet: Permanent Black, 2004.

4 Paul Gary Werskey, 'British Scientists and 'Outsider' Politics, 1931–1945', *Social Studies of Science*, Vol. 1, No. 1, 1971, pp. 67–83.

5 Ibid.

6 Ibid.

7 Ibid., 71.

8 Ibid.

9 Ibid., 76.

10 Another major influence was Vannevar Bush's report to the American president titled, *Science, the Endless Frontier* (1945), where he advocated the autonomy of scientists from the political, economic and social realms. The report placed the public at the receiving end, in contrast to the pre-war period of science activism in the West, where the learned public actively engaged with science. See Aant Elzinga and Andrew Jamison, 'The Other Side of the Coin: The Cultural Critique of Technology in India and Japan', in E. Baark and A. Jamison (eds.), *Technological Developments in China, India, and Japan*, London: Macmillan, 1986, pp. 205–251.

11 Maurice Goldsmith, *Three Scientists Face Social Responsibility*, New Delhi: Centre for the Study of Science, Technology and Development, 1976, p. 24.

12 The Lysenko affair was the major case sited by them.

13 Paul Gary Werskey, op. cit., p. 76.

14 An initiative of the radicals at Cambridge consisted of about 80 scientists and the graduate students of Cavendish Laboratory and the Biochemical Laboratory in 1933–1944. Aant Elzinga, 'Bernalism, Comintern and the Science of Science: Critical Science Movements Then and Now', in Jan Annerstedt and Andrew Jamison (eds.), *From Research Policy to Social Intelligence*, London: Macmillan, 1988, pp. 87–113.

15 Paul Gary Werskey, op. cit., p. 78.

16 Ibid.

17 Ibid; Maurice Goldsmith, op. cit.

18 Paul Gary Werskey, op. cit.

19 Aant Elzinga, op. cit.

20 Ibid.

21 Ibid..

22 Paul Gary Werskey, op. cit., p. 81–82.

23 Ibid.

24 Ibid..

25 Ibid..

26 For a detailed discussion, see Dhruv Raina and S. Irfan Habib, *Domesticating Modern Science*, New Delhi: Tulika Books, 2004.

27 Dhruv Raina, 'Evolving Perspectives on Science and History: A Chronicle of Modern India's Scientific Enchantment and Disenchantment (1850–1980)', *Social Epistemology*, Vol. 11, No.1, 1997, pp. 3–24.

28 Although the entire gamut of sceptical remarks about western science and technology became ascribed to Gandhi and the 'Gandhians', there were a spectrum of standpoints on the matter in the 1920s and the 1930s, ranging from Rabindranath Tagore to J.C. Kumarappa and J.C. Bose to P.C. Ray. For a detailed discussion, see Deepak Kumar, *Science and the Raj*, New Delhi: OUP, 2nd Edition, 2006.

29 The comment about Gandhians is applicable in the case of Nehru as well. There was actually a range of positions available on western science and industrialisation since the early twentieth century, which eventually became known as the Nehruvian position. See Benjamin Zachariah, *Developing India*, New Delhi: OUP, 2005.

30 A.V. Hill was the biological secretary of the Royal Society of London. See Deepak Kumar, op. cit., p. 254.

31 According to Deepak Kumar, 'Hill had an excellent personal rapport with Bhatnagar, and he corresponded with more than fifty other Indian scientists. Hill came to India with no political views, but quickly put economic problems high on the agenda'. See Deepak Kumar, op. cit., p. 254.

32 Dhruv Raina, *Needham's Indian Network*, New Delhi: Yoda Press, 2015, p. 13.

33 Deepak Kumar, op. cit., p. 257.

34 Dhruv Raina, *Needham's Indian Network*, op. cit., p. 13.

35 Ibid.

36 Dhruv Raina, 'Evolving perspectives', op.cit., p. 9.

37 For detailed discussion, see Shiju Sam Varughese, 'The State-Technoscience duo in India: A Brief History of a Politico-epistemological Contract', in Andreas Franzmann et al. (eds.), *Legitimizing Science*, Frankfurt and New York: Campus Verlag, 2015, pp. 137–156.

38 Ibid, p. 143.

39 For a detailed discussion, see ibid; Shiju Sam Varughese, 'Where are the Missing Masses?' op. cit.

40 The details in this section are from M.P. Parameswaran, *Janakeeya Sasthra Prasthanam*, Thrissur: KSSP, 2008, if not mentioned otherwise. Insights from Isaac and Ekbal, op. cit. also has been used in this section.

41 The office bearers of the organisation were P.K. Korumaster, P.T. Bhaskara Panicker and O.P. Namboodiripad.

42 Only the publication of the first issue of the magazine could be accomplished by them.

43 Mathew Zachariah and Sooryamoorthy op. cit., p. 54.

44 Ibid.

45 Ibid., p. 55.

46 Dr. M.P. Parameswaran did his PhD from the Institute and joined the Bhabha Atomic Research Centre (BARC), Mumbai in 1966. Later, he emerged as a renowned people's science activist as the leader of KSSP.

47 There were seven such associations that included Sastra Sahitya Parishad (Telugu), Marathi Vidnyan Parishad, Kannada Vijnjan Parishad, Hindi Vigyan Parishad and Vinjana Tamil Valarchikkazhakam apart from the Malayalam wing.

48 K.G. Adiyodi, 'Satra Sahithyakaranmarkku Oru Sanghatana', in *Sastra Sahitya-Parishad Utghatana Smarakam*, 1962, p. 9; quoted in Issac and Ekbal, op. cit., p. 5.

49 The term scientific temper was first used by Jawaharlal Nehru in his *Discovery of India* (1946).

50 Mathew Zachariah and Sooryamurthy, op.cit., p. 55. Prof. Haldane was a renowned scientist who spent his later years in India after his retirement from University College, London in 1957.

51 Ibid. See also Thomas Isaac and B. Ekbal, op. cit., p. 21, for a reference to the Haldane's keen interest in the organisation.

52 Thomas Isaac and B. Ekbal also acknowledge the influence of SRS movement on KSSP, although the internal differences of the movement are not considered. See ibid, p. 20–21.

53 See Thomas Isaac and B. Ekbal op. cit.; Mathew Zachariah and Sooryamoorthy, op. cit.; T.M. Thomas Isaac, Richard W. Franke and M.P. Parameswaran op. cit.

54 As quoted in *Parishad Vartha*, October 1991.

55 *Sastra Keralam* (1969) and *Eureka* (1970).

56 Thomas Isaac and B. Ekbal, op. cit., pp., 10–11.
57 Ibid.
58 Ibid., p. 12.
59 Mathew Zachariah and Soorymurthy, op. cit., p. 61.
60 M.P. Parameswaran, op. cit., p. 45.
61 KSSP, *Parishad Pinnitta Muppathu Varshangal*, Kozhikode, 1993. A similar attempt to start Factory Science Forums but failed.
62 Ibid., 46–47. These scientists were active sympathizers of the Naxalite movement of the 1970s. Especially the speech of Dr. K.R. Bhattacharya (President, CSIR Workers' Association) in the national science conference organised by KSSP as part of its annual meeting had a strong influence on adopting the slogan.
63 M.P. Parameswaran, op. cit., p. 30.
64 Ibid.
65 *Parishad Vaartha*, October 2001, pp. 1–15.
66 Kuttanadu, known as the 'rice bowl of Kerala' is a major rice cultivation area situated in the southern part of Kochi of central Kerala. It is an area with a subtle ecosystem. In 1974, Thanneermukkam barrage was constructed across Vembanadu backwaters in Kuttanadu to control the salt-water entry into the area from December to June, in order to make possible rice cultivation in this period. It has been creating severe ecological crisis in Kuttanadu.
67 In the context of the first Stockholm conference on environment (1972), Parishad organised a meeting that deliberated issues of air and water pollution and protection of environment. M.P. Parameswaran, *Vikenthreekritha-Janadhipathyam Keralathil, 1956–1998*, Kochi: KSSP, 1999.
68 K.P. Kannan (ed.), *Towards a People's Science Movement*, Kochi: KSSP, 1979.
69 Ibid.
70 A. Vaidyanathan et al., 'People's Science Movements and Science Wars?' *EPW*, Vol. 14, No. 2, 1979, pp. 57–58.
71 Jaffrey et al., 'Towards a People's Science Movement: A Report', in KSSP, *Science for Social Activism*, Trivandrum, 1984, pp. 54–67. The increased interaction and cooperation of PSMs thereafter led to the formation of All India People's Science Network (AIPSN) in 1988.
72 Krishna Kumar op. cit..
73 Membership of Parishad increased in 1981–1982 to 5,859, and by the next year, it was increased to 7,319. There is a steady increase of its membership in the 1980s, which turned to be 37,653 by 1988–1989. See Mathew Zachariah and Sooryamoorthy, op. cit., p. 139.
74 See Shiju Sam Varughese, 'Where are the Missing Masses', op. cit.; Shiju Sam Varughese, 'A Brief History', op. cit.
75 For a detailed exploration of these perspectives in civil society, see Dhruv Raina, 'Evolving Perspectives', op. cit.; Aant Elzinga and Andrew Jamison, op. cit.
76 Shiv Viswanathan, 'Between Cosmology and System: The Heuristics of a Dissenting Imagination', in Boaventura de Sousa Santos (ed.), *Another Knowledge is Possible*, London and New York: Verso, 2007, p. 184.
77 Because of pressures from civil society, the project was abandoned in 1983. See Dr. E. Unnikrishnan, 'Silent Valley Prakshobham Thudangiyatharu?', *Mathrubhumi Weekly*, December 7–13, 2008, pp. 8–15.
78 See KSSP, *Parishad Pinnitta*, op. cit. for some references to this crisis faced by the organisation.
79 See ibid.
80 For a detailed discussion on this shift, see Shiju Sam Varughese, *Contested Knowledge: Science, Media, and Democracy in Kerala*, New Delhi: OUP, 2017.

Section II
Technology and culture

5 Electricity and urbanization in Madras, 1895–1930

Y. Srinivasa Rao

Electricity and urbanization

Electricity and expansion of Madras and other towns in Madras Presidency have had an inverse relationship. Initial governments' attempts of the electrification of towns in Madras Presidency was to provide urban convinces to Europeans consisting primarily of government officials and others who were settling in the presidency and local rich who were seeing it as a luxury. Until the era of mass production of electricity was started with the commencement of big centralized thermal power station in Madras in the first decade of the twentieth century in the urban areas, it was a luxury and non-essential commodity for the locals. In the third decade, the mass production of hydro-electricity had de-urbanized electricity and opened gates for both industrial modernization in medium and small towns and agricultural modernization in the countryside. In fact, the development of port towns as colonial settlements and towns in the hinterland as industrial and agricultural market towns was attributed to electricity. Electricity as an agent of transformations had transformed nights into days, extended workings hours and opened doors for new night culture. By the middle of the eighteenth century itself, Southern India had a few medieval and colonial towns. There were port cities, capitals of various native kingdoms and colonial settlements such as Madras, Coimbatore, Bezawada, Kurnool, Bobbili, Tanjore and Madurai. These bigger towns were surrounded by clusters of smaller towns and semi-urban villages containing large communities of merchants and artisans and permanent marketplaces.[1] The main reason behind transforming few adjacent villages into towns was to serve colonial administrative and commercial purposes. This included creating new spaces for administration of annexed territories, consolidating colonial power, perfecting the revenue administration, introducing new economic activities and better networking of the empire. Densely populated areas – where non-agricultural occupations were the main economic activities – offered themselves as natural candidates.[2] With the electrification, the very nature of these towns was changed. Those which had already became towns were occupied by the British, which acquired new modern characteristics

DOI: 10.4324/9781003241980-8

and transformed from medieval towns to modern, and those which were developed by the Europeans from the scratch were also endowed with the infrastructure of modern city.

The port towns, which were gateways of seaborne trade of the empire, the colonial government developed them first and inland towns later. As the development of commercial activity had gained momentum, ports have established oil power plants to address the question of power requirements. Later, electricity supported the commercial activity and enabled the extension of the hours of work which resulted in the extension of day. Similarly, inland towns such as Coimbatore and Bezawada, which were industrial and agricultural business centres, were also promising pockets for electrification. For the effective management of material, market as well as for consolidating its power, the colonial government had created a technological network consisted of communication, electricity and railways. Electricity had a special place in this network, i.e., it has powered the network as communications and railways were powered by it. All the urban centres in colonial India including big cities such as Bombay, Madras and Calcutta, apart from being satellites for the colonial economy were also embryonic markets of foreign electrical companies. Though the colonial purpose of the technological network was understandably different, it has connected the newly developed port colonial cities with countryside and brought regional culture closer which was a by-product.

In big cities, electricity had performed multiple functions. It powered modern massive public transportation systems like tramways and railways, public utilities like modern hospitals, parks, fairs, business firms, banks, colleges and government offices. Madras town which was developed as the commercial and administrative centre and big in size was facing lighting and water supply problems. Hence, when the electrification process was started in Madras Presidency, the government gave priority to coastal towns, market centres and foreign settlements.

Apart from street lighting in towns, electricity opened the way for urban-based commercial products industries in coastal towns and food processing industries such as rice, flour and oil mills in inland towns. This in turn contributed to further growth of towns and cities in these both areas. In other words, while electricity expanded the boundaries of urban areas, the urban growth in turn produced greater demand for electric power. Electricity had entered the towns of the presidency much earlier, but the commencement of the operations of the Basin Bridge Thermal Power station of the Madras Electricity Supply Corporation (MESC) from 1909 marks the beginning of large-scale town electricity. This mass production capacity of the MESC had offered opportunities to the local government to lit streets, electrified public buildings, parks and operated public utilities with electricity and led to the penetration of electricity as power of light and other domestic necessities.

It was only Madras which had the ability to offer the required power demand to the big centralized electricity production and supply. In the rest of the towns, private firms which installed power plants for their business purposes also supplied excess power to the towns nearby. Within the first decade of its entry into the presidency, there was considerable number of power plant installations. 121 power plants were installed in the presidency, which was notified under the Indian Electricity Act (IEA), 1887. But this new technology also brought in new problems. Natives of the presidency, who were not educated about the way new technology works, lost lives due to electric shocks which were something new. Despite continuous inspections by the officials of the Public Works Department, electrical accidents were continuously reported[3] and they were increasing as the electricity continued to penetrate.

Small towns were developing slowly serving the needs of the colonial government. Yet, the colonial government was slow in providing electricity to towns as it was not seeing it as a governmental obligation or social necessity. Neither the native people saw it as an immediate societal necessity. Prior to the commencement of massive hydro-electric projects in the 1930s, electricity in the smaller towns was an industrial and public utility but not a domestic necessity. Though the government realized the potential of electricity for industrialization and revenue generation as a result of which government constituted Hydro-Electric Surveys of India (HESI) in 1919, until the fruition of the initiated hydro-electric projects in the 1930s, the government left electricity to private and local public bodies completely and it remained a licensing agency. From 1900 to 1930, the government was playing safe, leaving electricity generation and supply to the local bodies and private agencies. It was the colonial necessity that pushed the British Indian government to introduce, invest, manage and maintain the electricity generation and distribution systems. It went on with a massive electricity propaganda campaign as it was under pressure of providing economic justification of massive electricity projects to vindicate the position of the government. In the smaller towns of the presidency, where there was less power demand, private companies began to explore possible power generation sites for using water sources to generate electricity. Government too, though waited till such power generation was proved economical, in the second decade itself, had made attempts to conduct scientific surveys for generating power through hydro-electric systems not though thermal or oil engines. To supply power to the inland trade town, Coimbatore, schemes were prepared to use water sources of the Siruvani River. In 1913, a hydro-electric scheme was sanctioned to investigate power generation possibilities from this river for Coimbatore as well as at the Pykara falls for supplying Ootacamund and neighbourhood.[4] By 1918, this scheme became a reality and nearby towns such as Nilgiri Hills, Podanur, and Coimbatore were supplied with electricity and some schemes. Towns both in Tamil- and Telugu-speaking areas away from the water sources were given to private firms which used oil engines to generate power. Trichinopoly, Madurai and Karaikudi in Tamil Nadu and Tenali in Guntur district and Bezawada town were electrified.[5] From the 1920s onwards, municipalities began to show interest in installing power plants to supply power in their respective areas.

Table 5.1 Details of private and municipal stations prior to the hydro-electric era

S. No	Name of the undertaking	Location	Motive power	Capacity in KW	Est. Year	Fuel used
1	Madras Electricity Supply Corporation	Madras	Steam turbines	15,500	1909	Coal
2	Ootacamund Municipal Electrification Scheme	Ootacamund	Received from Karteristation	500	1923	Hydro-electricity
3	Bellary Municipal Electrical Works	Lake Nallacherum	Crude oil engine	177	1927	Crude oil
4	Kanadukathan Electric Supply Corporation	Kanadukathan	Oil engines	119	1924	Crude oil
5	Meenakshi Electric Supply Corporation	Devakotah	Internal combustion semi-diesel engine	100	1924	Liquid fuel
6	Cochin Electricity Supply Scheme	Kalvetti Lane-British Kochi	Semi-diesel crude oil-engine	85	1924	Crude oil
7	Electric Power Supply Company	Coimbatore	Oil engines	220	1924	Liquid fuel
8	Trichinopoly–Srirangam Electric Supply Corporation	Trichinopoly	Diesel engine	2–100	1924	Liquid fuel

Source: Army Department Correspondence to provincial governments on 14 December, 1927.

In urban areas, apart from the local bodies installing electrical power plants, private agencies with licenses in areas where municipal concerns were not able to provide power provided electricity, serving the needs of expanding urban areas beyond Madras. From the second decade of the twentieth century onwards, semi-urban towns slowly started accepting electricity and government apart from acting as an agent of providing public facilities, most of which were to work with power be it a public office for fans and lights, be it a hospital where apart from lights and fans most of the medical apparatus work with electricity, it was also acting as an agent of providing license to the private agencies which were interested in providing power for commercial gains. Government encouragement for lighting schemes increased the number of installations in the *mufassil* areas. Spending by the government on installing plants and maintenance were consistently on the rise from 1910 onwards. By 1925, the government spending on electrical installations was increased to 21 lakhs. Urban electrical installations by the municipalities continued to rise. Towns like Bellary, Cochin, Bezawada and Ootacamund started to provide supply to town folk.

Government's disinterest in investing in electricity systems in some towns of the presidency where electricity seemed to have less demand or not considered as a driving force of economy was justified from the point of the economic rational. In smaller towns of northern and non-delta areas of the presidency where agricultural and industrial production were still determined by rainfall, the local demand for electricity was quite dismal. This lack of interest from the natives and lack of interest from the government had created some sort of regional inequality in electricity within the presidency, between the south and north of the presidency. Massive hydro-electric power stations in the south which covered both rural and urban areas of south and large-scale thermal power station in Madras laid the domination of south over north in the presidency. The northern areas were only having oil engines and small- and medium-scale thermal stations were coming up in the 1940s. This had caused regional power inequality. Thermal power stations in northern towns of the presidency were not operationalized until the end of the colonial rule and hydro power stations have only came up after independence. There was also rural–urban divide in electricity. Even after the commencement of massive power stations in the 1930s, where de-urbanized electricity was a domestic necessity, in rural areas, electricity was an economic necessity instead of public utility in domestic sphere.

Tramcars: a modern transport

Transport of humans and goods from one place to other and problems related to it are more sociological and economical than purely scientific in nature.[6] But during the colonial rule, relocation of various transports technologies in 'peripheral towns' was also political, cultural and environmental. For instance, railways introduced by the British Indian government had served political and economic agendas of the colonialism and it had caused massive deforestation as well. Though in course of time, it was found to be acceptable but in the initial stages, train transport covering long distances and tramcars running in towns caused a debate on cultural, ecological and geographical suitability. Nonetheless, the relocation of electric traction and electric light has announced the birth of a modern city and became its integral parts by the turn of the twentieth century in colonial India. Though the railway's origin could be attributed to transportation of goods, the stimulus came from passenger traffic in England by the 1840s. Commercial electric trams were tested in Virginia and Richmond in the mid-1880s, spread to European towns despite aesthetic of overhead wires and supporting poles inhibiting their development.[7] Roughly in a decade, electric traction was introduced into colonial peripheries like Madras, Calcutta and Bombay not as a choice of the colonized but as a colonial convenience and as business opportunity to the western electric corporations. In an economic sense, it was cheap, fast and convenient compared to the traditional means of transport. Whether the tramcar transportation was necessity of the colonized or the colonial is debatable, but the expansion of the towns justified its introduction. Creating markets for the newly emerging electrical corporations, conveyance to citizens and a bit of revenue to the local government and profits for the foreign private companies, tramcars were an integral part of Madras until it was closed in 1953. Services and revenue allowed cities to expand geographically and changed the image of electric energy from luxury to necessity.[8]

Madras was considered the Queen of the Coromandel, as a trading post of the colonial government. According to S. Muthaih, the foundation for modern India was laid with the birth of Madras during the Age of Trade.[9] From its birth in the seventeenth century, within 100 years, by 1871, the city had covered an area of approximately 27 square miles with a combination of urban areas and rural patches. Its expansion in the first half of the twentieth century was quite sluggish. By 1931, only two more miles were added to the city. From then on, the emergence of colonies in the suburbs outside the jurisdiction of municipality had transformed it into greater Madras.[10]

Yet from the last decade of the nineteenth century, towns of the periphery were getting technological systems. In Madras, Calcutta and Bombay, the tramcar system was the first mass passenger commuting technology. It was later joined by motor cars which were elite luxury and motor buses in the 1930s. Along with the tramcars, the motor bus transport system had played a significant role in the development of greater Madras. Prior to the entry of electricity-run electric tramways, commuters were transported by the animal power-driven transport system, i.e., bullock carts and *Jutkas* (horse carriages). In the west, where some of the towns developed trams running on the rails fixed in the middle of the road were pulled by a bunch of horses. With electric traction, there was a conversion of horse trams to electric trams. But in Madras, *Jutka* was a single cabin wagon that was pulled by a pair of horses that ran on the tar road. Therefore, the entire system in economic, technical and cultural sense was a new experience to the people of Madras. Tramcars, which were operated with electricity in Madras city, completely dominated other means of transport. First, the tramcars and later bus transportation systems have challenged the traditional animal-power-supported transport systems. After the tramcars, it was the same company which introduced organized buses in 1927 in Madras. It introduced a fleet of buses but closed after three years. By this time, unorganized local private operators were already running buses supplied by Simpson's & Co. from 1910.[11] By the 1930s, the bus transportation system was challenging the tramways.

Though it took some time for the public bodies to benefit from the technologies comes under public utilities in colonies, local bodies such as municipalities and panchayat boards were involved in power generation and distribution. There were also local business firms which acquired concessions where public bodies did not come forward. Despite these two Indian agencies, the British firms, not only in the business of electricity but also in relocating, installing, maintaining and managing foreign technological systems, maintained an undeclared monopoly and domination. Despite the period of providing licenses to the private companies was truncated with the emergence of the Public Works Department from the 1860s, the domination of the foreign technological corporations was guaranteed as there was no competition from the native Indians entrepreneurs whatsoever. Both Crompton & Crompton that owned thermal power station and Hutchison & Co. that owned electric tramways in Madras have acquired licenses roughly for a period of 50 years. Yet during the first half of the twentieth century, given the experience in technology and management, foreign electrical companies enjoyed some sort of guaranteed monopoly in the colonial administrative towns which had the great possibility of expansion that offered greater marketing space have guaranteed maximum profits to foreign firms. The starting period of license started roughly in the 1890s, the time when large-scale technological systems – telephone, telegraph, hydro- and thermal electricity generation stations and supply systems – were making entry have opened up business opportunities to British technological corporations to involve in direct business and many more corporations of other European countries and North America to export technologies to newly expanding peripheral towns. British business concerns who had understood the potential of transport businesses in the fast-expanding city eyed to explore the possibilities by introducing the new transport systems which had already proved their potential to be a convenient urban mass transport in metropolis. Madras city is the first in the East where electric tramways were introduced. In 1892, Hutchison & Co. started the Madras Tramway Corporation which started its operations in 1895. It was renamed as Madras Electric Tramways Limited in 1904 and it continued its operations until 1953. The tramways were the single line system constructed on the conduit system, i.e., the live conductor carried underground. This was soon replaced by overhead lines as underground electric lines were troubled by rains and floods.[12] Consistent attempts were also made to speed up the service, to replace the old single track with double and to replace the old open-type bogies with the bigger closed-corridor cars.[13]

The company had its own power generation stations at various points in the city. It was through the MET's tramcars, the public, for the first time, had experienced the power of electricity, which not only propelled the tramcar but also provided lighting inside the car. The MET used oil engine generators to produce electricity. The power station had two Burnley horizontal engines of 200 h.p. each and two William central valves– one of 200 h.p. and the other 100 h.p. These were used to drive electric generators constructed by Electric Constructing Company Limited (UK). There was also a 20 h.p. Bellis &Morcon Electric Company equipment set for lighting purposes.[14] By the beginning of the twentieth century, the MET had 39 motor cars and two trailers, and the average number of cars was 35. The line equipment was the overhead trolley system throughout, with centre poles along the double track portion and side poles on the single track. Its route mileage was 9 miles and two and half furlongs (furlong is a unit of length equal to 220 yards), 2 miles and 6 furlongs of which was double track. The tram route covered Customs House, Egmore, Royapettah Police Station, Barber's Bridge and Kandappa Mudali High Road.[15] There was a two-minute service on all stations except at the Custom House and Barber's Bridge.[16] At the beginning of 1902, there were 39 tramcars and by the end of 1927, there were 88, but only 63 were in operation covering a distance of 2,009,447 miles and carrying 1,0,084,266 passengers.[17] Based on the increasing passenger traffic, in 1915, tram lines were extended by adding 5 more miles now the tramways was covering 12 miles in the city. In five years, another 96 yards of double track was built. Bogy and trailer cars were manufactured in the city itself. Increasing extension of line had necessitated the addition of new service bogies such as testing, road metal carrier and watering car in the rolling stock.[18] This clearly indicated that there was a visible growth in the number of cars, which also indicated that the city folk accepted this new transport.

As soon as the MESC, the biggest thermal power station in colonial southern India owned by Crompton & Crompton, had commenced its operations, the power generation system of tramways was altered. The engine for lighting purposes was kept unused because of the supply obtained from the Basin Bridge power station of the MESC in the year 1909.[19] Once the centralized thermal power was made available by the MESC, the isolated stations generating electricity through generators were made as standby systems or else completely removed. In fact, the opening of the MESC furthered the tramway's domination over all other modes of transport and it emerged as one of the most favoured massive transportation system that was readily accepted by the Madras populace. Table 5.2 shows the growth of the tramcar system in the city.

There was a constant growth in the number of cars, distance covered and in the number of people transported. This indicates the ready acceptance of the new and imported technology by the people of Madras as the best means

Table 5.2 Tramways development from 1902 to 1910

Year	No. of tramcars	No. of passengers transported	Distance covered in miles
1902–03	39	6,290,610	944,336
1903–04	40	7,670,832	971,083
1904–05	49	7,802,226	1,070,848
1905–06	NA	NA	NA
1906–07	60	9,579,832	1,178,492
1907–08	NA	NA	NA
1908–09	60	12,215,101	1,218,532
1909–10	63	13,056,166	1,227,609

Source: *Reports on Administration of Madras Presidency 1902–1910.*

of urban transport. Tramcars, running through streets of Madras, were a source of excitement and wonder to the people. From 1895 to 1925, tramway maintained its domination as the sole passenger commuting system, but it began to lose its sheen due to financial losses. This downward trend began from 1919 itself. The average daily services decreased from 69 to 65 cars and mileage fell from 1,582,439 to 1,400,278 miles. But the passenger traffic continues to grow. In 1924, tramways carried 18,425,185 commuters but in the year 1925, the number had come down to 15,510,007.[20] However, in the year 1925, another new alternative transportation technology, i.e., motor buses, appeared in the scene. Ironically, it was the MET which introduced 12 buses in the same year to compensate financial losses of the tramcar transport system. Domination of the MET over the urban commuter transportation by monopolizing the tramways from the 1890s and later in bus transportation from the 1920s became a debatable issue in the local media. When the MET decided to run buses, *Swaraj* and *Andhra Patrika* were suspicious of the intensions of the company. In a context where local entrepreneurs have already invested in this urban transport business, allowing MET to enter motor-transport business for which it generates necessary capital from England instead of Madras did not seem necessary. Both these newspapers argued that allowing MET to run buses with its foreign capital would not do any good and it would dominate the local private bus operators and raised questions about the unequal competition.[21]

Though tramcars attracted quite a huge number of people, it also faced criticism. Contemporary newspapers of the presidency were very critical of tramways companies' method of operations, exploitation and practice of racism. *Andhra Patrika*, a Telugu daily, was critical of the way trams cars were operated in Madras. In 1901, reporting frequent accidents, it criticized the company for running tramcars through congested and narrow lanes in Triplicane and Chintadripet market close to inhabitant houses which forced children of these area to lockup inside rooms for fear of being run over. It also suggested that drivers of the cars should drive slowly through congested streets, use whistles to warn passengers, use vacuum-breaks, get the service of cow-catchers and use engines that were used in railways of long distance.[22] In 1919, *Andhra Patrika* reported another accident where a finger of a lady was smashed and suggested that separate accommodation should be provided to ladies inside the car, not in front seat.[23] *Swadeshimitran*, a Tamil newspaper, reported on the racial discrimination practised by the tramway in its daily operations.

> A boy of 18 years was run over by the tram car at Chintadripet. Witness says that the drivers are indulging in fast driving in crowded quarters where accidents of this kind reported are imminent. The drivers are indulgent and obliging to Europeans and stop the car at the pleasure of such passengers and pay regard to their convenience in all possible ways, and if the car is over crowded the drivers generally ask a native passenger to vacate his seat for the sake of European passenger. Towards natives, drivers are insolent and they do not stop the car wherever the native asks them to stop and do not even patiently wait till they get into or alight from the car, but hastily set the car in motion while the native passenger has one foot on the foot board of the car and the other on the ground.[24]

Another newspaper, *Desabhaktan*, complained:

> The authorities of the company (tramway) are busy fleecing money from the people and do not care to secure the conveniences of the latter. They raise the fares as they liked and it is said that they have the support of the Madras Corporation and the Government of Madras. Why did not those who sanctioned an increase in fares insist on the tramway authorities securing the conveniences of the passengers? The government and corporation would realize the bug nuisance on the tramcars only if they travel in the cars. Once medical men were of opinion that the poisonous bite of the bug leads to many diseases and, yet the health officers of the corporation did not attend to the matter. The tramcar should be cleaned at least twice in a week with hot water and insecticidal solution before use.[25]
>
> Apart from the racism and poor maintenance, the extravagant expenditure on the construction of tramway tracks was faced criticism. In Madras, to construct an 18 miles tramway line between Tambaram and Beach, the government had incurred Rs. 1.75 lakhs, The *Vira Bharathi*, from Madras, had questioned the rationality behind such huge investment by the railway while it was laying-off poor labour working with it. It further argued that this sort of investment was being made to benefit iron and cement factories of England.[26] This new transportation system had demanded large number foreign technicians who were to be imported from England.

The MESC was economically powerful. It dictated terms to the colonial government and made laws on behalf of the government. It had control over other companies through having more shares. The MET and MESC's, both were British companies, capital was generated in Europe. There were also a few Indian shareholders. But they had nothing to say in the company's affairs. Though MET and MESC were different companies, some of the directors of the MESC were directors of other companies. The MESC held all ordinary shares in the MET. These two companies complimented each other in the Madras town in maintaining domination in business. The commencement of the MESC, which supplied power to the MET for running tramcars, supplemented its growth. While the MESC supported the MET's growth, in turn, the MET supported the MESC which was not paying its dividends in the initial years. There were also plans for merging these two companies together. In Madras, these companies had faced no competition at any time in their half a century of existence in the only biggest southern colonial city – hiked power charges when they wished to and exploited the population of the city.[27] These two companies were liked by the English and for the English. Hence, one could safely argue that British companies had largely benefited from the urban electrification. In the 1930s, ties between these two companies strengthened further. In 1930, both companies together formulated an advisory committee to give advice on promoting relationship between consumers and companies. The advisory committee consisted of agents of the two companies as ex-officio chairman and one member each nominated by the Government of Madras, Municipal Corporation of Madras, Commissioner of Police, Madras Chamber of Commerce, South Indian Chamber of Commerce and Trade Association of Madras. However, these companies had also made it clear that the advisory committee's recommendations and suggestions would be considered and treated as advisory only and it had no authority to enquire into the general administration of the two companies.[28] The Electric Tramway and Supply Corporation Workers Union which has been dominated by communists equally posed challenges to the might of these companies. But the company had full support of the government from the beginning to the end of the British rule. When the workers union sought government intervention into some of their grievances, the government showed no interest and further it was of the opinion that grievances were not genuine.[29] But persistence of the union in solving the workers' problems in both tramways and other railways, in the following year, the management heeded to demands of Madras tramway workers: increasing wages and dearness allowances and a scheme of sickness insurance was proposed to be introduced.[30] Apart from fighting for their own rights, the tramway workers joined anti-British strikes and revolts. They have joined the general strike call given by the Communist Party of India when Royal Indian Navy ratings in Bombay revolted in 1945. The Tramway Workers union held meeting under the auspices of the All India Trade Union Congress and passed resolutions condemning the use of India as a base for operations in Java and Indo-China and called upon Dock workers to refuse loading of ammunition and material into ship destined to East Asia.[31]

> Despite the racism, unsuitability, poor maintenance and foreign capital, the trams were symbols of the technological progress of the West and its economic and political power. The trams, car barns, powerhouses, rails and overhead wires were highly visible evidences of the capital investments and civic improvements promoted by colonial government and private entrepreneurs and of the role of electricity. By 1953, however, the progress of motor bus transport system eroded the viability of the tram cars system and tram transport was closed on April 12, 1953. During its lifetime, although it faced criticism on grounds such as being unsuitable to Madras city conditions (narrow and congested streets), racism and economic exploitation, it comforted and moved thousands of commuters of the city. Most importantly it served as one of the earliest examples of the power of electricity.

Electric light: symbol of new culture

The history of the evolution of lighting technology in Madras city or probably that was the case with every city that was began to develop after the British took interest seems to show only two phases: oil lamps lit with kerosene and electric lighting. This is not to ignore that there was debate on whether to install arch lighting in Madras went on for some time. Most of the colonial officers were suggesting the government or the city administration bodies to opt for electric light over arch lighting. Unlike Western Europe, colonies had less opportunity to use gas lighting which was the second phase of the evolution of lighting technology in Europe. Neither India had the necessity of importing gas lighting which needed huge investment as the gas lighting was provided through centralized gas supply stations in Europe. Except the native lighting systems, the installation of foreign lighting systems was expensive and maintenance was considered as luxury, and the Madras Municipal Corporation collected lighting-tax whenever Madras city was hit with an epidemic.[32] But in European electric trams, electric lighting was an accepted feature of urban life in 1900. Though still in competition with gas and arc lighting systems for long, Europe and North America cleared the way for supremacy of the incandescent lamps because they were safe, convenient, reliable and clean.[33] Until the beginning of the second decade of the twentieth century, even the political and educated community did not see electricity as a public necessity in Madras. But since these towns were being developed by Europeans, despite opposition, electricity made inroads into public life and with its entry, and towns in colonies began to acquire the characteristics of the Metropolis. Light marked the social gulf between the leisured classes and the working population, but also the difference between the metropolis and province.[34] This is also true with the metropolis and peripheries. By the time electric lighting entered into Madras, first colonial administrative city in India, European cities made enormous progress in both domestic and public lighting systems, while towns in the periphery were being lit with oil and Kerosene lamps, most of which were designed locally by the 'native' craftsmen.

Madras, as the sole colonial capital of South India, received lighting system of various kinds as it was accommodating English people for a long time. Native urban rich also started installing lighting systems. Madras streets were lit up with kerosene oil lamps until 1910. Though it could be assumed that European lamp lights could have been introduced into Madras from the very beginning, local and foreign oil lamps both were in use. While the domestic space of the locals was lit by locally made native lamps, domestic space of the Europeans, and office and public space of the government were lit by imported lamps of different variety and quality. There were ordinary, patent, Kiston and Washington lamps which were serving the city. By 1857, the city had some 200-kerosene oil-lit public lamps which gradually increased to about 6,500 by the time electricity was introduced. Though at the end of the first decade, mass electric lighting came into existence, the government was preparing ground for the introduction of electric lights in the cities at the beginning of the twentieth century by preparing legal mechanisms to administer the technology and control its usage and exploitation. Neither the Bengal presidency nor the Madras Presidency had produced legal mechanisms with which this new technology could be administered. There was lack of such specialized legal mechanisms for governing electric light and traction, which was considered necessary for the entry of foreign electrical corporations such as Crompton & Crompton which expressed interest in entering the business of generation and distribution of electricity in Madras. The Secretary of Madras Government was critical of the provisions of Electricity Act XIII of 1887, the objective of which was protecting persons and property from the risks of electricity supply. In the place of the old protective act, he suggested a new liberal and progressive act like the Imperial Act that was governing railways objective of which was to 'regulate and facilitate'.[35] But the local business community represented by Chairman of Madras Chamber of Commerce opposed the moves of the government to introduce the imperial electricity act and demanded that the company should have its registered office in India and be given only 42 years of license. Though giving contracts to companies cancelled from 1864, yet in urban areas building public utilities was handed over to foreign private companies.

In 1906 itself, there were electric lights and fans installed in the govern office space at Fort St. George, Government House, Guindy and Government Post Office. These were lit by power generated by the generators and oil engines. In the same year, with the success of the North American experience and with the emergence of the massive centralized power supply system, the Madras municipal corporation had understood the necessity of mass electricity production and distribution for domestic, office and public purposes and it had granted license under the provisions of IEA 1887. Commencing its operations, the Basin Bridge in April 1909, it began to provide electricity to the local government for various governmental and public purposes. With its main station at the Basin Bridge, initially, it had two sub-stations one at Egmore and Mylapore. In the central parts of the city, the company supplied 440 volts of Direct Current (DC) to motors and 220 volts of DC to lights and fans. Consumers in the outlying areas: Mambalam, Santhome, Adayar, Saidapet, Kilpauk and Royapuram were supplied with 440 volts of three phase Alternate Current (AC) and 250 volts of single phase AC. In 1934, an extension was sanctioned to supply power to Tambaram from Kilpuak and the service was commenced in 1937 to Tambaram, and its neighbouring areas Sembian, Thiruvatriyur, Alandur and Vallivakkam were served with bulk supply at 500-volt AC. The supply systems were high- and low-tension distribution networks that comprised around 275 miles of underground cables and 140 miles of overhead lines.

This rose up to 10,000 in 1932, controlled by several street electric lighting substations.[36] Introduction of electrical street lighting (incandescent bulb) systems did not totally decimate the kerosene oil (lamp) street lighting system. Yet, it is quite visible that the earlier was dominating the latter slowly by replacing the oil lamps with electrical incandescent bulbs. In 1917, there were 4,374 oil lamps lighting streets of Madras; this went down to 1,805 in 1922. In 1930, all streets in Madras were installed with electric lighting and covered a total length of 302 miles. But there were still 50 oil lamps existing in Mambalam by the turn of the third decade.

As shown in the table, the consumers, in the town, have almost doubled or tripled every year. Half of the power demand of the Basin Bridge station was met by the textile giant Benny & Co. which was also the managing agent for the greater part of the 42 years of license, M&S M Railway Workshop, the South Indian Railway suburban electric lines, Madras Corporation street lighting and water works and Madras Port Trust. The rest was consumed by the general domestic and industrial sphere of the city.

Table 5.3 Increase of consumers, load and units generated by the MESC in 1909–1947

Year	Number of consumers	Peak load in KWs	Units generated in millions
1909	2,539	2,129	8.2
1925	5,445	4,380	16.0
1933	9,197	11,000	37.0
1940	15,912	15,400	60.5
1947	19,600	20,200	81.93

Source: *Madras Information*, Vol. 1, No. 20, 1947.

With the commencement of the MESC in 1909, it had become the main provider of electricity to industrial works, government establishments and to the centralized street lighting system. Street lighting as a public facility on the one hand and electric light at individual houses and shops on the other hand gave a new look to the streets of the city. There was an increasing demand from both government and domestic sectors in the city. Within the urban limits itself, the thermal electricity gained popularity and there was rapid expansion of consumer base for the MESC. It was not only getting popular in the urban limits but also in *mufassil* areas. The number of consumers in 1912 was 1,316 and within five years, i.e., by the end of 1917, this number climbed up to 2,529 consuming 5.1 million units.[37] The standard of supply was improved and the consumption of energy increased by over 20%.[38] Electric street lighting that was started in 1910 had gradually gained momentum. In 1912, only two localities were lit up with electric lights, but within five years, there was 24 miles of street lighting with over 2,400 lights. In *mufassil* areas, private plants were 79 in 1913 and they rose up to 167 by 1917, most of which were installed in the pre-war period. In 1928, electric bulbs with higher candle power were substituted for bunches of lamps of smaller candle power. Due to this, there was some decrease in the use of electric simple lamps. In another decade, it had increased to three-fold and reached 10,231 lights by the year 1932.[39] In the same year, schemes were approved with a cost of Rs. 12.75 lakhs to install more substations, approved by council for municipalizing the entire lighting system. These expansions were planned to be met by funds raised by fixing prices for advertisements on street electrical lamp posts. Towards the end of the third decade, there were three substations at Royapuram, Permabur and Thyagarayanagar. Another 13 substations were added to improve the street lighting in use. In 1938, two high-tension bunks were erected to complete the high-tension ring mains for guarding substations against interruptions. The capacity of the substations was increased based on the increasing demand of power consumption. The Perambur substation capacity was increased from 50 to 100 KVA.[40]

In the early years of the twentieth century, electricity was a costly 'new thing' but was opened to public and had become a public amenity. Electricity was considered as panacea for basic urban problems and as one of the important means in the management of urban space. Lighting public spaces such as streets, parks, fairs, churches and temples were an important feature of urban life. An overhead streetlight during night time was an indication of a 'modern city'. During pre-electricity era, in Madras, religious places were illuminated with only candles and oil lamps. However, with the advent of electricity, new 'electric-holy cross' or 'electric star' came up on the top of the church. Incandescent bulbs were installed on top of the temples and inside as well. The 'holy crescent' on the *Gumbad* (dome) of Masjid was accompanied by an electric bulb. Electrification of temples, churches and masjids added new 'electrified' mood to religious festivals. During festival times, complete illumination of these places, installation of electric religious symbols on top of these religious structures, and birth of 'electric deities' (lit gods' photos and frames) in and around the place gave new festive and aesthetic experiences. Electricity supply at the prominent religious temples like the Tirumala Thirupati temple was started by a private company in July 1929. In 1936, throughout the city, 265 ornamental electric lights were installed at the places of public worship. Maintaining lighting systems at public parks was shouldered by the government. In 1916, the Government spent Rs. 36,000 to maintain electric lighting systems at People's Park, Napier Park, Robinson Park and Loane Square.[41] Ornamental posts with electric bulbs were installed at important bridges, around temple tanks and in certain parks.

Illumination at parks, fairs and religious festivals brought a whole new 'night culture'. It also had important commercial dimensions. According to Schivelbusch,

> night life includes this nocturnal round of business, pleasure and illumination. It derives its own, special atmosphere from the light that falls onto the pavements and street from shops, cafés and restaurants, light that is indeed to attract passers-by and potential customers.[42]

Similarly in Madras, shops owned by British companies, which lit up the surroundings with electric bulbs, drew people towards them. Electric light in these shops was used to symbolize the new business culture. Electricity had almost become an inevitable source for the emerging factories of Madras. While the arc lamp, which was bigger in size and worked with batteries provided lighting outside the location of the firm, electric incandescent lights illuminated the indoors and electric fans provided ventilation and comfort. Before the emergence of well-structured marketplaces, public spaces such as parks were not only places of pleasure and entertainment but also home to business fairs during night time. These business fairs, which promoted night-time business and culture, received electricity from individual foreign firms. Prior to the starting of Basin Bridge, business companies 'owned' light. In 1901, a five-day Madras Fair, held at the People's Park, was illuminated with electricity supplied by the famous watch-marketing company Messrs P. Orr & Sons. Alfred Chatterton, who had installed his booth in the fair, had bought power from P. Orr & Sons. They used small oil engines and dynamos to generate electricity.[43] In 1903, R.G. Piggott, an electrical inspector of the Madras Government, inspected the P. Orr & Son's firm's premises and reported on some of power generation technology used: (1) a machine giving about 6 amperes at 220 volts for working of 6 arc lamps; (2) a machine giving about 40 amperes at 140 volts for working of ten electric fans and 16–25 candle power incandescent lamps; and (3) another low voltage dynamo for electroplating.[44] This inspection also intended to examine the safety measures taken in the wiring of the premises. In all emerging towns, private companies, which were considerably big in size and work, also began to install dynamos for power generation.

This raised the safety question of public in cities. These companies were to be controlled by the IEA. Clause 3 (b) of the IEA guarantees the protection of public from the risk of electricity. The clause necessitates any person who was intending to use electricity in a place likely to affect the public to give notice at least one week before to the commissioner of police in a presidency town. Section 4 of the Act empowered the Governor-General in Council to make rules for the protection of the person and property from injury by reason of contact with or proximity of apparatus used in the generation and supply of electricity.[45] The local government was quite strict on ensuring protective measures in installation of devises which work with electricity at public and private establishments.

The act under appendix 1 rule 34 stipulates instructions for the restoration of the person suffering from the shock should be fixed in a conspicuous place. If these measures were not taken, licenses for the installation of dynamos were rejected. In 1911, I. H. Cardew, Electric Inspector, disallowed the license as Indian Aluminum Company Ltd., Triplicane, which requested for the permission to use the plant, did not fix danger-warning boards at the dynamo site.[46] Though the company had met all rules and regulations of the IEA, it had failed to meet the clause on safety. However, powerful private companies like Benny & Co. found that section 30 of the modified IEA 1910 which necessitates the permission of the Police Commissioner if any alterations were carried out was discriminatory and anti-industry. At many instances, stricter implementation of the IEA rules and regulations had brought in conflicts between bureaucracy and industrialists. Later in 1914, the government issued order on the question of using power for the purposes other than the sanctioned. According to paragraph 3 of the order, the consumer permitted to use power for domestic purposes should only extend the usage on the same system and consumer acquired permission should only extend the usage on the same system. This rigidity was found to be commercially impractical by the MESC officials, including J. T. Jones. However, with the commencement of the Basin Bridge, these private firms and industries getting supply from the MESC' station have shut down on their own.

Alongside, the number of lamps lit by oil came down drastically. Entry of electric lighting thus came in conflict with already-existing kerosene, gas lighting and battery-based arc lighting. However, in the battle of systems, electricity as a better lighting agent eclipsed already-established technologies despite the various strategies of businessmen involved in those technologies. In USA, centralized gas lighting systems in New York went in dire state when electric lighting system was introduced. The electric arc lighting system was threatened by the incandescent electric bulb. A similar situation emerged in Madras Presidency as well. The inefficiency of the older lighting technology prompted many residents of the senate road, Presidency College and Buckingham Canal, to complain against the existing situation during the night time. When the Madras Corporation planned to illuminate the Chepauk Road and Marina Beach, E.P. Richardson, an engineer of the

Corporation of Madras, recommended the introduction of incandescent bulbs in the place of gas lights and arc lights. He continuously opposed the installation of arch lights which he believed were not economical. In his correspondence to the engineer of the Madras Corporation, he wrote:

> Personally I do not advocate arc lighting for long service economical work. At home (London), arc-light is superseded by incandescent lighting, preferably with tungsten filaments, single bulbs from 100 to 1000 C.P. are readily obtainable and this kind of lighting is much cheaper than arch lighting which now cheaply used for exhibitions, special places, large shop fronts, important congested parts of big stations and so on. Do not go in for the Tantalum filament, they consume more energy and have shorter life than the Tungsten.[47]

The introduction of the new electric system in Madras generated opposition as well as support. In the first half of the twentieth century, the city's economic position did not permit the administration to go for luxurious electric lighting technology. But it was true that the people of Madras wanted the city to be lit with better lighting system. The corporation received complaints from the people about poor lighting and inconveniences during night time. Roads, government offices and public places which were lit up with lanterns, slowly lit with electric arc lights and incandescent bulbs. Government had received public complaints regularly on the existing conditions on streets and roads. Due to this public demand, the Corporation of Madras drafted a scheme for the electrification of Chepauk Park.

With the advance of large-scale thermal electricity, though production costs of electricity did not come down, there was an increase in public demand for electricity. The government took up the programme of electrification of government offices and institutions such as Kings Institute, Guindy; Lunatic Asylum, Kilpauk; Government Opthalmic Hospital and the Royapet Hospital. Technical and science institutions such as Guindy Engineering College and Veterinary College were supplied with electricity. In 1914–1915, the Consulting Architect had prepared some 260 draft plans for electric installations in the above-mentioned public facilities. This continuous electrification of public utilities furthered the process of urbanization. Factories established in the suburban areas extended the boundaries of cities. In the *mufassil*, the MESC did not extend its network as it was conscious of the costs of single consumer extensions. Hence, the other private players erected plants to provide power.

It is interesting to note how electricity intersected with another element of scientific modernity introduced by colonialism: modern science education and medicine. Developments in modern medicine had a very direct relationship with electricity. In a modern hospital, with sophisticated medical technology, electricity was crucial for general maintenance and for surgical operations. As a source of lighting, electricity helped to achieve precision in conducting operations. In Madras Presidency, during the pre-electricity era, while the day light was the source of light during night time, apart from lanterns, moon light was one of the main sources of light during night time. Along with the office space, medical institutions were in the list of the priority for electrification. In 1906 itself, ophthalmic and maternity hospitals were electrified. The doctors in the hospitals, mostly Europeans, explained the necessity of electric light in the hospitals. Emphasizing the necessity of lights in the hospital, Surgeon General of the Maternity Hospital, Madras, wrote that 'light in all areas is quite necessary on a moonless night passages are black darkness and so were the varandhas of the pupil's blocks'.[48] At the turn of the second decade of the twentieth century, among the public facilities, government had undertaken large-scale electrification of medical establishments: Royapuram Hospital, Lunatic Asylum in Kilapauk and Kings Institute of Guindy. In 1914, in Madras city, Consulting Architects' section of the Public Works Department scrutinized many applications for installation of electric plants. Largest among them was the power plant installation at the Royapuram Hospital. In the following year, while 237 estimates were scrutinized by the electrical inspector, again it was the Royapuram Hospital with the largest of the estimates with a budget of Rs. 24,460. Apart from providing power for lights and fans, in large hospitals, it was also used for driving pumps (water supply), lifts, sterilization and powering other medical appliances; new electric lifts and cranes, imported from London, were installed in the Royapuram Hospital in 1916 itself[49] and at a medical school at Coimbatore in 1927.[50] The X-ray room and dark rooms in the hospitals were also equipped with electric lights. Electricity also helped in the process of instructions in modern medical science in the presidency. Permanent classrooms and practical rooms in medical colleges came up with the advent of electricity. Earlier, lanterns were used in the lecture halls of the medical colleges. With the advent of electricity, they were replaced with the new 'magic lanterns' – electric lights.

European doctors, accustomed to cold weather conditions, faced difficulties in working under humid conditions. W.B. Bannerman, Surgeon of the Government Ophthalmic Hospital, Madras, in his official correspondence to the Secretary of Madras Government in 1914, complained that

> there is only one fan in the centre of the examination room and it contains two examination couches around which surgeons, assistants, students and nurses congregate and the air from the single fan is quite insufficient to reach both tables.[51]

Such lack of proper working conditions and institutional facilities for European doctors had affected the pace of the growth of medical knowledge. There were no well-equipped permanent practical classrooms in hospitals or in colleges in the 1930s. Sprawson, Surgeon General of the Government of Madras, who understood the importance of advanced medical education, requested for additional electrical installations in the Pathology Department of Madras Medical College for holding classes in morbid histology. He also pointed out the recommendation of the Sadler Commission (constituted in 1917 to look into the problems of Calcutta University) which suggested transforming the existing laboratory into a permanent classroom.[52]

Electricity also had its impact on other scientific education in general. The Presidency College, established in 1841, started offering subjects such as Botany, Zoology, Chemistry and Physics. In the pre-electricity era, classes and scientific experiments were conducted with the help of lanterns. Professor Herbert Spencer Duncan, the principal of the college (1920–1925), in his correspondence with the Director of Public Instruction, explained the need for electricity in classrooms, laboratories and in the library. This correspondence contained the specific requests of the professors of Botany, Zoology, Geology and Physics. According to the professor of Botany, during the night time, electricity was necessary because 'observations on the growth and behaviour of plants will sometimes have to be made at night in connection with the plant physiology. Plant material will have to be fixed at certain definite hours in the night'. For the professor of Zoology,

> lights are required in the Zoology laboratory because the work has sometimes to be done late in the evening and before the daybreak. Wall plugs are necessary because the incubator in the embedding bath can be most satisfactorily worked with electric current.[53]

This illustrates the point that for the impartation of advanced scientific education, electricity was becoming indispensable. Many scientific instruments, used in conducting experiments needed electricity. Concurrently, electricity transformed the method of lecturing, extended the hours of classes and gave comfort to teachers and students in the classrooms.

Thus, electricity acquired new functions and roles day by day. Increasing popularity from various corners of the cities and towns for variety of new necessities provided continuous demand for electricity supply companies. The trend was the same throughout the presidency. To take advantage of the situation, electricity supply companies and municipal licensees kept changing the unit prices, volts and standards. They had no commonality in the way they operated in the presidency. Under these circumstances, the government saw the necessity of evolving common rules and regulations for the operation of electricity business firms.

Notes

1 Dharma Kumar, 'South India', in Dharma Kumar (ed.), *The Cambridge Economic History of India*, New Delhi: Orient Longman, 2005, p. 357.
2 S. C. Misra, 'Urban History in India: Possibilities and Perspectives', in Indu Banga (ed.), *The City in Indian History*, New Delhi: Manohar, 2005, p. 1.
3 *Reports on the Administration of Madras Presidency* (RAMP), 1914–1915, p. 93.
4 RAMP, 1913–1914.
5 RAMP, 1917–1918, p. 81.

6 J. D. Bernal, *The Social Functions of Science*, London: George Routledge & Sons, 1946, p. 373.
7 Jonathan Coopersmith, *Electrification of Russia*, New York: Cornell University Press, 1992, p. 75.
8 Ibid., p. 77.
9 S. Muthiah, 'First City of Modern India', *Seminar*, Special Issue on Chennai, No. 535, 2004, p. 15.
10 Rao Sahib C.S. Srinivasachari, *The History of the City of Madras*, Madras: Varadachari &Co., 1939, p. 287.
11 S. Muttaih, *Madras Discovered*, Madras: East-West Press Ltd., 1987. p. 213.
12 Rao Sahib C.S. Srinivasachari, *The History*, op. cit., p. 301.
13 Ibid.
14 RAMP, *1902–1903*, p. 63.
15 RAMP, *1902–1903*, p. 63.
16 Ibid., p. 63.
17 RAMP, *1926–1927*, p. 56.
18 RAMP, 1919–1920, p. 75.
19 RAMP, *1909–1910*, p. 67.
20 RAMP, *1925–1926*, p. 137.
21 *Andhra Patrika*, 23April 1925; and *Swaraj*, 27 April 1925.
22 The English papers examined in the Secretariat of Fort St. George and thevernacular papers examined by the translators to the government of Madras for the month of October 1901, Andhra Patrika, Madras, 16th 1901.
23 *Report on English Papers Examined by the Criminal Investigation Department Madras* and *Report on Native Papers Examined by the Translators to the Government of Madras for the Fortnight ending the 15.9.1917.* (Hereafter REPECID-M, RNPET-GM). The Colonial Government maintained this special branch to keep an eye on anti-colonial news items in Tamil, Malayalam, Telugu and Kannada.
24 REPECID-M, RNPET-GM, p. 307.
25 *REPECID-M, RNPET-GM, for the week ending 11.10.1919*, p. 1551.
26 REPOICID-M &RNPET-GM, for the week ending 19 May 1931, p. 162.
27 Industrial Commission, Minutes and Evidences, 1916–1917, Vol. III, Madras, Bangalore, and Calcutta, Superintendent of Government Printing, India, 1918.
28 Government Order. Number 21W, Public Works Department, 3.1.1930 (from here on GO. No. PWD)
29 Confidential Fortnightly Reports of Public (General) Department, 24 October 1944
30 Confidential Fortnightly Reports of Public (General) Department, 25 October 1945
31 Confidential Fortnightly Reports of Public (General) Department, 24 December 1945.
32 Rao Sahib C.S. Srinivasachari, *The History of the City of Madras*, op. cit., p. 282.
33 T. K. Derry and Trevor I. Williams, *A Short History of Technology*, New York: OUP, 1960, p. 634.
34 Woflgang Schivelbusch, *Disenchanted Night*, Oxford: Berg, 1988, p. 142. (translated from German by Angela Davies)
35 G.O. No. 2933, PWD, 1900
36 Woflgang Schivelbusch, *Disenchanted Night*, op. cit., p. 299.
37 RAMP, 1914–1915 and 1918–1919.
38 RAMP, 1915–1916, p. 83.
39 RAMP, 1931–1932, p. 76.
40 RAMP, 1937–1938, p. 82.
41 RAMP, 1916–1915, p. 49.
42 Wolfgang Schivelbusch, *Disenchanted Night*, op. cit., p. 142.

43 GO No 3509, PWD, 19.12.1901.
44 GO No. 180 W, PWD, 24.1.1903.
45 GO No. 2998 PWD, 5.10.1901.
46 GO No. 1347W, PWD, 16.10.1911.
47 GO No. 1006W, PWD, 22.7.1912 and GO No. 152W, PWD, 30.1.1912.
48 GO No. 1255W, PWD, 20.6.1913.
49 RAMP, *1915–1916*, p. 83.
50 GO. No. 2005W. PWD, 10.9.1927
51 GO No. 1991Electricity, PWD, 20.10.1914.
52 GO No. 1701W, PWD, 7.6.1930.
53 GO No. 298W, PWD, 11.3.1926.

6 Academic engineering and India's colonial encounter

Bengal Engineering College, Sibpur, a historical perspective[1]

Suvobrata Sarkar

One of the leading English dailies recently reported the following:

> An engineer or a tech graduate will never be out of a job, right? Well you're only 40% right. New data indicates that hotel management graduates are far more likely to be placed than those trained in architecture, engineering or even technology.[2]

A career in engineering is always lucrative to the Indian youths. Why this profession is experiencing hardships in recent years? To find the answer, as a student of history, I went back to the past to study the evolution of engineering education in West Bengal. Currently, the Indian Institute of Engineering, Science & Technology (IIEST), Sibpur, and the Jadavpur University, Kolkata, are famous for engineering education in India. The IIEST is the progeny of Bengal Engineering College, one of the oldest in the country (established in 1856). But due to nationalist connection, Jadavpur University (earlier the National Council of Education) grabbed all the limelight of the academic community, so far as historical study is concerned.

Academic engineering

The term 'academic engineering' describes the teaching of engineering within a university or college of higher education; it differentiates an institutional teaching framework from the broader assimilation of engineering working practices by the method of apprenticeship or pupillage. The growth of academic engineering, both in terms of student numbers and variety of courses, profoundly influenced the structure of what we might call 'practical engineering'[3], the status of engineering as a profession searching for recognition within society, and the corporate relationship between the administrators, engineers, and places of higher education. The emerging discipline of academic engineering in the late nineteenth and early twentieth century was by no means uniform and unvarying across the range of institutions of higher education. One might wonder why a college of engineering was opened at

DOI: 10.4324/9781003241980-9

Roorkee (1847), and subsequently at Calcutta (1856), at a time when Britain itself did not provide academic training to engineers except for military purposes. This chapter highlights some of the relevant issues of academic engineering and its colonial linkages in the context of Bengal Engineering College, Sibpur. Although started as the College of Civil Engineering, at the Writers' Building, Calcutta, in 1856, this chapter focuses on the history of college during 1880–1947.The year 1880 was a major milestone. As in that year, the College was restored to its former status of independent existence after an uneventful 14 years (1865–1879) under the administrative control of the Presidency College. The Workshops of the Public Works Department at Sibpur were attached to the College for practical training of the students. Even the Government of Bengal advised the University of Calcutta to revise the existing regulations to enable a candidate to obtain a License in Engineering or the Degree of bachelor's in engineering in either of the two branches—(1) Civil Engineering and (2) Mechanical Engineering.[4]

Early diffident steps

Though the College was founded for training Civil as well as Mechanical Engineers, in actual practice, however, no proper arrangement was made for giving training in higher Mechanical Engineering of the standard of the degree course. In November 1884, a Parsee student, Sarabji Shavaksha applied for permission to take up the Mechanical Engineering course. He also expressed the reason behind his joining the Sibpur College to pursue this branch of engineering, as he could have easily learned Civil Engineering in Bombay. At this, the Principal communicated his inability to teach Mechanical Engineering course 'up to the university standard' to the higher authority and proposed that 'a properly qualified European should be appointed, upon whom would devolve the duties of superintending both Engineer and Apprentice students whilst engaged in the shops'.[5] Ultimately, it was decided in July, 1885, to appoint E. F. Mondy, Professor of Physical Science, also as Professor of Mechanical Engineering. It was further decided that this arrangement would be applicable only when a class of at least six students was being formed for the study of Mechanical Engineering. No class could however be formed.

Here, the attitude of the British Raj towards the technological calibre of their subject people is significant. The Superintendent of the Workshops, Sibpur Engineering College, was apprehensive as he found a great many of the native students consider manual labour distasteful and endeavour to avoid it. To stop such tendency, he introduced rigorous rules for annual examinations which he hoped 'will convert the idlers into earnest workers…'[6] Such negative attitudes remained throughout the colonial rule about the techno-scientific proficiency of the Indians. Of course, there were loopholes in the system, which were pointed out by several bureaucrats. In one such instance, Sir Alexander Pedler, Professor of Chemistry at the Presidency College and later appointed as the Director of Public Instruction,

Bengal, remarked, '...the present constitution of the Engineering College, and the course of studies pursued in it, require to be thoroughly overhauled. There is no doubt that the College is, for some reason or other, not attractive to students...'[7] His recommendation to the students was to look for the private fields of employment where the modern, practical, and adaptable worker alone was valued.

Sibpur Workshops, as has been mentioned earlier, were under the control of the PWD and run on a commercial basis. The College could not, therefore, utilize the Workshops solely for instructive purposes. When the advice of F. J. E. Spring, Engineer-in-Chief of the East Coast Railway and a member of the Board of Visitors of the College, was sought, he submitted a memorandum on the subject in January 1893:

> It must be ever borne in mind that the problem with which we are dealing, when treating of Sibpur College matters, is not one which solely concerns the well-being of the College itself, but that it is a problem which is intimately bound up with the entire technical education pyramid of which, in Bengal, Sibpur is, of rather ought to be, the top stone...Now when I pass for the handing over the shops to the College authorities, I do so far two principal reasons; because I am convinced from a careful study both at home and out there of the now fairly well understood subject of technical education that until the shops shall have become strictly and solely educational and no longer commercial shops, they will fail in efficiency for educational purposes.[8]

In May 1894, the Government sanctioned the transfer of a part of the Workshops to the Education Department and placed a grant of rupees 50,000 at the disposal of the Principal of the Sibpur Engineering College to erect new shops for it. The Workshops were finally transferred to the Education Department in April 1897.[9] An important decision aimed at promoting the study of engineering as a career was taken in September, 1891, and all appointments in the Upper Subordinate Grade of the PWD of the Government of Bengal were reserved for being filled up by the graduates of the College.[10] By this time E. W. Collin, under the direction of the Government of India prepared a report on the arts and industries of Bengal, which contained certain comments and suggestions in regard to the Sibpur College.[11] The establishment of a special class for the training of Mining Assistants was one of them. The Directorate of Public Instruction, Bengal, also suggested that an alternative course for Mining Engineers should be introduced into the university curriculum. It was decided accordingly that successful candidates would receive 'diploma of qualification as Mining Engineers issued by the Principal of the Sibpur College and countersigned by the Superintendent of Mine'.[12] Subsequent things, however, shaped slowly. The proposal took another 14 years to see the light of day. On the recommendation of a Committee set up by the Government in 1903 to consider the question of providing facilities for the training of qualified Managers and assistant

Managers for the Mines in India, a Mining Department for the College was sanctioned in 1905. Another Committee under the Chairmanship of the Commissioner of the Burdwan Division was appointed for finalizing the curriculum of mining instruction with reference to the practical training of the students. The Mining Classes were opened in February 1906.[13]

In April 1893, the then Principal Civil Engineering College, J. S. Slater wrote to the Director of Public Instruction about the 'liberal loan' from the Messrs Martin & Company, Calcutta-based Engineering firm, of certain electric machines with a view to begin Electrical Engineering at the College. The Principal stated that this would enable him to demonstrate the student's various electrical equipment which were hitherto unavailable to him due to high cost. As this branch of engineering was in an underdeveloped state due to unavailability of trained manpower, efforts were made to install small electric lighting plant at the College and train interested students in Electrical Engineering.[14] Thus, a full practical course in Electrical Engineering was introduced at the Civil Engineering College in 1895 and in the same year the proposal for a complete electric lighting installation was initiated.[15] The early installation work was supervised by Dr. P. Bruhl, Professor of Physics of the College, and helped by some senior students of the Apprentice Department. Three students who had gone through the new electrical course passed out during 1898–1999, and the Public Instruction in Bengal reported, 'all obtained well-paid appointments at once'.[16]

In 1902, the post of Professor of Mechanical and Electrical Engineering was created.[17] As a measure of developing the Workshops, the Principal sent up a proposal for the introduction of electric-power drive into the Shops. The substitution of the old steam-power drive by the modern electric drive was necessary from the point of view of training, as the major shops of those days, viz., The Calcutta Tramways Company, the Cauvery Falls in Madras, the Railway Workshops at Jamalpur, Lillooah, Kharagpur, etc., had already decided to install electric drives. The proposal was accepted in 1902 and the new electric-power plant was inaugurated in March 1904, by Sir Andrew Fraser, the then Lieutenant-Governor of Bengal.[18]

Quest for technological knowledge

During these 20 years (1880–1900) of its existence, the Civil Engineering College, Sibpur, produced several efficient engineers. To name a few who graduated in the late 1890s and made their mark are Surendra Kumar Basu who cleared B.C.E. in 1880 after the College opened at Sibpur. The degree B.C.E. was replaced by B.E. (Bachelor's in engineering) in the year 1885. Basu was the famous author of the popular textbook, *Building Materials and Construction*. Aannada Prosad Sarkar (1883) rose to the position of the first Indian Chief Engineer (Irrigation), Government of Bengal. Anukul Chandra Mitra (1887) became famous as a practical engineer of great skill. He oversaw the construction of the 'Victoria Memorial'. Girish Chandra Das (B.E. in 1891) became the Chief Engineer of the Light Railways under

Messer Martin & Co. He was also the Engineer-in-Charge of construction of High Court at Calcutta. Another engineer Amar Nath Das (1895) joined the Indian Service of Engineers. Later, he rose to the position of Chief Engineer of Bengal. Another distinguished student of the period is Benimadhab Mitra (B.E. in 1900) who made his mark as an authority on roads construction.[19] But the combination of engineering and entrepreneurship was extremely rare in those days. The only notable exception was Sir Rajendra Nath Mookerjee (1854–1936).

In the year 1902, the students of the college started publishing a monthly journal in Bengali, *Shibpur College Patrika*. Their intention was noble— 'Can we improve the existing state of our agriculture and industry with the help of science? To learn the basic scientific principles, we need a command over the English language. But one feels comfortable in his own mother tongue while learning a complex subject like science. Alas we don't have such comfort. Hence only the English-educated have the entry-pass to the wonder world of science. Thus, the mass, without the English-skill, miss the chance to get the flavour of modern science. Unfortunately, our artisan class, the flag-bearer of our industry, belongs to this section. The transmission of scientific ideas, in simple Bengali, among them is the main purpose behind our humble effort...'[20] This journal published a wide variety of popular articles on science and technology and the contributors were mainly students of the College. In one such instance, the author narrated the marvels of electricity:

> The main gossip of today's Calcutta is the electric tramways! This modern mode of transport is not horse-driven or human-driven, not even steam-driven—but electricity is the prime mover for tramcars. One become perplex while considering the vast range of activity of this new technology! In every aspect of our life electricity is now an obedient servant to mankind. Electric-stove for modern cooking (as substitute of fire), electric-fan during the hot summer season, and electric-lamp for illumination in the night—everywhere is electricity. To run the machines of a factory, electricity plays the pivotal role by producing hundreds and thousands of horse-power. If one wants to talk with a friend mile away from him, the remedy is telephone—another offshoot of electricity. One cannot think without electricity in modern times.[21]

On another occasion, the journal observed:

> To man the Public Works Department, Municipalities and District Boards, the college was first established at Sibpur. So, the general conception about the college is that its main intention is to train the engineers, overseers, and sub-overseers. To some extent it is true as the favourite destination above 80 % of the alumni of the college to these departments and bodies. It does not imply that the college only entertains the job-seekers. The choice for government job, over entrepreneurship,

among the youths is not the result of the instructions of this college. It is the common tendency and result of the overall national degradation of our society. The college courses are ideal for the potential entrepreneurs and industrialists. But unfortunately, those who come here for training in engineering, come only when they failed in general education and with an intention for quick employment. So mostly they are job-seekers, not would-be entrepreneurs and industrialists.

The demand for Mechanical Engineering has been increasing speedily. The reason behind this is that our educated-folk become aware of the benefit of entrepreneurship and industry. Compared to other engineering colleges of India, Sibpur provides a thorough training in engineering and the courses are also of advanced character. Manufacture of engineers for the Public Works Department is not its main goal—but, the students of this college have every potentiality to become a successful engineer-entrepreneur.[22]

This venture however did not last long and with the departure of the editors (Atul Chandra Bagchi and Upendranath Kar) from the College, the publication was discontinued in 1908.

A scheme to impart Industrial Chemistry at the Sibpur College was worked out and submitted to the Government saying 'instruction should be given in the scientific principles underlying the chief industries of the country, and the facilities should be afforded for research work' in 1907. It was practically a post-graduate scheme for students with science degrees, except those who had passed the intermediate examination in engineering. It also proposed to offer admission of occasional students with a practical knowledge of some industry and wanted to study the theory underlying that industry. According to the proposal, the course of instruction was divided into two: (I) technological chemistry and (II) dyeing and chemistry of dyeing.[23] Although the Bengal Government was optimistic, the Secretary of State was sceptical about the demand for men with such advanced training in India and asked the Bengal Government to justify its proposal.[24]

The Government of Bengal tried hard to convince that the objective of the proposed courses is to train men for direct employment in the different cotton mills, engineering firms and manganese concerns, or 'who will at least be in a position to proceed to a further course of practical work in actual manufacturing concerns such as is necessary to make them experts in their respective subjects'. Several Indians also thought such advanced level of training was not the call of the time and instead suggested to confine it to the artisan level.[25] But this time, a determined Bengal Government argued:

> ...Applications are also received occasionally by the Principal of the Civil Engineering College for chemists for iron works, engineering firms and manganese concerns. Such applications must be refused owing to the absence of any properly qualified men, since the chemistry hitherto

taught at Sibpur and the other Bengal colleges has not been such a nature as to fit students to take up analytical work of a special kind. The new classes, however, will be able to turn out men suitable for such appointments...Indians with an enthusiasm for the commercial development of the country on Indian lines are founding and will no doubt continue to found business of their own in preference to seeking employment with European firms; and as the dyeing industry, unlike cotton industry, is one which can be profitably conducted in small factories, and with a moderate capital, it may confidently be anticipated that in the immediate future private enterprise will provide a large field of employment for students who have taken the course of dyeing under proposed scheme.[26]

Finally, sanction was given for an experimental period of three years with a clause that the continuance of the classes would depend upon the practical results obtained. The inauguration of the classes originally fixed for November 1909 had to be postponed for a year. Rajendra Nath Sen, who had been selected for the post of Professor, was pursuing his training at the University of Leeds. In the meanwhile, steps were taken to equip a laboratory and to purchase a large stock of chemicals and apparatus. Sen was eventually appointed for three years as Professor of Tutorial Chemistry at the College and the classes were started in January 1911.[27] Although the number of students declined in the subsequent years, a few of them were successfully employed in various industries and started their own business.[28] The Government of India felt that the paucity of students might be due to some extent of public ignorance and hence suggested the local government to encourage the public through advertisements, not only in Bengal but in other parts of India as well.[29]

In the early twentieth century, an important step in the form of State Technical Scholarship was taken for the Indian students to pursue technological studies in Europe or the United States for the betterment of Indian industry. It was decided to exclude engineering from the scope of such scholarship and the mining industry, in the case of Bengal, was found to be the most favourable field.[30] The Government of India declared: 'There is an urgent demand in this province (Bengal) for persons who had some training in mining; and a scheme for utilizing the Sibpur Engineering College as a centre where such instruction should be imparted...'[31] Accordingly it was decided to send few selected persons to Europe for a special training course in mining. The Government of India encouraged the local governments to consult the leading commercial firms, both European and Indian, in the task of selecting scholars. But the European firms were generally hostile to the scheme. They were apprehensive about the idea that 'natives of India trained in business methods and capable of competing with them on their own ground in the capture of the European markets'. The Government of India believed a sound commercial education, on the lines of Manchester and Birmingham University, would benefit the emerging Indian entrepreneurship to compete in various modern business ventures. The scheme of State Technical Scholarships would be beneficial for prospective Indians as it enabled them to proceed to various foreign countries for advanced technological training.[32]

In 1909, the Bengal Government recommended 'two outstanding students', Girindra Nath Dutt and Jitendra Nath Das Gupta, for the State Technical Scholarships in the field of Mechanical Engineering. But the Government of India rejected both the proposal claiming that engineering was 'expressly excluded from the scope of technical scholarship' and asked for alternative proposals.[33] At this, the Bengal Government recommended one Manmatha Nath Baysack in the field of textile chemistry and the proposal was readily accepted. But this didn't go without debate. Finally, in the very next year the Government of India announced:

> ...electrical and mechanical engineering should be considered as falling within the scope of the technical scholarship scheme, as the instruction available in these subjects in this country is far less advanced than the procurable in civil engineering, while the demand for qualified electrical and mechanical engineers to take charge of factories and works is steadily on the increase.[34]

Academics and infrastructure: frailty and failures

In those days, Electrical Engineering mainly dealt with electrical power generation, transmission and utilization. With the multiplication of the number of electrical power houses, and with the increase of its industrial usage, the demand for men trained in the use of electrical machinery and acquainted with the fundamentals of this technology began to grow and the contemporary engineering colleges started upgrading their facilities for this branch of engineering. Sibpur Engineering College was also not an exception.[35] The training in Electrical Engineering, in early twentieth century, was confined almost entirely to the apprentice (or subordinate) stage at the Sibpur College. Students in this branch first had to go through the ordinary four years' course for PWD overseer students, and then specialize in electrical work for some 15 months. In the first four years, they spent much time in the various mechanical workshops and took such subjects as steam engine and applied mechanics. Thus, huge time was devoted to Civil Engineering and other subjects of little subsequent use to them in later life.[36] Mechanical Engineering was not a favourite destination of the students as the course was meant to train the electrical men. W. H. Everett, the then Acting Superintendent of Industries, Bengal, was all praise for the Plant:

> In the running of plant the students get an excellent training, both with direct and alternating machinery, including testing of engines and boilers. They have to keep a regular log daily, and work out the costs for each month. I know of no college in England where such a full training

is given in actual running, the plant being in actual use for power and light and about 12 hours daily. The students also get much experience in wiring; and several 1 ½ H.P. motors have been made by them. They go through a simple course of electrical testing, as well as lectures and class work.[37]

When the College was steadily developing into an institution of high standard in Sibpur, it was suddenly decided to shift it to another site on the ground of un-healthiness of the place. Meanwhile, the advisability of creating a Technical Institute in Calcutta itself was brought to the notice of the Government from various quarters.[38] The creation of Technical Institute of Calcutta was advocated on the ground that in all European countries in which technical education had reached considerable degree of development, such education had almost entirely confined to industrial centres.[39] The proposal of abolishing the Civil Engineering College at Sibpur aroused a considerable opposition in Calcutta. One Surendra Nath Roy moved a resolution in the Bengal Legislative Council to the 'effect that the proposal to abolish the Sibpur Engineering College be dropped' in February 1914.[40] Finally, in 1918, the Government of Bengal decided to retain the Civil Engineering College at its present site, Sibpur.

In the meanwhile, the Government of India had deputed E. H. Atkinson and Tom S. Dawson to enquire about the state of technical education in the country with special reference to their industrial applicability. They visited various industrial concerns, both private and government owned, and prepared a huge report.[41] The general assumption of the employers (i.e. various industrial firms) in Bengal, as mentioned in their report, was that the educated Bengalis were averse to manual labour as a regular employment even after their training in a technical institute. Their other major observations include: 'There is at present no demand with a university training in mechanical and electrical engineering...the university type of education unfits a Bengali for practical work, and although an engineering graduate would have a more complete scientific education he would be far behind the technically-trained student in practical ability and less valuable to an employer'.[42] Instead, they suggested for training as an apprenticeship on a living wage, work with their hands, and observe factory hours and rules. Such training would enable the Bengali job-seekers to convince the employers as there was a huge opportunity for the men with practical knowledge in mechanical and electrical engineering.

Despite the hindrances created by the proposal to shift the Sibpur College, B. Heaton, the then Principal of the College, proposed a three-year Overseer Course in Mechanical and Electrical Branch. The Director of Public Instruction, Bengal, was also in favour of creation of a Professorship in Mechanical Engineering who would confine his attention to Electrical Engineering with the addition of the electrical portion of Physics.[43] In January 1914, the Government of Bengal sanctioned the proposal.[44] In the session 1912–1913, the Electrical Engineering Department was separated from the

Mechanical Engineering Department. The Mechanical Engineering Department received great impetus in 1914 when the newly appointed professor C. A. King put forward a scheme for the development of the Department by improving the courses of studies and all sections of the laboratories including Prime Movers & Workshops by procuring some essential equipment from abroad and also by preparing a large number of small equipment and models in the College Workshops.[45] During the period of uncertainty as to the future of the College, some minor projects were taken up. A motor mechanic class was opened in 1907 and was abolished in 1910. A class to train selected telegraphers was also opened in 1909. A Dyeing class was opened in 1910 and abolished in 1916, as it failed to attract students.[46]

Technology, education and imperialism between the World Wars

Thus, a change in the official attitude was around the corner as priorities were shifting from agriculture to industry, hence to technical education. The First World War (1914) exposed India's industrial backwardness and her dependence on other countries for a variety of commodities like machines, prime movers, and most importantly technical skill. As an important source of manpower and materials, mostly war-related, combined to effect a change in the British attitude preferring India's industrialization.[47] The most significant outcome of this change in attitude was the appointment of the Indian Industrial Commission (1916) which was constituted to study the question of India's industrialization and explore the possibility of state participation in it. It was in this connection that the Commission directed its attention to the various aspects of techno-science including education, research, and development.[48] Its findings revealed several lacunas in the existing state of technological development and the government approach to that. In order to remove these drawbacks, the Commission emphasized the necessity of closer coordination amongst the technologists, government departments, universities and institutes, and the industries. It wanted to relate technical education and research to the actual industrial needs of the country.[49] However, the changes suggested in the report were never translated into practice. In fact, the recommendations suffered from inherent contradictions. The needs and interests of the foreign masters and their subject people were mismatched. So, the recommendations were bound to be doomed.

The Indian Industrial Commission discussed the engineering curricula, method of teaching, etc., and recommended substantial changes. The Commission visited the Sibpur College during 1916–1917. The then Principal, B. Heaton, although expressed dissatisfaction over the Bengali character in terms of technological proficiency, but was also optimistic that '…a Bengali student, whose tendency is more towards philosophy and the abstract ideas found in law, we must take every possible means to bring his mind from the clouds, from thoughts of the abstract to the practice of the concrete'.[50] His proposal was to retain the Thomason College, Roorkee, for higher civil engineering course and convert the other three (Sibpur, Madras and Pune) into 'mechanical and electrical engineering colleges and into mining colleges where facilities could be offered for advanced courses of a university type'.[51]

In pursuance to the policy recommended by the Industrial Commission, the Mechanical and Electrical Engineering Classes of the College were being developed independently of the Calcutta University and it was intended that the University would in due course recognize these courses for awarding a Degree in Mechanical and Electrical Engineering. It was thought the ideal method of training Mechanical Engineers was to combine workshop practice and technical instruction as closely as possible.[52] Since the Mechanical and Electrical Engineering and Mining Classes were separated from the Civil Overseer classes four years ago (1914) the students who applied for these two branches had increased enormously. The Bengal Government sought to raise the strength of the faculty by appointing 'an Assistant Professor of Mechanical Engineering and a first-class Mechanical Draughtsman'.[53] This time, the local government had a valid reason as it wrote 'A further important consideration is that the classes in Mechanical and Electrical Engineering and Mining are rapidly increasing both in importance and popularity. There is an ever-increasing demand for well-trained Mechanical Engineering for the development of the industries...'[54] Both the proposals were sanctioned in 1919. The Principal of the Sibpur College selected Purna Chandra Ganguli, a B.Sc. of Glasgow and Professor of Mechanical Engineering of the Bengal Technical Institute, for the post of Assistant Professor 'on probation for six months'. Ganguli was selected due to his nine years of teaching experience as In-Charge of the Mechanical Engineering Department and his knowledge of the Bengali students and of local conditions.[55]

One of the major reforms undertaken during the period was to re-organize the various Departments of the College, which was suggested by the Mookerjee Committee. This Committee under the Chairmanship of Sir Rajendra Nath Mookerjee was formed in February 1919, to coordinate the Apprenticeship Training Scheme of the Eastern Bengal Railway Workshops at Kanchrapara with the Mechanical Engineering Classes of this College.[56] The Committee recommended that the lower classes in the Mechanical Engineering Department of the College should be given up and instead a Technical School at Kanchrapara should be established. The Sibpur College should be utilized for the final training of the Mechanical Engineers, after they had completed the full course of apprenticeship at Kanchrapara or other large Workshops in Bengal. In accordance with these recommendations, the Board of Control for Apprenticeship Training was created in 1921.[57] The improved courses received Government sanction in the next year and with this reform, the College was mainly concerned with the training of persons to occupy higher position in the engineering industries of the country.

During the deputation of W. H. Everett, Professor of Mechanical & Electrical Engineering, B. C. Gupta had overseen the Electrical Engineering Department and Power Installation. Jitendra Nath Chakravarti was appointed at the Department as demonstrator to the special class for training Subordinate Telegraphists of the Government of India. It was said that both were acclaimed for their teaching efficiency.[58] The Indian School of Mines at Dhanbad was built by one of the ex-students of the College, Anath Bandhu Bhattacharya under Messrs Martin & Co. in 1926. So, the admission of students into the Mining Classes was stopped and mining was finally closed at the Bengal Engineering College in 1929.[59] To utilize the space vacated by the Mining Department, it was proposed that a course of instruction in advanced metallurgy might be introduced, but without effect.

The next important event in the history of the College was the introduction of the Degree Course in Mechanical Engineering and the first batch of Mechanical Engineering graduates, six in number, passed the Degree Examination of the Calcutta University in 1932.[60] There was some consequence of the Great Depression upon the development of the College. It was due to financial stringency, that Degree Course in Electrical Engineering could not be introduced along with the Mechanical Engineering. It was only in 1935 that the Degree Course in Electrical Engineering came into existence.[61] The Degree Course in Metallurgy was introduced in 1939–1940.[62] Dr. A. H. Pandya, a renowned engineer of international repute, took over as Principal of the Bengal Engineering College in 1939 and he was the first Indian Principal of the College.

During the First World War, the Testing Laboratory of the College was utilized for testing of war materials. During the Second World War, many technicians for the Munitions Factories, Armed Services and the Technical Wing of the Air Force were trained up in the College Workshops which served as a combined centre under a special technical training scheme. The Second World War affected the College adversely, as entire energy and attention was diverted towards producing war technicians. Temporary barracks and classrooms were built in 1940, and training commenced in the next year. During this entire academic year, no admissions were made to the first-year class to make room for the technicians![63] It did not take long to realize the massive potentialities of techno-science for the successful conduct of a modern war. As the war raged, it once again badly exposed India's technical and industrial backwardness and her dependence on others for a large variety of goods and commodities irrespective of wartime requirements.[64]

Engineering education and industrialization

The Bengal Government constituted a committee under the Chairmanship of F. Rahman in 1943 to enquire amongst other things the prevailing state of technical and industrial education in Bengal and also to suggest a comprehensive scheme so as to serve the needs of the prospective industries of Bengal.[65] In regard to the Bengal Engineering College, it was pointed out by the committee that the Workshops and Laboratory arrangements at the College were inadequate, and for successful training in higher engineering these defects should be removed. The committee further observed that apart from the degree courses in Civil, Mechanical, Electrical and Metallurgical Engineering, which were already in existence, arrangements for starting new courses, such as Ship-building, Aeronautics, etc., should also be made, and that this College should impart training in higher engineering only, leaving the Diploma and the Apprenticeship courses to the Calcutta Technical School.[66]

Soon after the outbreak of the Second World War, various committees had been constituted to meet with its different, especially technical, requirements in India, and along with it came the realization of the magnitude of the task of reconstruction.[67] But for India, with divergent interests of the rulers and the ruled, the post-war reconstruction was rendered more difficult. The importance of applied science and technology and necessary manpower for fulfilment of the task were, however, keenly felt. Meanwhile, industries in the country were expanding and their needs as well as the needs of the Government for higher technical personnel to meet various increasing activities became urgent. In its report on post-war educational development of the country, the Central Advisory Board of Education remarked, 'In view of the recent expansion of industry and the likelihood of further development after the war it is necessary to plan immediately a comprehensive system of technical education at all stages'. It also emphasized:

> ...the necessity for ensuring that in the University technological departments ample facilities are provided for research directed towards the solution of practical industrial problems and that the need for testing the results of laboratory experiments on a commercial scale is not overlooked. It is not less important that technological degree courses generally should be made more practical than they are at present and that the students should spend a considerable part of their time throughout the course, and not merely at the end of it in works and factories.[68]

The Bengal Engineering College which was one of the few institutions for producing the higher engineering personnel had to keep pace with these expanding activities. A 'Committee for the Development of Higher Engineering and Technical Education in Bengal' was accordingly formed in 1945 under the Chairmanship of the DPI.[69] The Committee submitted its reports in two parts: one containing the 'Immediate Plan' and the other 'Five Year Plan'. While submitting its report under the 'Immediate Plan', the Committee confined their recommendations to higher engineering training and the training of supervisory personnel, and aimed at increasing the number of qualified engineers and supervisors in the more important and immediately necessary branches of engineering, improving their quality and standard and also providing facilities for advanced instruction and research.[70] In the opinion of the Committee, these improvements were urgently needed in the light of post-war expansion. Government accepted these recommendations and the Calcutta University, after consulting the Faculty of Engineering and the Board of Studies in Engineering, framed new curricula and revised the Regulations relating to Degree Examination in various branches of Engineering in 1946.

The objective of the 'Five Year Plan' prepared by the Development Committee was the ultimate attainment of converting the Bengal Engineering College into a most modern and well-equipped institution comparable to the best institutions of the world, with a greater range of degree courses and facilities for the post-graduate and research work. But the scheme materialized almost ten years later, only after independence.[71] The collective adoption of the 'Immediate Plan' and the 'Five-Year Plan', by the Committee, was with a view to fostering close co-operation and collaboration between industries and the College.

Conclusion

This essay has delineated the process of the colonial transplantation of a modern institutional structure to produce technical knowledge in the late-nineteenth and early-twentieth-century Bengal. The Bengal Engineering College, Sibpur, sandwiched between colonial-tapping and nationalist antagonism, fulfilled long the demand for technical manpower in the Eastern India. The Sibpur College, compared to other three government engineering colleges, maintained a practical character of teaching from the beginning which in turn provided better industrial orientation among the students. But colonial constraints compelled this college to adopt a 'go-slow' policy in the field of advance technology, especially in mechanical and electrical engineering. Although it produced a large cadre of civil engineers, in the fields of mechanical and electrical sciences its performance was meagre. The result was as expected; among the alumni of Sibpur, with the rare example of Sir R. N. Mookerjee, very few turned to business. In the fields of mechanical, electrical or mining engineering, with rudimentary training, the students were mostly eligible for lower level technical jobs. Entrepreneurship was not there in the vocabulary of the Sibpur students.

Unlike the predominant historiography which highlights the 'centre-periphery' relationship and influence of PWD and Military on engineering education[72], this essay elucidates several positive points, in the case of Sibpur College and its struggle to transcend the barrier imposed by colonialism. On several occasions, the British bureaucrats claimed the superior quality of teaching, apparatus, workshop facilities at the college, which was sometimes better, or at least equal to their 'home' condition. Many demanded upgradations of training and courses, but without any effect. Almost all the latest engineering sciences—mechanical, electrical, mining, industrial chemistry, etc., were introduced immediately with their application in Western Europe, but on a limited scale. Sibpur was unable to attract always the best brains of Bengal. This is evident from the testimony of the Principal of the college in front of the members of Indian Industrial Commission (1916–1918) and from the writings of the contemporary students (1907).The favourite destination of the Bengali youths was legal profession, followed by medicine. But, despite several obstacles, the Bengal Engineering College survived, and became a premier seat of higher learning in the field of engineering and technology. As recognition of its almost 160 years of glorious existence, the IIEST, as the first of its kind, was set up by the Government of India in 2014 by converting the erstwhile Bengal Engineering College, Sibpur.

Notes

1 Originally presented at the Silver Jubilee National Seminar on 'Contemporary Debates on Science, Technology and Nationalism in India', Department of Humanities & Social Sciences, IIT Guwahati, October 31–November 2, 2018. The author is grateful to the organizers for the opportunity and the participants for insightful comments and suggestions.
2 *Times of India*, 13 August, 2018, Kolkata edition.
3 Ben Marsden, 'Engineering science in Glasgow: economy, efficiency and measurement as prime movers in the differentiation of an academic discipline', *BJHS*, Vol. 25, No. 3, 1992, pp. 319–346.
4 Francis Spring, *Technical Education in India*, Calcutta: W. Newman & Co, 1887, p. 20.
5 S. F. Downing, the then Principal of the College, stated that, 'The subject laid down by the University authorities for the above examination, which, I with my present instructive staff, I am unable to teach, are as follows- (i) Mechanical Machine design, (ii) Design and workshop appliances, (iii) The Steam Engine (in its practical aspect)'.Proceedings of the Government of Bengal, General, Education, No. 3709, July 1885, West Bengal State Archives, Kolkata (hereafter WBSA).
6 *General Report Public Instruction in Bengal, 1884–1885*, Calcutta: Bengal Secretariat Press, 1885, p. 7.
7 *General Report Public Instruction in Bengal 1885–1886*, Calcutta: Bengal Secretariat Press, 1886, p. 75.
8 'Sibpur Civil Engineering College and its connection with the Sibpur Workshops', Note by F. J. E. Spring; *General Report Public Instruction in Bengal, 1893–1994*, Calcutta: Bengal Secretariat Press, 1894, p. 17.
9 *General Report Public Instruction in Bengal 1897–1898*, Calcutta: Bengal Secretariat Press, 1898, p. 9.
10 *Bengal Engineering College Centenary Souvenir*, Calcutta: no pub, 1956, p. 29.
11 E. W. Collin, *Report of the Existing Arts and Industries in Bengal*, Calcutta: Bengal Secretariat Press, 1890, pp. 3–7.
12 Proceedings of the Government of Bengal, General, Education, File No. 1485, May 1892, WBSA.
13 Home, Education, A. Proceedings, August 1907, Nos. 84–86, National Archives of India, New Delhi (hereafter NAI).
14 The following electric machines were received by the Civil Engineering College, Sibpur: Ferranti (120 Incandescent 16 C.P. lights, one Dynamo Motor, two Brush Lamps, one Brush and eight light Dynamos, one Gramme Excitor for the Ferranti Dynamo. General, Education, File No. 4C/22—1 & 2, Nos. 4–5, May 1893, WBSA.
15 PWD, Civil Works—Miscellaneous, April 1896, Nos. 1–7, NAI.
16 *General Report Public Instruction in Bengal for 1898–1899*, Calcutta: Bengal Secretariat Press, 1899, p. 115.
17 Home, Education, A. Proceedings, October 1901, Nos. 80–84, NAI.
18 *General Report on Public Instruction in Bengal for 1903–1904*, Calcutta: Bengal Secretariat Press, 1904, 30.
19 *Bengal Engineering College Centenary Souvenir*, op.cit., pp. 26–35.

20 *Shibpur College Patrika*, Vol. II, No. 1, 1903, p. 4.
21 *Shibpur College Patrika*, Vol. I, No. 8, 1902, pp. 231–232.
22 *Shibpur College Patrika*, Vol.1, No. 2, 1907, pp. 25–29.
23 The course in technological chemistry would include—(A) lectures and laboratory work in organic chemistry and applied chemistry; and (B) theoretical and
   practical work in the following branches of mechanical and civil engineering,
   heat engines, principles of machine design, drawing of building and machines,
   and manual training, carpentry, and fitting. Home, Education, A. Proceedings,
   November 1907, Nos. 47–50, NAI.
24 Home, Education, A. Proceedings, March 1908, Nos. 8–13, NAI.
25 Raja Peary Mohun Mukherjee, member of the *Association for the Advancement
   of Scientific and Industrial Education for Indians*, commented, 'A graduate who
   has gone through a three years' course of special study and obtained a diploma
   will, I am assured, except not less than Rs. 100 per mensem from his employer,
   and very few of the smaller industries on which the Government of Bengal lays
   so much stress can hope for a margin of profit after paying so much'. Home,
   Education, A. Proceedings, October 1908, Nos. 110–111, NAI.
26 H. C. Streatfeild, Officiating Secretary to the Government of Bengal, wrote to his
   counterpart at the Government of India, Home Department (Dated: July 1,
   1908). Ibid.
27 General, Education, File 4C/81—1, Nos. 61–65, November 1913, WBSA.
28 The final examination was held for the first time in 1912 and two candidates
   cleared it out of three. One obtained employment at the Tata Iron & Steel Works
   while the other remained at the College for research work. One occasional student started a dye factory in Calcutta and became very successful, while another
   employed at Dr. K. C. Bose's technical lab. Ibid.
29 General, Education, File No. 4C/9—12, No. 253, July 1914, WBSA.
30 Home, Education, A. Proceedings, October 1903, Nos. 14–18, NAI.
31 Home, Education, A. Proceedings, October 1904, Nos. 91–124, NAI.
   Three students were selected for advanced training in Mining at the University
   of Birmingham. A short introduction about the first three scholars who received
   the State Technical Scholarship:(1) S. C. De was trained at the Bihar School of
   Engineering and obtained a Survey Standard Certificate. He served the Birbhum
   Coal Company and Nandi Coal Company at the capacity of Mining Surveyor;
   (2) Asoke Bose, son of great geologist P. N. Bose, was studying Geology at the
   Presidency College when he was selected for the scholarship. He already received
   practical and theoretical training in mining at the West Baraboni Coal Company
   which belonged to his father; (3) P. K. Majumdar was the Manager at the estate
   of the Maharaja of Gidhour. In that capacity, he established a mica-mining
   industry there and thus gained a practical knowledge of mining. Ibid.
32 Home, Education, A. Proceedings, July 1907, Nos. 134–143, NAI.
   Many times, these Indian students encountered difficulties in the foreign
   countries due to the lack of sufficient information. In 1907, the Secretary of State
   appointed a Committee to enquire the position of theses scholars in England. In
   one such instance, this Committee helped one Gopal Bhaduri, who came to
   England to study Electrical Engineering without much information, to locate
   ideal institution for this branch of engineering. Home, Education, A. Proceedings,
   October 1908, Nos. 27–35, NAI.
33 Home, Education, A. Proceedings, June 1909, Nos. 145–152, NAI.
34 Home, Education, A. Proceedings, October 1910, Nos. 69–70, NAI.
35 The Bengal Government sanctioned a contingent grant for the Electro-Technical
   Laboratory of the College in 1911 under the head 'Provision for additional
   teaching staff for Electrical Engineering Department'. General, Education, File
   No. 4C/37, Proceedings Nos. 221–222, June 1911, WBSA.

36 *Report on Public Instruction in Bengal for 1910–1911*, Calcutta: The Bengal Secretariat Book Depot, 1911, pp. 35–37.

37 Ibid, p. 41.

38 J. G. Cumming, *Technical and Industrial Education in Bengal, 1888–1908*, Calcutta: Office of the Superintendent Government Printing, 1908, p. 31.

39 The following subjects were recommended to include at the proposed Technical Institute: 'Mechanical and Electrical Engineering, in both the higher and lower grades; Civil Engineering, up to and including the Overseer grade in any case; Textile subjects, and Commercial subjects'. *Proceedings and Report of the Committee Approved to Advise on the Creation of a Technical Institute for Calcutta and Allied Subjects*, Calcutta: Bengal Secretariat Book Depot, 1912, p. 7.

40 *Calcutta University Commission Report 1917–1919*, Vol. III, Calcutta: University of Calcutta, 1919, p. 94.

41 E. H. Atkinson & Tom S. Dawson, *Report on the Enquiry to bring Technical Institutes into closer touch and more practical relations with the Employers of Labour in India*, Calcutta: Superintendent Government Printing, 1912.

42 Ibid, pp. 89–90.

43 General, Education, File No. 4C/59—1, Proceedings Nos. 50–53, March 1913, WBSA.

44 General, Education, File No. 6E/19—1, Proceedings No. 61, January 1914, WBSA.

45 *Bengal Engineering College Centenary Souvenir*, op.cit., p. 54.

46 *Report on Public Instruction in Bengal for 1915–1916*, Calcutta: The Bengal Secretariat Book Depot, 1916, pp. 19–23.

47 Jagdish N. Sinha, *Science, War and Imperialism*, Leiden & Boston: Brill, 2008, pp. 27–9.

48 *Report of the Indian Industrial Commission*, Calcutta: Superintendent Government Printing, 1918, pp. 109–110.

49 Ibid, p. 95.

50 Evidence of Mr. B. Heaton, *Indian Industrial Commission, Minutes of Evidence 1916–1917, Vol. II, Bengal & Central Provinces*, Calcutta: Superintendent Government Printing, 1918, pp. 249–250.

51 Ibid, p. 253.

52 Government of Bengal, General, Education, File 4—C/3 1–2, Nos. 1–6, April 1918, WBSA.

53 Government of Bengal, General, Education, File No. 4-C-11 (1), Nos. 11–15, June 1919, WBSA.

54 General, Education, File No. 4-C-11 (2), No. 16, June 1919, WBSA.

55 *Report on Public Instruction in Bengal for 1920–1921*, Calcutta: Bengal Secretariat Press, 1922, p. 58.

56 *Report on the Bengal Engineering College for the Quinquennium 1917–1918 to 1921–1922*, Calcutta: Bengal Secretariat Book Depot, 1922, p. 6.

57 The Upper Subordinate Classes were abolished after long years of existence. The last link of the College with the PWD was severed, and no more students were sent up for the examinations under the Joint Technical Board after July, 1922. The control of examinations in the improved Mechanical and Electrical Engineering courses was transferred to the newly created Board of Control for Apprenticeship Training from March, 1922. Now the various Departments of the College were thus recognized:(1) Civil Engineering Classes, (2) Mechanical & Electrical Engineering Classes, and (3) Mining Classes. (2) & (3) were formerly Apprentice Department. Ibid, pp. 7–9.

58 *Report on Public Instruction in Bengal for 1923–1924*, Calcutta: Bengal Secretariat Press, 1924, p. 27.

59 Arun Kumar, 'Bengal Engineering College (Sibpur): A Study in Historical Perspective', in Chittabrata Palit and Arun Kumar (eds.), *Science and Environment*, Delhi: B. R. Publishing Corporation, 2013, p. 17.

60 The successful candidates were Brojendra Chandra Bagchi, Satipati Bhattacharya, Kumudranjan Chaudhuri, Bijoygopal Dutt, Ahmed A. N. Kaliruddin and Ambikacharan Mukhopadhyay. *Report on Public Instruction in Bengal for 1932–1933*, Calcutta: Bengal Secretariat Press, 1933, p. 19.

61 The first batch of students appeared at the Degree Examination of the Calcutta University in 1936. The successful degree holders were 3 in number: Ramendra Nath Bandopadhyay, Sibapada Chattopadhyay and Md. Mahibul Majid. *Report on Public Instruction in Bengal for 1937–38*, Calcutta: Bengal Secretariat Press, 1938, p. 34.

62 *Report on Public Instruction in Bengal for 1939–1940*, Calcutta: Bengal Secretariat Press, 1940, p. 7.

63 *Bengal Engineering College, Calendar for 1940–1941*, op.cit., p. 14.

64 Jagdish N. Sinha, 'Science and Globalization: Indian Experience in the Second World War', *IHR*, Vol. XXXIII, No. 2, 2006, pp. 134–160.

65 *Bengal Engineering College Centenary Souvenir*, op.cit., p. 40.

66 *Bulletin Bengal Engineering College: One Hundred Twenty fifth Anniversary Issue*, Calcutta: no pub, 1981, p. 13.

67 N. Das, *Plans for a Better Bengal: Progress of Post-War Reconstruction Planning in Bengal from 1ˢᵗ January 1944 till 31ˢᵗ October 1944*, Alipore: Superintendent Government Press, 1944, pp. 32–35.

68 *Post-War Educational Development in India: Report by the Central Advisory Board of Education*, Delhi: Manager of Publications, January 1944, p. 41–7.

69 *Report of the Committee for the Development of Higher Engineering and Technical Education in Bengal*; quoted in *Report of the Director of Public Instruction, Bengal, 1945–1946*, Calcutta: Bengal Secretariat Press, 1946, p. 9.

70 Ibid, p. 10.

71 The Calcutta University approved new regulation for the degree of Master of Engineering in Civil, Mechanical, Electrical and Metallurgical Engineering in 1953. The first batch of post-graduate students came upon in the year 1956. During the same time Doctorate programme were also introduced to enhance the research activities of the College. *Bulletin Bengal Engineering College*, op. cit., p. 14.

72 A recent example, John Black, 'The Military Influence on Engineering Education in Britain and India, 1848–1906', *IESHR*, Vol. 46, No. 2, April-June 2009, pp. 211–239.

7 Of geologists and water-diviners

The quest for groundwater knowledge in mid-twentieth-century India

Kapil Subramanian

Despite the emphasis on centrally managed large-scale surface irrigation schemes in the historiography,[1] privately owned groundwater pumps account for two-thirds of Indian irrigation today. Interwar India was a global pioneer in the use of tube-wells for irrigation and such was their key role in the Green Revolution that some have called the latter a tube-well revolution; indeed as I've argued elsewhere, the technology central to the Green Revolution was tube-wells rather than the widely hailed new seed varieties.[2] Activities in pursuit of groundwater knowledge increased manifold during the mid-twentieth century and in this essay, I shall explore the practice, practitioners and politics of that knowledge in mid-twentieth-century India. I shall argue that knowledge about groundwater became increasingly important, if heavily contested concern.

I shall begin by exploring the employment of water-diviners at the highest levels by interwar and postcolonial Indian governments. While the practice received sharp criticism from colonial Indian politicians including the conservative Congress Party premier of Madras C. Rajagopalachari, it barely received political comment in postcolonial India, even as Nehru himself was a well-known enthusiast of a famed diviner.[3] I shall argue that this had in part to do with the inescapable irony, in the colonial era, of a self-proclaimed scientific empire imposing what was seen as superstition on natives; science was central to the nationalists' claim to the universal.[4] Equally, the criticism of state employment of diviners had to do with the push for Indianization[5] (the diviner in question was English) which was a key political theme of the last quarter century of colonial rule.

As I will show, one reason for the employment of diviners was that groundwater science was a science of observation and deduction rather than one of precision and quantification; it was unable to offer the firm answers farmers and governments demanded. This was even as the Geological Survey of India became increasingly involved in providing advice about groundwater for irrigation, municipal and military use in

DOI: 10.4324/9781003241980-10

the interwar years. This engagement intensified through the war and the Nehruvian era, the latter in large part due to a massive American-aided programme of drilling to explore India's groundwater resources. The story of groundwater science makes for an important case study of American technical aid to postcolonial India; two of the operational agreements under the Indo-US Technical Cooperation Agreement of 1952 had to do with groundwater.[6] This essay explores the shift from British to American expertise at the GSI in the 1950s. The shift was driven by the formidable expertise developed by the United States Geological Survey (USGS) and private firms while exploring the American west in the early twentieth century, by the availability of surplus technical manpower at the USGS and by the departure of the British from the GSI due to sympathy for Indianization and fears of political interference.[7] Besides the general increase in groundwater activities, the transition also qualitatively transformed the GSI; the American-aided All India Groundwater Exploration Project was as much a massive engineering and logistical exercise as it was a scientific one, thus presenting new kinds of challenges for the Survey. I shall likewise illustrate how the impact of the project lay as much in training skilled drillers who would go on to drive India's Green Revolution as it lay in producing geological knowledge and groundwater geologists.

Water-divining and the state

Founded in 1949, the Rajasthan Underground Water Board was the first provincial institution dealing exclusively with groundwater. Its members included the Union Minister for Agriculture, the State Premier, its Chief Electrical Engineer, a professor of botany, its Director of Agriculture and other civil servants. The only member with expertise in water issues was an individual referred to as Paniwala Maharaj.[8] Born Jivaram Vyas, the water-diviner was about 50 and reputed to have acquired his powers through years of yogic practice. Described as having "X-Ray sight", sitting in a room or travelling in a car he would see "a cloud of haze in the depths of the earth" and indicate the quality and quantity of water with "mathematical precision". His methods, which included moving his hands across a map to pinpoint sources of water had apparently been successful across northern and western India, baffling "men of science who were inclined to scoff at this mumbo jumbo of psychic powers".[9] In addition to a handsome salary, allowances and charges,[10] the central government also took over the maintenance of his ashram from the Saurashtra government who had previously employed him.[11]

This was not the first time that Indian governments had employed diviners. In the late 1930s, the Madras Public Works Department rejected the offer of a Dutch diviner's services deeming the existing services of a Mr. Pinto adequate. Pinto's own employment was the source of some discomfort; the conservative Congressman and Premier of the Presidency

C. Rajagopalachari did not want "this water-divining superstition to be part of a government communiqué" as

> The employment of water-diviners is a mere device to get rid of the uncomfortable feeling in the mind when we make a choice without reason. The government cannot be a party to such expenditure, but villagers and officials may indulge in spending private money on such luxuries. I would advise a decisive step putting an end to this practice.[12]

The last water-diviner with a status comparable to the yogi's was Major C.A. Pogson, of the Maratha Light Infantry who went on to become something of a legend in the global water-divining world. Pogson presented a paper on water-divining at the Bombay Engineering Congress in 1923, going into the (Anglo-German) history of water-divining. Explaining the phenomena in terms of radiation emitted by metals and water that human "galvanometers" could detect, he also related how he learnt the skill from his father in south west England.[13] The summer and (failed) monsoon of 1925 saw Pogson rise to prominence. Faced with famine and drought, the Bombay government emphasized irrigation from wells, as colonial Indian governments had often done during past droughts. Wells were seen as the key to quick results (unlike large surface irrigation schemes), and as they were largely constructed through private efforts, an emphasis on wells enabled the government to be seen to be doing something while spending little money. It also diverted attention from the failure to spend on canal irrigation which many had called for but was an expensive proposition in the Deccan.[14] Arguing that peasants were reluctant to spend money on wells due to the high incidence of failures, the Revenue Member of the Bombay Council, having apparently learnt of Major Pogson through the newspaper, decided to employ his services for famine relief.[15] Through that summer, Pogson indicated water sources in Tarapur, Cambay state, Baroda state, Sholapur and Ahmednagar, as well as holidaying in Mahabaleshwar where the hill station and its club requested his services.[16] A well at a site recommended by Pogson for a club in the hill station of Materhan was inaugurated (complete with Parsi prayers) in May 1925.[17]

In August that year, the government moved to appoint Pogson full time as the Water-Diviner to the Government of Bombay. The appointment, carrying a princely salary with expenses, was met with strident opposition in the provincial legislature whose session had to be extended by a day to accommodate all speakers. In a speech which the *Times of India* called his most fiery that session, Mulund Rao Jayakar of the Swaraj Party satirically suggested that the government also appointed a fortune teller to foretell his party's fate. Moulvi Rafiuddin Ahmad protested "in the name of Oxford and Cambridge and of science and of Huxley and Herbert Spencer" and said that he wouldn't be surprised if the next appointment was a Surgeon General for Faith Healing. Reminding the house that the Surgeon General had called the Ayurvedic and Unani medical systems as superstitions, he argued that by placing faith in an "English quack", the House would lose the right to go to the people and call for Western science to take the place of Eastern superstition. Calling upon the government to draw upon the belief in the civilization "which they had been preaching to Indians" he warned,

> At this rate, you may have Colonel Alcott tomorrow and all sorts of spiritualists…I am not for encouraging supernatural powers. It is a bad principle.[18]

One critique in the *Times of India* proclaimed that any virtue in the superstition stemmed from the psychology of the individual rather than "from the Deity, from Satan, from affinities and sympathies, from electric currents or from passive qualities of organo-electric force". The consensus of modern scientific opinion, it argued, was reflected in studies at the University of Bristol and the Royal Sanitary Institute and was on the side of those who voted against the appointment.[19]

Government members of the House pointed to Major Pogson's success rate and the confidence villagers reposed in him. Examples were given of Scotland, where water-divining was apparently a regular trade, as well as of Australia and South Africa where governments employed diviners. Calling himself a scientific man, the Home Member argued that Pogson's skill was "no magic". Supernatural powers were merely those that were not possessed by all and not at all extra-scientific; civilization had robbed man of several powers and water-divining was no different from the keen sense of smell possessed by hounds. Major Pogson had merely scientifically systematized his powers. Divining merely implied a process of deducing from facts or guessing from physical experiment; a purely scientific exercise no different from the Finance Member deducing what income and expenditure figures to enter into a budget. Appropriating the enthusiasm of science, he appealed,

> If in the past years people did not carry on experiments, where, Sir, would be your steam engines, and railways and steamships? Who ever thought 50 years ago of wireless telegraphy? The progress of science has been so remarkable in recent years that it seems to me he would be a rash person indeed who would protest that it was impossible for a person to develop his physical powers so as to make it possible for him to locate scientifically underground supplies of water. What is this, Sir, to the many experiments which have been made in recent years about the electrical forces of the universe? I should strongly recommend to this honourable House therefore, to boldly make this experiment for the sake of the inhabitants of the Deccan and other dry tracts of this Presidency.[20]

Calling this "scientific colouring" as nothing short of pure superstition, the opposition insisted that it was impossible to calculate the Major's success rate. Indeed, the government admitted that of 126 spots located in Ahmed-nagar by Pogson, sufficient equipment had only been available to bore ten holes of which six had borne water.[21] As boring was expensive, in general, the sites selected were those recommended by multiple experts (say the celebrity diviner, water supply engineers and perhaps the local diviner); so even if the numbers indicated success, it would be impossible to attribute it to Pogson alone. Pogson's publicized achievements in Bombay were largely confined to the hill stations and even there the record was far from stellar. Besides Mahabaleshwar where club members were invited to feel the twitching of the dowsing rod (it is unknown if a well was actually sunk), in 1923, he indicated a site for a club in Materhan, where a (very expensive and specialized) well was finally inaugurated in 1925. Another spot he located for the town's municipality failed to strike water at an expensive 120 feet around the same time; optimists recommended continued digging while others acknowledged failure.[22] C. M.Saptarishi, of the draught-hit Ahmed-nagar district, said in the legislative council that the Deputy Collector in Ahmednagar had characterized Mr. Pogson's work there as nothing short of humbug. Critics argued that rather than the government having found a diviner, it was the diviner who had found a simpleton in the government.

Some suggested a geological survey while others questioned whether any expertise was at all required; the Bombay Public Works Department had a high success rate with boring wells and the few failures did not justify any expense. The government claimed that Pogson's utility was proven by how he had been besieged with requests from peasants– again, betraying how a diviner made for excellent publicity of government efforts. After a protracted debate, the resolution to employ Major Pogson was put to vote and carried by a single vote.[23]

As Major Pogson's appointment came up for renewal the next year, the government was well prepared with a statistical dossier attesting to his success, but the opposition alleged obfuscation by the selective presentation of data and by a confusing categorization of failures (into heads such as "abandoned by owner" and "destroyed by blasting") that made it impossible to calculate success rates; Pogson could also merely predict the presence of water and not its permanence or quantity. The very definition of a successful well was up for debate; the government defined it as one on which a *mhote* could be worked for four non-consecutive hours a day in a 12-hour day-light period, while an opposition member claimed that successful irrigation required at least eight hours of working. The testing of Pogson's success with a Mansfield water finding machine which the government had previously termed as useless was also criticized; the government responded that experience had meanwhile been gained in the use of the machine. Yet again, the government managed to pass a resolution ensuring Pogson's employment till the end of the financial year in 1927.[24]

Pogson next came up for discussion in the council only in March 1928 when it was noticed that his salary had been moved from the "votable" to the "non-votable" section of the budget; this was even questioned later in the British Parliament.[25] The government responded that as a military man appointed by the Secretary of State, his salary was protected. The Government of India had argued that "Persons and not merely posts are protected"; the fact that he was employed by Bombay in his personal capacity did not matter. Through some bureaucratic creativity, the government had managed to get away without a discussion in March 1927; the House President had to affirm that the government had acted in good faith. This time, Indian members called Pogson a "*Jagirdar* of the service" and condemned the way he had glided out of their control.[26] If this was indeed a move by the government or the Major Pogson's friends in high places to protect his employment, the impact was short-lived; in 1929, council members were openly calling his position a sinecure and Pogson's employment fell prey to Depression-era retrenchment later that year. Ever entrepreneurial, Pogson had advertised extensively in the *Times of India* through the 1920s, which not only gave him space to write several articles about water-divining but also strongly supported his appointment through the years.[27] In the late 1930s, the Major wrote to the Colonial Office, offering his services for a rumoured new Jewish settlement in Tanganyika.[28]

These colonial-era examples demonstrate that the Indian polity had developed a critique of water-divining early on and that their employment was far from uncontroversial. It is thus curious that there was little debate on the employment of Paniwala Maharaj in independent India; a parliamentary question in 1951 even suggested that he be employed to find oil.[29] Even UNESCO did not find it necessary to comment on Paniwala Maharaj, though their reports note his membership of the Board.[30]

Perhaps the contrast may be explained by the stark irony in Major Pogson's case of a colonial government imposing a seemingly superstitious practice on natives; as one Bombay politician put it, "East has become West and West has become East". The debates over Pogson also had as much to do with Indianization of the services as with questions of science and superstition; every new European appointment in the era was endlessly debated in provincial councils. Indeed, some like D.P. Desai, while expressing disbelief that the "agnostic government" believed in supernatural powers, objected to the appointment merely on the grounds that Indian water-diviners were available at no expense; divination "was the special monopoly of the east" and there was no need to turn to the west. Others suggested that some Indians could be trained in each district by Pogson. The government defended itself by admitting that Indian water-diviners were quite good, but argued that they could only locate springs, not *underground* springs. Mr. Pogson also drew excellent graphs besides indicating quantity;[31] this façade of science made western water-divining superior to the Indian practice.

In postcolonial India, no less than Nehru himself was a big enthusiast for the yogi's skills, motoring down to Faridabad to explore 2 square miles of land with Paniwala Maharaj to find water for a refugee colony. The GSI only stepped in to drill at the spots suggested by the seer.[32] In the words of the man who was in charge of settling the colony,

For a community of 40,000... we needed a vast quantity of water...
The Government engineers said that there was no sub soil water in the
region... it did not seem to be a practicable proposition to build a town
there, the engineer said... Nehru... told me ...about the water-diviner...
The Army Colonel...scoffed. The Colonel and I, however walked over
the 3,500 acres of land along with the Paniwala Maharaj and from
time to time he would stop and stamp his foot in a particular way and
say "Here you have water". So I got hold of a drilling rig and a crew
and drilled ten holes in ten places pointed out by the diviner...Up to
this day there is no other source of water for this large industrial town
with all its big industries except these tubewells... in an area where we
were told scientifically by engineers that there was no water and a town
could not be built.[33]

As late as the 1970s, Paniwala Maharaj's experience was quoted in an otherwise sceptical article as one of the few acceptable examples of the harnessing of extra-rational abilities by the state.[34] Water-diviners continue to
be used in Rajasthan and the rest of India, though their relationship with
the state may have turned adversarial.[35]

Groundwater geology and the GSI: a useful science?

While the GSI had dealt with matters concerning groundwater since the late
1800s, only starting in 1921 did these activities reach a scale justifying a
separate section in its annual report.[36] Over the next five decades, the GSI
provided services to industries, municipalities, railway stations, military
installations and aerodromes across India on groundwater issues.

A typical groundwater investigation report would start with a general
examination of the geology of the area, noting the various rock formations
and their properties. It would go on to discuss the possible underground
water bearing formations, the sources of recharge and the possibilities of
their development. As the GSI had not surveyed most of India, often even
the general geological features mentioned in a groundwater report would
be the product of a quick observation of visible features and a deduction
of the underlying geological formations. This methodology was heavily
dependent on the knowledge and experience of the individual geologist;
reports were peppered with phrases like "probably" and "in his opinion".
The characterization of the movement and flow of water through porous
rock formations was also speculative, descriptive and non-quantitative; in
fact, the only occasion in the early years a formula was devised to calculate
recharge, it was important enough to be mentioned in the GSI's annual
report, though the required data were unavailable. In the Bombay debate
on water-divining, a government member had said that water-divining was
not a science like physics or chemistry the truth of which could be demonstrated by experiment but one like astronomy whose truth could be judged
by observation. Of groundwater geology in India, it could be similarly said

that it was less an exact quantitative science than it was a system of observation and deduction.

The appointment of Paniwala Maharaj and the associated press coverage of geologically improbable finds of water in the Rajasthan desert led the GSI to publish in 1950 the first of a series of bulletins on groundwater, authored by a remarkable Cambridge-trained geologist called John Bicknell Auden who was the last European to join the permanent cadre of the GSI in 1926 and the last to leave in 1953. Known for his passion for Himalayan geology (he was part of the Karakoram expedition immortalized in Eric Shipton's *Blank on the Map*) as well as his work on the Bihar–Nepal earthquake, he was the poet W. H. Auden's brother and married to the granddaughter of the nationalist W. C. Bonnerjee.[37] Beginning in the 1930s with advice for the United Provinces' Ganges Valley State Tubewell Irrigation Scheme which pioneered the use of tube-wells for irrigation in India[38] and its emulators, Auden acquired much experience in groundwater issues and headed the Engineering Geology and Groundwater Section of the GSI after the war.

Auden's 1950 bulletin on the groundwater resources of Western Rajasthan concluded with a section on water finding methods, quoting the USGS's O. E. Meinzer (considered the father of modern geohydrology) extensively to critique water-divining. Admitting that geologists could not claim 100% success in locating water, Auden argued that at least their methods were based on the best science at any given point of time. Water-divining on the other hand was shrouded in mystery with successes accepted as proof of magical gift and failures having a curious habit of remaining concealed. Auden gave some examples where military airfields had sought the advice of both diviners and geologists; perhaps the reason, as he failed to recognize, lay in the cautious prescriptions that geologists often made when compared with diviners. In the case of Rajasthan, for example, he recommended investigations "requiring a team of geologists and geophysicists working over a period of years". A water-diviner at an airfield on the other hand had predicted the depth and yield of a well to a precision which was "apt to mislead the gullible and simple but renders any experienced person immediately suspicious". That officials were often inclined to trust water-diviners more than geologists was exemplified in Coimbatore (Madras) where of 22 sites recommended by a geologist, 18 bore water, 2 were barren and 2 were actually abandoned at the advice of a diviner. Even this scathing critique had to admit that a significant proportion of water-diviners might just be people with good practical experience of terrain. Such was the trust in diviners that Auden offered to set up a system to test the abilities of "water witches" and determine the extent to which their site selection consisted of fraudulent guesses dressed behind a "pseudo-quantitative facade".[39]

A crucial source of groundwater data came from existing wells: their daily yield, the diurnal or seasonal fluctuation in the water level, the daily consumption of water, the rock strata visible on the walls of the well and the quality of the water. These data became all the more important with increasing methodological sophistication; for example, it was crucial for the plotting of water table contours or isochlor lines (salinity contours). These data were infrequently maintained by local bodies and so usually obtained from a survey of the wells and local conversations; local people being "illiterate and unobservant", what was brackish to some was sweet to others. Sometimes, it was felt that interested parties wilfully misled geologists to get a government well sited in an area advantageous to them. Data from the few extant boreholes were even less reliable as the recorded description came from drillers with little knowledge of geology.[40] The need for data meant that the advice most often given was to drill test bores before proceeding; such prescriptions seemingly defeated the very purpose of seeking expert advice as test bores were fairly expensive.

This uncertainty conflicted with the needs of engineers and politicians for quick, firm and favourable scientific advice about the viability of their favoured groundwater irrigation schemes, making for much tension. In 1951, much to the GSI's chagrin, the central government's Agriculture Ministry requested Auden's services at a short notice of 48 hours to save a tube-well project in North Gujarat after a Finance Ministry administrator alleged that in the absence of geological knowledge, the project was a speculative exercise.[41] In its earlier report on the scheme, the GSI had noted that the terms of the contract for drillers prepared by the government were impracticable; this was why only a single company had bid for the scheme. The company later managed to get the contract controversially modified to terms which would have interested the original competitors; Nehru had to make a statement in parliament as it emerged that the company was connected to the son-in-law of K.M. Munshi, his Agriculture Minister.[42] After much prevarication in its various publications on the issue, the GSI admitted that some element of speculation was inevitable. "No amount of deductive reasoning on still manifestly inadequate data" could take the place of actual exploratory tube-wells which other firms had envisaged. It noted that if the primary consideration was growing more food, the development of groundwater resources was crucial, even if some areas proved barren.[43] Knowledge of groundwater resources could thus come only from actual development of those resources.

One way to improve the precision and sophistication of geological recommendations was the application of geophysics, and the Geophysics Section was slowly emerging in the late 1940s as the most prestigious department in the GSI. One of the GSI's first publications resulting from the application of geophysical methods to groundwater problems was a study of the complex volcanic Deccan Trap region in West-Central India published in 1956 which noted that the scientific method thus far in vogue to find water had a "hit or miss" quality to it; by close examination of a village, it was possible to locate a small region with good prospects of striking water, though much was left to chance. This "naturally gave a handle" to dowsers who could not only pinpoint exact sites for wells but also predict the quality and quantity of water. The authors had little faith in the Mansfield water finder which had the semblance of a scientific instrument and consisted of a magnetic needle and a coil of wire packed in an attractive box with a well-illustrated leaflet promising water and treasure. It was felt that there was indeed a need for instrumentation, but that the appropriate device was a resistivity meter; a study of electrical resistivity could help gauge the nature of the underground strata and water. The Deccan Trap study had mixed results. There was some controversy in the literature over which precise rock formation was the major source of water in the Traps; the study vehemently claimed that it was weathered zones rather than cracks or fissures, perhaps because weathered zones could be detected with a resistivity meter while cracks and fissures could not. Even so, the spasmodic nature of weathering made the interpretation of resistivity data unaided by geological observation difficult. The report concluded that the methods of geophysics were uneconomical as the scattered villages made equipment transport expensive.[44]

While the GSI does not appear to have used the resistivity meter on any significant scale for groundwater studies in the 1950s, other water finders did. One particular case is of great interest in demonstrating how experience could substitute for academic training in geology. Poona resident D. G. Limaye, a graduate in agriculture, began work in the 1930s with the Deccan Agricultural Association to help farmers site wells, eventually setting up a very successful private practice called The Geophysical Explorers. Previously reliant on the Mansfield Water Finder, having observed a resistivity meter imported by the nearby Central Water Power Research Station, he had a resistivity meter built by a local technician in the 1950s, though the instrument itself appears to have been sparingly used. Besides aiding farms and industries, he was involved in prestige projects like the siting of tubewells to supply Chandigarh, the new capital of postcolonial Indian Punjab. He also made contributions to the scientific study of groundwater in the Deccan Traps which were continued by his son S. D.Limaye who joined the family business with a Ph.D. from the Indian Institute of Technology, Bombay.[45]

Enter the Americans

The growing interest in groundwater irrigation spurred by the success of the Ganges Valley tube well scheme, the military's need for water supply in new installations such as airfields built during the Second World War and the post-war plans for dams and reservoirs (the siting of which would require groundwater expertise) spurred the demand for groundwater knowledge and an Engineering Geology and Groundwater Section was created in the GSI in 1945. Headed by J. B. Auden, it was the Survey's first non-regional technical section.

While there were plans to recruit foreign experts to the new section as early as 1946 to deal with the sheer volume of dam related work even as some officers were in training in Australia and more were expecting to go to US Bureau of Land Reclamation, it was initially felt that the section could cope without new recruitment even if it sent a few groundwater officers to the USGS for training.[46] However, by the late 1940s, the GSI's groundwater branch came under increasing pressure. In 1950, it requested the services of G. C. Taylor Jr., a USGS geologist to assist with some specific investigations and to design a long-term programme of assistance to the GSI in groundwater geology and hydrology under Truman's Point Four Programme. Taylor stayed on in India until 1955 to head the programme he designed and was replaced by P. H. Jones whose Indian service ended in 1957 when long term US technical assistance to the GSI came to an end, though there were several short-term projects later.[47]

In 1953, within a couple of years of the Americans' arrival, the last formal link of the GSI to Britain was severed with the departure of the J. B. Auden, the last British geologist still in Indian service. Then at the brink of becoming the director of the GSI, Auden opted to resign and pursue opportunities elsewhere in the developing world, including a long stint with the FAO, later writing that he left partly as he felt that an Indian should head the GSI and partly fearful of "headaches" in recruitment and state representation.[48]

The USGS had developed a formidable expertise in groundwater matters. While Auden and others at the GSI were essentially generalists who were developing expertise in water issues as a result of regular assignments, as early as 1903, the new Hydrographic Branch of the USGS had been split into a Division of Hydrography (for stream gauging), a Division of Hydro-Economics, a Reclamation Service and a Division of Hydrology (for groundwater). In addition, a 1916 act authorized funds for discovering, developing and protecting desert water supplies. In the same decade, Western Congressmen also managed to appropriate large funds for a programme of exploratory drilling. The Division of Hydrology was headed from 1911 by Oscar Meinzer, acknowledged as the father of modern geohydrology. Faced with the persistent question of "What is the safe year-by-year yield from groundwater supplies", hydrology in the US took an increasingly quantitative turn from the first two decades of the twentieth century and Meinzer developed many new methods.[49] In India, there was a great preference for American expertise in part due their domestic experience with groundwater issues; when asked by the UK Trade Commissioner in New Delhi why the government was not purchasing more tube-well equipment from the British rather than the Americans, Dr. Piplani, a Joint Secretary in the Ministry of Agriculture bluntly replied that there was no area in the UK which had tube-wells.[50]

On his first visit, Taylor assisted with studies in the Purna Valley, in Coimbatore, in Neyveli (on the problem of flooding of lignite mines) and Kutch (to develop water supply for the new port) and in Meerut (on tube-well irrigation). Besides assisting with the regular programme of the GSI, he helped formulate a massive American-aided project to sink over 2000 tube-wells in India.[51] Other USGS personnel who served in India in the 1950s included A.A. Garett who was sent in response to the GSI's request for assistance with quantitative methods and P. H. Jones, Taylor's replacement who helped train GSI personnel for the All India Groundwater Exploration Project which I will detail in the next section. By 1957, the USGS had trained 60 GSI personnel, prepared 25 technical reports for the Government of India and co-authored 45 studies of various localities with the GSI besides providing administrative guidance.[52]

Project elephant

In 1950, the Government of India had contracted Parson-Johnston-Brush International, an Anglo-American consortium formed to execute post-war tube-well projects in India, to carry out a groundwater reconnaissance survey in Madras, UP, Orissa, Bihar, West Bengal and Saurashtra. The contractors' report recommended exploratory drilling. A committee of geologists and the Technical Cooperation Administration (the organization set up to coordinate American aid) staff had also recommended a nationwide survey of economic groundwater supplies for irrigation, while a detailed project had been proposed by Auden and Taylor in August 1952. With the signing of the 12th Operational Agreement of Indo-US Technical Cooperation in 1953, the All India Groundwater Exploration Project was born. It envisaged the drilling of 350 holes in 15 "soft rock" regions of the country to prospect for water.[53] For the first time, organized groundwater investigations not in support of any particular project would be conducted in India. The project was one of the largest exploratory drilling programmes in the world and was to be described in later decades as the "forerunner of modern concept introduced in India in the search for ground water resources with a multi-disciplinary scientific approach".[54]

The Ministry of Food and Agriculture was to be the general supervisor. A contractor was to be selected for the drilling and equipping of the tube-wells, but it was felt that the project being of a research nature with works scattered all over the country, it was best handled departmentally;[55] the bids received were also deemed "disadvantageous to the Government of India". The Operational Agreement was amended to allow the Ministry to do the job departmentally with technical consultancy and engineering services to be provided by a contractor. An Exploratory Tube-wells Organisation (ETO) was set up for the purpose in the Ministry in 1954 headed by a Chief Engineer and containing four divisions, each headed by an executive engineer, with most personnel being on deputation from other government agencies. The Board of Management for the project consisted of the Agriculture Secretary, the Tube-well Projects Administrator (an Agricultural Ministry position created to manage the tube-well projects of post-war reconstruction schemes, the Five Year Plans and the Technical Cooperation Mission), the Finance Secretary and the Irrigation Adviser to the Government of India.

A new Groundwater Exploration Section of the GSI was to provide technical support for siting the boreholes and for recording the strata encountered (well logging).[56] The new section, which consisted of 24 geologists, a geophysicist, a well logging technician, a geophysical assistant, six electrical logging operators, an instrument operator, four surveyors and three chemists was distinct from the Engineering Geology and Groundwater Section which was to continue local investigations.[57] The GSI had only five officers working on groundwater when the Engineering Geology and Groundwater Section was set up in 1945;[58] this was five times the number that was initially appointed to work exclusively on the All India Groundwater Exploration Project, reflecting the growing importance of groundwater.

The consultancy contract was awarded to the Los Angeles-based Ralph M. Parsons Company, one of the members of the aforementioned Anglo-American consortium. "Project Elephant" as the company called it was a massive engineering and logistical enterprise and in its final report to the ministry, the company blamed management difficulties for the project running five years instead of the planned three. Fifteen heavy-duty drilling rigs were procured from the US. The trucks to carry the rigs were received in complete disassemble to save on customs duty and the company contracted to assemble them was only experienced with light vehicles. Pumps were slow to arrive. Systematic maintenance protocols were hard to establish and there was much damage to the equipment from causes such as poor lubrication and failure to fill radiators. The inexperience of the ETO meant new operating procedures had to be developed from scratch; the organization developed along the lines of a Public Works staff, which according to the company was inadequate for research projects requiring immediate on the spot decisions. Frequent reference to headquarters hundreds of miles away resulted not only in lost time but also in the loss of uncased wells.

The company quickly developed differences with the GSI; for example, they differed on the intent of the operational agreement; while the contractor held that the term "economic supplies of groundwater" necessitated discontinuance of drilling once supplies adequate for a well were encountered, the GSI called for determination of the water yielding capacity of the most important aquifer as well as the testing of all aquifers to determine their importance. Conflicting instructions were frequently issued as a result; in the contractor's words, many deep holes were drilled in search of information "not readily associated with the development of economical ground-water supplies". The GSI's use of unskilled labour for taking samples often led to inaccuracies and well logs were sometimes taken from improperly washed holes. The procedure for the selection of gravel for packing well shrouds was also a point of controversy between Parsons and the GSI which was never resolved.[59] Nevertheless, considerable groundwater reserves were located in Gujarat, the Narmada Valley, Orissa and the Jaisalmer desert, and possibilities for the same were indicated in UP, Bihar, Punjab and West Bengal.[60]

Of the 25 American staff working with the company in India, only two were hydrogeologists. Three were engineers, one was a stores officer and the rest were concerned with the practice of drilling (five drilling superintendents and 14 drillers).[61] Amongst the project's accomplishments, the contractor noted that many drillers had been trained who would be adept at exploration for water, oil or minerals. Perhaps more than the scientific data obtained, this was the project's true accomplishment. For example, Gurnam Singh, a Punjabi farmer's son joined the ETO as a driller and served in various parts of India, gaining experience in drilling holes deeper than he had previously encountered. He resigned after 12 years, working briefly for the National Coal Development Corporation and as a driller at Bird and Company in Calcutta where an ex-ETO engineer was manager. Quitting in 1970, he bought a small drilling rig in a Calcutta scrap market and set up as a drilling contractor in Punjab in the wake of the Green Revolution. Gurnam Singh and Company, today run by his family has clients ranging from various industries and provincial government departments to the World Bank.[62]

Conclusion

Knowledge of groundwater became increasingly seen as necessary in mid-twentieth-century India and a variety of means were deployed to acquire the same, from water-diviners and quick geological investigations to a massive programme of exploratory drilling. The employment of water-diviners by the colonial state was driven as much by a desire to be seen to be doing something (popular) during droughts as it was by a desire for meaningful drought relief. The employment of European water-diviners by the colonial state attracted much criticism from Indian politicians, both on the grounds of it being superstitious as well as in an attempt to push Indianization. In postcolonial India, however, the seemingly superstitious practice attracted little comment from politicians, and Nehru himself was an enthusiast for dowsing; much to the GSI's chagrin. However, groundwater geology was a science of observation and deduction rather than one of precision and quantification and was very dependent on data from existing groundwater development such as wells; it was thus no match for the seemingly precise and definitive pronouncements of diviners.

The importance placed on groundwater knowledge was reflected in the increasing number of geologists employed in groundwater work and in the large programme of American aid in support of groundwater exploration in India. Like in many other technical fields, this expansion in groundwater-focussed geological activities was accompanied by a mid-twentieth-century shift in India from British to American expertise. This shift was driven by an expansion in the GSI's work and the consequent personnel shortages, the availability of USGS personnel (as well as American aid) and the formidable groundwater expertise the USGS had developed in exploring that country's varied terrain. The departure of British experts was driven by sympathy towards Indianization policies as well as fears of changing postcolonial employment cultures and political interference. The expertise developed in India as a result of this exchange with the US lay as much in engineering, logistical management and skills such as drilling as it lay in geology.

With the completion of the American project in 1959, the groundwater work of the GSI suffered something of a setback. The Groundwater Exploration Section was wound up and some officers were deputed to help ETO finish up and execute the second phase of the project. On their return, many geologists trained in groundwater work were diverted to other branches of the GSI to fulfil urgent industrial targets in support of the Second Five Year Plan. Much follow-on research from the project however continued, and it was realized that large-scale exploration was not an end in itself but a precursor to longer term assessment of the new reserves. In particular, the exploration programme failed to provide answers to questions like how much water could be safely withdrawn without detriment to the hydrological balance and how irrigation would affect the water table. The Planning Commission's study group to assess the raw material position of the country for the Third Five Year Plan also envisaged an ambitious programme for groundwater investigation.[63]

While it did much in the 1960s, the GSI's approach was somewhat desultory and it found itself unprepared to answer the increasingly specific and complex questions demanded by policy makers in the wake of the Green Revolution groundwater boom. Perhaps the most important reason was that while the cadre of groundwater geologists had increased manifold, the numbers clearly weren't enough; in 1970, B. B. Vohra, the Agriculture Secretary estimated that at the then extant rate, the GSI's 80 groundwater geologists would take 30–50 years to survey all of India. As the ETO typically took up drilling work only once an area was prospected by the GSI, this was a serious bottleneck. Vohra made the case for all groundwater work to be vested with the ETO which could also serve as a single point of contact for credit agencies such as the World Bank which had begun to lend massively to India for groundwater development and demanded increasingly complex data to appraise loans. Vohra saw the Agriculture Ministry as the right home for the new body for the "simple reason that agriculture is the greatest beneficiary of groundwater and therefore most interested in its development", leaving unsaid the fact that in reporting to the beneficiary department, experts might lose their independence. The ETO became the Central Groundwater Board in 1970, and the Groundwater Wing of the GSI was merged with it in 1972.[64] Thus ended an interesting mid-twentieth-century chapter in groundwater exploration at the Geological Survey of India.

Notes

1 Important works on colonial Indian irrigation schemes include Elizabeth Whitcombe, *Agrarian Conditions in Northern India*, Berkeley, CA: University of California Press, 1972, Ian Stone, *Canal Irrigation in British India*, Cambridge: CUP, 1985, Imran Ali, *The Punjab Under Imperialism*, Princeton, NJ: Princeton University Press, 1988, M. Mufakharul Islam, *Irrigation, Agriculture and the Raj*, New Delhi: Manohar, 1997, David Gilmartin, 'Scientific Empire and Imperial Science: Colonialism and Irrigation Technology in the Indus Basin', *The Journal of Asian Studies*, Vol. 53, No. 4, 1994, pp. 1127–1149. For an overview,

see Elizabeth Whitcombe, 'Irrigation' in Dharma Kumar (ed.), *The Cambridge Economic History of India*, Cambridge: CUP, 2005, pp. 677–736. For a literature review, see Rohan D'Souza, 'Water in British India: The Making of a 'Colonial Hydrology'', *History Compass*, Vol. 4, No.4, 2006, pp. 621–628. On dams, see Daniel Klingensmith, *One Valley and a Thousand*, New Delhi: OUP, 2007 and Rohan D'Souza, 'Damming the Mahanadi river: The emergence of multi-purpose river valley development in India (1943–1946)', *IESHR*, Vol. 40, No. 1, 2003, pp. 81–105. Also see Daniel Headrick, *The Tentacles of Progress*, New York: OUP, 1988.

2 Amongst others, the phrase "tubewell revolution" was used by Robert Repetto, *The "second India" revisited: Population, poverty and environmental stress over two decades*, Washington, DC: World Resources Institute, 1994, p. 35. For a history of the tubewell in India including its role in the Green Revolution, see Kapil Subramanian, *Revisiting the Green Revolution: Irrigation and Food Production in Twentieth Century India*, Unpublished Ph.D. thesis, King's College, London, 2015.

3 Perry Anderson, 'After Nehru', *London Review of Books*, Vol. 34, No.15, p. 31 begins his controversial essay by challenging the standard notion of Nehru as secular and rational to a fault.

4 On science and nationalism during the period, see especially Gyan Prakash, *Another reason: science and the imagination of modern India*, Princeton, NJ: Princeton University Press, 1997, pp. 159–226, Deepak Kumar, 'Reconstructing India: Disunity in the Science and Technology for Development Discourse, 1900–1947', *Osiris*, Vol. 15, 2000, pp. 241–257, and David Arnold, *Science, Technology and Medicine in Colonial India. The New Cambridge History of India III: 5*, Cambridge: CUP, 2000, pp. 129–210.

5 For an excellent account of the slow process of Indianization of the technical services in the interwar years, see Aparajith Ramnath, *The Birth of an Indian Profession*, New Delhi: OUP, 2017.

6 Operational Agreement No. 6 provided for a large tube-well irrigation programme, while this essay engages with Operational agreement No. 12 which was to aid groundwater exploration. See John Perkins, *Geopolitics and the Green Revolution*, New York: OUP, 1997 and Nick Cullather, *The Hungry World: America's Cold War Battle against Poverty in Asia*, Cambridge, Ma: Harvard University Press, 2010 for accounts of American aid to Indian agriculture.

7 On the employment of Indians in the GSI in the nineteenth century, see Deepak Kumar, *Science and the Raj*, New Delhi: OUP, 1995, pp. 187–188.

8 Anon., 'Underground Water Board for Rajasthan', *Times of India*, 11 December, 1949.

9 Anon., 'Yogi sees into bowels of earth', *Times of India*, 8 October, 1950.

10 Anon., 'Escape of witnesses in Laik Ali case: Questions in parliament', *Times of India*, 6 March, 1951.

11 Anon., 'Yogi sees into bowels of earth', *Times of India*, 8 October, 1950.

12 G.O. Ms. 2725 (PWD) dated 22-11-1939, Tamil Nadu State Archives.

13 David Clarke, *Britain's X-traordinary Files*, London: A&C Black Business Information and Development, 2014, p. 94.

14 On droughts and groundwater irrigation, see Kapil Subramanian, *Revisiting the Green Revolution*, op. cit. On the call for canal irrigation in Bombay in 1925, see Anon., 'Bijapur Needs', *Times of India*, 10 September, 1925.

15 Anon., 'Bombay Council: Government's famine policy', *Times of India*, 8 March, 1925.

16 Anon., 'Cambay State: Prospecting for water', *Times of India*, 10 April, 1925, Anon., 'Water divining at Mahabaleshwar', *Times of India*, 23 May 1925, Anon., 'Co-operation: The movement in Sholapur', *Times of India*, 4 September, 1925 and Anon., 'Current Topics', *Times of India*, 26 August, 1925.

17 Anon., 'Materhan Notes', *Times of India*,1 May, 1925 and Anon., 'Materhan Notes', *Times of India*, 1 June, 1925.

18 Anon.,'Locating water supplies: Debate on water divination in Bombay Council', *Times of India*,10 August, 1925 and *Proceedings of Bombay Council*, 8 August, 1925.

19 Anon.,'Fact or Fallacy: Tales of water divining', *Times of India*, 20 August, 1925.

20 Anon.,'Locating water supplies: Debate on water divination in Bombay Council', *Times of India*,10 August, 1925 and *Proceedings of Bombay Council*, 9 August, 1925.

21 Anon.,'Locating water supplies: Debate on water divination in Bombay Council', *Times of India*, 10 August, 1925 and *Proceedings of Bombay Council*, 9 August, 1925.

22 Anon., 'Materhan Notes', *Times of India*, 1 May, 1925 and Anon., 'Materhan Notes', *Times of India*, 1 June, 1925.

23 Anon.,'Locating water supplies: Debate on water divination in Bombay Council', *Times of India*, 10 August, 1925 and *Proceedings of Bombay Council*, 9 August, 1925.

24 A *mhote* is a traditional water lift. See *Proceedings of Bombay Council*, 27 July, 1926 and Anon., 'Major Pogson's appointment extended to March', *Times of India*, 28 July, 1926.

25 See *Hansard*, 29 March, 1928.

26 Anon., 'Bombay Council: Mr. J. C. Swaminarayan's motion negative', *Times of India*, 6 March, 1928, Anon., 'Adjournment motion on Bardoli situation', *Times of India*, 8 March, 1928 and *Proceedings of Bombay Council*, 5 March, and *Proceedings of Bombay Council*, 7 March, 1928. A *Jagirdar* refers to a feudal landholder.

27 See, for example, Anon., 'The Water Diviner', *Times of India*, 10 August, 1929.

28 Frank Shapiro, *Haven in Africa*, Jerusalem and New York: Gefen, 2002, pp. 65–66.

29 *Lok Sabha Proceedings*, 7 May, 1951.

30 See, for example, UNESCO, Arid Zone Programme, Development of the Rajasthan desert, India Paris, 11 September 1951 UNESCO/NS/AZ/50 http://unesdoc.unesco.org/images/0014/001485/148500eb.pdf

31 *Proceedings of Bombay Council*, 7 August, 1925.

32 Anon., 'Delhi Diary', *Times of India*, 24 July, 1949.

33 Sudhir Ghosh, *Gandhi's emissary*, London: Cresset Press, 1967, pp. 237–238. Faridabad today incidentally also houses the headquarters of the Central Groundwater Board which is dedicated to the scientific management of India's groundwater resources.

34 V. Subramaniam, "Dror on Policy making" *Indian Journal of Public Administration*, Vol. 60, No. 1, 1970, pp. 84–96.

35 See Trevor Birkenholtz, *The Politics of Groundwater Scarcity: Technology, Institutions and Governance in Rajasthani Irrigation*, Unpublished Ph.D. Dissertation, Ohio State University, 2006. During my own visit to the Board's office, the disdain for water-diviners was demonstrated when thus far uninterested geophysicists took a keen interest in explaining to me everything that was wrong with the craft as soon as dowsing was mentioned. They refused to believe their board could have ever employed such charlatans, though they admitted they could not explain the practitioners' high success rates.

36 L. L. Femor, *General Report of the Geological Survey of India for the Year 1921*, Calcutta: Geological Survey of India, 1923.

37 B. P. Radhakrishna, *J.B. Auden: A centenary tribute*, Bangalore: Geological Society of India, 2003.

38 For history of the Ganges grid and its globally pioneering efforts in tube-well irrigation, see Kapil Subramanian, '*Rural Electrification on the Ganges Canal: The Colonial Origins of India's Tubewell Revolution*', *forthcoming* or Kapil

Subramanian, *Revisiting the Green Revolution: Irrigation and Food Production in Twentieth Century India*, Unpublished Ph.D. thesis, King's College, London, 2015, pp. 86–91.

39 J. B. Auden, *Bulletins of the Geological Survey of India Series B: Engineering Geology and Groundwater, No. 1: introductory report on the groundwater resources of Western Rajasthan*, Calcutta: Geological Survey of India, 1950.

40 B. G. Deshpande, *Bulletins of the Geological Survey of India Series B: Engineering Geology and Groundwater, No. 4: Groundwater resources of alluvial tracts of Gujarat*, Calcutta: Geological Survey of India, 1954.

41 M. S. Krishnan, *General Report of the Geological Survey of India for the Year 1951*, Calcutta: Geological Survey of India, 1954.

42 Anon., 'Sinking of tube-wells', *The Hindu*, 8 March, 1951.

43 M. S. Krishnan, *General Report of the Geological Survey of India for the Year 1951*, Calcutta: Geological Survey of India, 1954.

44 B. G. Deshpande and S. N. Sengupta, *Bulletins of the Geological Survey of India Series B: Engineering Geology and Groundwater, No. 8: Geology of Groundwater in the Deccan Traps and the application of geophysical methods*, Calcutta: Geological Survey of India 1956.

45 Interview with S. D.Limaye and consultation with his papers.

46 Anon., *A Five Year Plan for the GSI*, Calcutta: Geological Survey of India, 1946.

47 G. C. Taylor, *Historical Review of the International Water-Resources Program of the U.S. Geological Survey 1940–1970*, Washington, DC: US Government Printing Office, 1976.

48 B. P. Radhakrishna, *J.B. Auden: A centenary tribute*, op. cit.

49 Robert Follansbee, *A history of the Water Resources Branch of the US Geological Survey: Volume 1, From Predecessor Surveys to June 30, 1919*, Washington, DC: US Government Printing Office, 1994.

50 J. N. McKelvie (UK Trade Commissioner) to F. Doy (Commonwealth Relations and Export Department, Board of Trade), 1 February 1949, Oriental and India Office Collection, British Library File: IOR/L/E/8/7431.

51 For details about this massive project and how the Anglo-American consortium came to be, see Kapil Subramanian, *Revisiting the Green Revolution*, op. cit. pp. 139–146.

52 G. C. Taylor, *Historical Review of the International Water-Resources Program of the U.S. Geological Survey 1940–1970*, Washington, DC: US Government Printing Office, 1976.

53 Anon., *Final Report: All India Groundwater Exploration Project*, Los Angeles, CA: Ralph M. Parsons Company, 1959.

54 T. Charlu and D.K. Dutt, *Groundwater Development in India*, New Delhi: Rural Electrification Corporation, 1982.

55 Anon., *Tubewells in India*, New Delhi: Ministry of Agriculture, 1955.

56 Anon., *Final Report: All India Groundwater Exploration Project*, Los Angeles, CA: Ralph M. Parsons Company, 1959.

57 M. S. Krishnan, *General Report of the Geological Survey of India for the Year 1953*, Calcutta: Geological Survey of India, 1958.

58 Anon., *A Five Year Plan for the GSI*, Calcutta: Geological Survey of India, 1946.

59 Anon., *Final Report: All India Groundwater Exploration Project*, Los Angeles, CA: Ralph M. Parsons Company, 1959.

60 Anon., *Symposium on Groundwater in India*, Calcutta: Geological Survey of India, 1963.

61 Anon., *Final Report: All India Groundwater Exploration Project*, Los Angeles, CA: Ralph M. Parsons Company, 1959.

62 Gurnam Singh, *Merian Pairhan Mera Safar*, Chandigarh: Shabadkar, 1999.

63 Anon., *Symposium on Groundwater in India*, Calcutta: Geological Survey of India, 1963.

64 G.O. Ms. 1938 (PWD) dated 14-12-1973, Tamil Nadu State Archives.

8 From battlefields to homes

Oil's imperial and quotidian life in colonized and independent India

Sarandha Jain

For long, scholarship on technology in South Asia has entertained a dichotomy between big "conquest" technology such as the railways, dams, and the telegraph, and small "everyday" technology, such as sewing machines, typewriters, and bicycles. The former are tied up with imperial politics, characterized by a narrative of subjugation.[1] The latter are said to be embraced and appropriated by subjects in their everyday-life worlds, given vernacular meanings, thereby breaching colonial politics.[2] Discussing the ubiquitous and pedestrian nature of small technologies, it is argued that they became deeply embedded in social practices of ordinary Indians and lost their meaning as colonial impositions. Such a notion attributes a coherent and singular politics to both, as if big technological networks did not constitute everyday experiences, and household items were always outside the logic of empire.

Ritika Prasad questions this rigid separation of conquest and everyday technologies through her history of prosaic experiences of the railways: how this infrastructure became an inextricable part of the day-to-day lives of colonized subjects, even if they were not traveling on it. She constructs a history of how various aspects of the railways impacted people and how they responded to this: daily routines of railway travel; altered landscapes; new ways of scheduling, measuring, and organizing; allied industries; tracks made on rivers, forests, and land leading to changes in agriculture, housing, etc.; diseases and contagion brought by railway travel; and military control and surveillance.[3] This didn't just haul the state and big technology into people's humdrum affairs and refashion different aspects of their existences, but also made the state and big technology penetrable by concerning them with the mundane. By reaching into the mundane, neither the state, nor big technology remained untouched by it.

I present a narrative on oil showing that big and small technology worked in tandem, and was not detached in their operation, politics, or genesis. Imperial politics and the everyday life-worlds of common people were brought to interact and shape each other in profound ways through technological encounters provided by the telegraph, dams, and suchlike. Big technologies weren't just "conquering". On the other hand, bicycles

DOI: 10.4324/9781003241980-11

allowed state actors to penetrate deep into territories hitherto untouched, and dependence on items like sewing machines and typewriters increased dependence on the colonial state and trade, thereby augmenting imperial power. This division is put into question also by something like kerosene, a poor household's everyday artefact, but at the same time, a product of a networked infrastructure created by big capital, regulated by the state, yet fully appropriated by the people – so much so, that it is hard to wrest it out of their control without banning it today. Petroleum products were organized by big technology and imperial politics (albeit haphazardly), but simultaneously permeated into the daily lives of common people via small everyday technologies like cars, mills, lamps, etc. This had a double effect: bringing the colonial state into people's private affairs, but also upsetting state politics and government with people's commonplace aspirations and actions.[4]

Undoubtedly, petroleum products were used as imperial tools for filling imperial coffers and for fuelling colonial networks. But by turning into a daily necessity for common people, oil cannot be understood as an imperial tool alone. Its chemical properties have the capacity to generate several kinds of substances (asphalt, petrol, gas, plastics, pesticides, paints, etc.), making its political and social possibilities versatile. Aside from its physicality, the socio-economic context in which it emerged is equally significant in shaping its politics. The oil industry was assembled in a haphazard manner that ended up ascribing to it the capability to shape more than imperial possibilities, as it was not entirely top-down or steered by the state. It was assembled by diverse chance incidents, individual desires, local conditions, and debates at local levels, which intermeshed with colonial calculations and motives that reined in this networked product and manoeuvred its development. In addition to its production being troubled by politics outside state control, its consumption too could not be entirely disciplined by the colonial government. Although the widespread use of petroleum by common people was a calculated move, promoted by the industry and the state, once it became a household item it was no longer just a colonial instrument. Colonized subjects exercised some agency over this product, beyond the command of the state, and expressed a great demand for it, altering the meanings attributed to it by the state. The production and use of this substance were shaped partially by local people, giving its receivers partial influence over its politics.[5]

This begs us to question the unitary politics of control attributed to colonial expansion. Given that imperialism in India was disjointed, run by the East India Company (EIC) until 1857, not the British Crown, and differently practised in time and space, can we speak of one imperial ideology or one imperial state, especially before 1857? While these questions are important to complicate colonial politics, it is equally vital to neither romanticize native agency in the face of colonial power, nor chaos within the empire. Deepak Kumar asks pertinent questions regarding this, in relation to technology: to what extent was the technological context responsible for colonialism and to what extent did colonialism drive technologies to do its work? He argues that inconsistencies in the imperial project were marginal to upturn its broader politics, and despite variations, a basic character of colonialism can be seen in formation. Even though neither colonial ideology, nor the state were fully developed or driven by a singular and organized logic, science, and technology in colonized India cannot be seen as benign and driven only by curiosity, Kumar asserts. Technology of that time was ontologically linked to the colonial context it emerged in, and served its purpose. The colonial government, amorphous as it was, had goals to do what it did, even if spontaneous and ad-hoc. From those actions and objectives, we can glean their policy. Kumar admits that in many cases provincial governments made on-the-spot decisions, but the ultimate authority rested in London.[6]

Archived documents about petroleum in colonized India point to something similar: although local bureaucrats were not always following the script of imperial revenue collection or imperial science, and local administrative sovereignty played a noteworthy role in composing the British Empire, the institutional frameworks these bureaucrats and experts were limited by and the socio-political contexts they were embedded in were essentially imperial, which defined their broader politics. This circumscribed oil's politics – what it was to become, how it was to be used, made, taxed, etc. Several arguments and frustrations ensued because not every officer was deferent, which make for important stories about defiance, distributed sovereignty, contradiction, and confusion within the colonial bureaucracy. These *trouble* but do not *upend* colonial politics. Tussles amongst private oil companies and bureaucrats over leases and finances hint at the absence of an ideology of extraction and domination. But eventual decisions point to it, as leases *were* signed, finances *were* managed and extraction *did* happen, albeit after protracted debates, and not out of an overarching ideology, but for individual profit. The *effect* this had was colonialism. The condition of their possibility was also colonialism, as they were able to extract and sell oil by shaping labour practices in particular ways and by disregarding local rights.[7]

Numerous petroleum products saturating society today, hence, are derivatives not only of crude oil, but also of colonialism (with all its contradictions), without which, oil might not have been produced and used in these exact ways.

Narrating the gradual infiltration of people's private lives by oil, this chapter discusses how oil assisted the state, but also escaped it. Being a slippery and unknown substance, it was hard to control or even understand. But gradually an industry and science for it were developed to deploy it for colonial government. Nevertheless, its versatility accredited it with the possibility of being mangled and used in umpteen ways, many of which were undisciplined and illegal, such as adulteration. Even being demanded for legal uses by common people, from the state, as an entitlement, took it outside state control because its mass use could neither be curbed nor managed beyond a point. Through oil's ordinary uses, we can observe the complications in colonialism, and its more pedestrian face.

The many faces of oil

"Starting with the oil can and extending to the oil tanker, global trade of crude had become a worldwide system".[8] A substance "left largely undisturbed and satisfying no identifiable need" for the greater part of history, got refashioned in a way that more and more uses for it radiated outward to create a paradigmatic shift in the social, economic, and political life of the world at large. How does oil, from being nearly useless, become the lifeblood of society and a "critical actor" in shaping world politics?[9]

Oil use began in India in the latter decades of the nineteenth century. It was being imported since the 1860s and locally produced since the 1880s. Before the twentieth century, petroleum's primary use was kerosene for lighting, and occasionally it was used as a lubricant for machines. It was widely distributed for this purpose across the globe in tin cans.[10] By the 1880s, in India, kerosene was used in lighthouses on the Coromandel Coast at a "considerable saving of expenses", requiring less attention and providing far better light than coconut oil, which had been the primary resource for this activity thus far. Its use spread to other lighthouses soon.[11] Uninterested in using oil as fuel or lubricant, as was being done in several other nations, at this time the British Indian Government was resistant to the idea of putting oil to uses other than lighting. Reports revealed that Indian oil was heavy and better as a fuel and lubricant than as kerosene, for which light American oil was considered superior. In the eyes of the British Administration this was a "great falling off from expectations", and reason enough to reconsider granting leases to companies sequestering oil. What purpose would all that petroleum serve if it couldn't be used for illumination? Why should the government grant concessions for developing this industry if it couldn't serve the most important purpose? These questions were being raised by bureaucrats even when the railways expressed a need for fuel.[12]

This was the reason that oil had not been envisaged by the British Indian Government as a means of propelling the empire. It was seen as a subsidiary substance and could not displace coal as the global fuel. Only when *fuel* was developed out of oil did it become a colonial infrastructure for spreading the empire's tentacles. This shift in the significance of oil – from illuminant to lubricant to fuel – is what gave it a completely new identity and meaning. Alongside, with the abundance of petroleum in multiple aspects of people's lives, Indian society began to depend on it more and more. Its use for illumination, which initially raised the value and importance of crude oil globally, turned peripheral over time, in comparison to its other uses. Why did its fuel status raise its importance for the empire? Armies and governments both depended on energy for expansion, penetration, and control. Fuel was, thus, essential for colonialism, and also for trade. As petroleum products started proving to be more promising fuels than coal, they raised colonial hopes and created new dreams and means for more efficient expansion and trade (better ships, railways, weapons, and factories), and deeper penetration into the interiors (via cars and motorbikes). However, the British Administration took time to act on this.

In the early 1880s when the Punjab oil fields were being explored, the M.D., Chemical Examiner of Punjab, wrote to the local government informing them of the various uses oil could be put to for the government's benefit. His note said that in the USA and Russia, petroleum was being used to prevent oxidation of metals, lubricate machinery, fuel steamers, run heaters, prepare paints and varnishes, tan leather, substitute turpentine, cure skin diseases, and water-proof buildings.[13] The USA and Russia were already making manifold uses of this substance and that was catching on rapidly in other parts of the globe, making oil the most utilized product and the most important discovery of the nineteenth century. In the 1890s, the principal use for oil in these countries was in manufacturing.[14]

When the Public Works Department inquired into the use of oil as fuel in the early 1880s, it was told to refer to the journals *Engineer*, *Engineering*, *Herapath* and *Scientific American*, which carried articles about such research. Government officials discussed methods for burning liquid fuels, some of which had already been patented. They debated the advantages of liquid fuel over coal in metallurgical operations, as there was less waste of heat, it required less space, air, and time, produced less smoke and greater speed, had less weight, and the fire from it was more controllable. It was concluded that this information must be circulated, as in Burma (now Myanmar) and Assam, much oil had been drilled out but officials did not know what to do with it.[15]

By 1898, petroleum was being used as a lubricant in the railways in large quantities. Castor oil had been the standard lubricant but petroleum lubricants were cheaper. Most of them were being supplied by Standard Oil in the USA.[16] The shipping and the railway industries were also discovering the prospects of using oil as a fuel and not just a lubricant. The media routinely reported the triumphs of such experiments in Europe and the promises they held for the expansion of empires.[17] Steamers were gradually shifting to oil from coal.[18]

Petroleum now oiled not just the empire's machine but also found its way into the ordinary lives of people through machines they used and substances it was becoming a vital ingredient of, such as waxes, paints, medicines, and so on.[19] As cars made their way into India, so did petrol. It was categorized as "dangerous petroleum" and the Indian rules made its storage and use by private car owners prohibitive. At that time, not more than 40 gallons of dangerous oil could be transported or stored at one time. Discussions between officials of Bombay and Bengal in the early twentieth century, regarding the import and storage of petrol for cars, aimed at relaxing the rules to promote the trade of petrol and cars. Seeing that their use had increased tremendously, the Home Department wrote to local governments to draft rules according to the central guidelines. These recommended that

(a) 60 gallons could be stored in one place at one time under a special licence; (b) storage should be in large tanks; (c) precautions must be taken by car owners while storing this petrol in separate buildings constructed for this purpose and isolated from the dwelling area.[20]

Oil's usefulness for a multitude of activities was not a natural evolution or simply an outcome of its chemical properties, but also the result of a calculated process. Petroleum products were deliberately made to replace other substances used to make various household products such as medicines or bottles. Moreover, completely new products were being created from petroleum, and much research was being dedicated to inventing new uses for it, to make investments in it viable. Oil's many uses made its industry profitable. It was made to infiltrate several aspects of life, as science and technology sponsored by the state and industry consciously looked for ways in which it could replace other materials and become an integral part of people's lives.[21]

These developments, portraying the usefulness, accessibility, and affordability of petroleum, mirror a change of heart in the British Indian Government which grew abuzz with excitement about the possibilities of this new fuel. It took a few decades and many successful experiments for it to shed its hesitation over this mysterious substance. While the army was already using petroleum in small proportions for the preservation of leather hides, lubrication, illumination, etc., the navy now wanted to extend complete control over India's oil production and imports, claiming that naval use of fuel to serve the empire was the topmost priority and its other uses were secondary. The navy's need was being portrayed as a national need and it was deemed justified if consumers of wax, kerosene, etc., suffered from a diminished supply.[22] Oil's importance for warfare and conquest had been realized by the naval forces before the First World War, and it was recognized as a "strategic tool for ensuring global power".[23]

Crisis of shortage during the wars

Global use of petroleum grew by 50 per cent during the First World War (1914–1919).[24] Several new products with oil as a crucial component were invented for warfare. Armies and oil corporations were now inextricably bound to one another, with governments as the third partner, in the act of extracting and burning petroleum for national expansion. Peculiarly, petroleum tied civilian culture to military life and battlefield activities as no other substance ever had.

Militaries and navies claimed first rights over oil. A shortage of petroleum for civil and commercial purposes ensued. Railways, oil companies, and all agencies concerned with the supply of oil were asked to cooperate as the pinch of shortage and rising prices was felt across India.[25] Local governments wrote to the centre that all kinds of factories were on the verge of shutting down. The Government of Ajmer claimed that its minimum monthly requirement was 260 gallons of petrol and 12,500 gallons of kerosene. Actual requirements, the administrator wrote, were far more, but it was receiving even less than its minimum requirement. This led to a black market for petroleum. Rule 11J of the Defence of India Rules was imposed in some parts of the colony to prevent riots on account of unemployment resulting from factories closing down.[26]

Delhi Electric Tramways and Lighting Co. (DETLC) and Karachi Electric Supply Corporation complained to the government about the Anglo-Persian Oil Company cutting off oil supply as it claimed that the government had requisitioned its tanker steamer for carrying oil. They complained that this would force them to close down their operations and the cities they serve would be without power. DETLC's note to the government said that 70–80 per cent of its output was generated from diesel engines. With such severe shortage in the entire country and with cotton mills in Bombay expected to stop operations, large-scale unemployment was anticipated. Other companies also requested the government to not curtail supply, as they fed street lights, public buildings, hospitals, railways, and other services.[27]

This was a difficult time for curtailing oil supply for civilian needs, especially with a rising number of automobiles. Madras Presidency had approximately 3000–4000 cars plying in 1918. Local governments were amending laws for storage and transport of petrol to make life easier for car owners, while the central government was curtailing oil supply for them.[28] The Bombay Government complained that at a time when oil was needed to win battles, people in the city were wasting it for "pleasure drives". Conversion of railways and mills from coal to petroleum would have further increased the demand for oil, but was stopped midway.[29]

Oil companies were now in a powerful position as far as government decisions were concerned, because of the latter's dependence on them, and their partnership in the colonial endeavour. The Burmah Oil Company (BOC) wrote to the British Indian Government that the maximum supply of kerosene from Burma, if secured, would not exceed 140 million gallons. The company's experience and calculations showed that if supplies fell below this it caused administrative, political, and economic difficulties for the Indian state. It was essential, according to BOC, that it got maximum freight available for supplies to India. Freight congestion at Rangoon ended as a result. The Ministry of Shipping also chipped in by introducing a new distribution system in which sufficient tonnage was available.[30] Oil for war, oil for government, oil for mundane living – each activity was now carefully managed by the state and the oil corporation.

With oil saturating various aspects of life, the need for less regulation, easy accessibility, and affordability was expressed by some. Rules and taxes on it had to be rolled back. Over time, and especially once the First World War was over, the government realized that greater benefit lay in encouraging its use than in earning tax from it. Buses and lorries had become common on rural roads, ferrying goods to remote villages. In the late 1930s, British India had 175,000 vehicles plying on its roads.[31] Storing petrol in a special tank under controlled conditions specified by the government was turning out to be a hassle, and the need for readily available petrol in the market was felt by car users. Petrol pumps began cropping up in the 1930s.[32] The fledgling aviation industry in the 1920s and the 1930s was also encouraged, with concessions being granted to them with regard to petroleum.[33] Jawaharlal Nehru is said to have been the first member of the Indian National Congress (INC) to exploit air travel in his political campaign of 1936–1937.[34] Oil was important, therefore, not just for winning colonial battles but also democratic elections.

The Second World War (1939–1945) is sometimes referred to as the "war for oil".[35] In 1939, the Labour Department at the centre wrote to local governments that all oil exploration should stop during the War as companies were now expected to focus on increasing production of petroleum, not on prospecting.[36] The army, navy, and air force again lay claims on all supply, and their need was prioritized. Even though this need was not high enough to surpass supply, the armed forces demanded that all supplies be saved for their future need. Despite a sharp rise in military use of oil, its overall use fell drastically because of the blow that civilian and commercial use took, which shows how high the latter was.[37]

In 1942, the Department of Supplies at the centre wrote to the Controller of Supplies in Bombay, Madras, and Karachi that the time had come to bring diesel under government control, to reduce and regulate its consumption for ensuring long-term supplies. There was also a dire need for more space on ships for war-related goods, which could be created by reducing the space given to petroleum products for civilian and commercial consumption. The British Indian Government issued orders to regulate supplies and dispose stocks with all companies. State governments were instructed to restrict supply and consumption in consultation with electricity companies, factories, and other commercial users of petroleum. Kerosene for illumination had shrunk to half its earlier supply. It was feared that diesel engines would start using kerosene as fuel, as diesel was in even shorter supply, and this would threaten kerosene's supply for illumination even further.[38]

With the enforcement of curtailed oil supply, city governments began writing to their respective state governments about the situation under their jurisdiction. For each district, quotas were specified for different operations–irrigation, cotton gins, flour mills, electricity, etc. Frequent demands to alter these quotas were made.[39] Such correspondence recorded by the Punjab Government reveals the importance of petroleum even in small towns at that time.

In Nahan and Ambala, Achroo Mal and Sons, local distributing agents of Burmah-Shell, had apparently not supplied even 25 per cent of the fixed quota to the electricity department. The local city agencies requested the state government to ensure that tank wagons were allotted for carrying oil and its supply was resumed. The Dewan of Dujana complained that the supply of diesel had been cut to half and now electricity engines ran only till midnight. Likewise, the Khairpur Agency noted that diesel was being used for electricity, cotton gins, the press, flour mills, water works, and the drainage plant, but the hours of supply were now restricted to only six. In Bilaspur, people could not watch cinema as its engine ran on diesel. The demand for electricity in Sirmur was 150 units (1200 gallons of diesel) per month as against 75 units allotted to it. Various measures were taken to reduce consumption, such as stopping the powerhouse from operating during the day, switching off street lights when there was moonlight, and disallowing compound lights after midnight. In spite of this, the daily consumption of electricity required 30 gallons of diesel. According to this correspondence, there were roughly 12,000 oil engine owners in the Sind Circle alone.[40]

Meanwhile in Nabha, people faced grave health consequences without electricity, and inconveniences were on the rise in hospitals, jails, offices, railway stations, etc. Life without electricity was now unthinkable in these towns; so much so that their health depended on it. Rural areas were relatively insulated from the seepage of oil. The Malerkotla agency, for instance, wrote that its rural areas didn't need oil as they could meet their requirement by using hand-, water-, and animal-driven mills, but powerhouses in towns required diesel. Chamba, being a hill town, was also free of this dependence, as its diesel consumption was negligible. The Merchia flour mills complained to the Controller of Supplies in Karachi that there had been a 60 per cent reduction in electricity supply since 1941. Owing to this, out of 12 mills, only four survived until 1943. Since these mills were bound by military requirements, they were compelled to continue operating. They complained of corrupt practices and a black market for diesel. Oil dealers were apparently making private sales by making bogus entries against allotments. Incidents of corruption were commonly reported now.[41]

Many local agencies gave the plea of military requirements to demand more petroleum. The secretary of Tehri Garhwal claimed that it supplied huge amounts of grain to the military, was a part of the War effort, and must be allowed 54 units of electricity per month as against the 34 units allotted. Similarly, the Government of Faridkot claimed that its flour mills were providing an essential service to the community and military. Although several local agencies protested against curtailing oil and electricity supply, there were many who wrote to the Punjab Government that they would be willing to introduce compulsory registration of diesel engines to assist the War.[42] Through the management of oil and addressing the problems it threw up, the state got pulled into governing small everyday activities, spaces, and relations, which affected both – the day-to-day lives of people, as well as the character and functioning of the state.

By 1945, local agencies were probably tired of shrunken supplies and a fresh slew of complaints began. A political member from Jind wrote that not only were oil supplies far below essential requirements, but were also not in proportion. His note explained that bona fide consumers were being starved while a flourishing black market for petroleum products was thriving for those who could afford to buy at higher rates. Consequently, the Karachi Fuel Oil Advisory Committee framed new rules for distributing quotas. These decided the sale and uses diesel could be put to (only fuel, no lubrication, preservation, etc.), industries to be granted quotas, and the deeds included in an "offence" regarding the purchase of oil.[43]

None of this mattered to the public, for a war taking place far away, over issues irrelevant to them. For instance, the Dewan of Loharu wrote to the government that a rich resident wanted to start an ice factory. Its engine would require 15 drums of crude oil and 45 pounds of ammonia monthly. The daily turn out would be 10 tonnes of ice that would meet the requirements of Bikaner and Jaipur. Ice, at this time, was being imported from Delhi with a permit. The government was getting requests for various new factories and trades, but refused them all.[44]

War shortages revealed that the demand for oil was uneven; and for a considerable section of the populace it was nil, as oil had not infiltrated into their lives, or its requirement was marginal. The relative insulation of these sections from the oil economy was, interestingly enough, flagged as a source of concern in many a conversation between government officials and oil barons from the beginning of the twentieth century. They insisted that if the use of oil was to be extended into more economic domains, then the empire's Indian subjects had to be made to move from a culture of restraint and shared consumption into one of private and excessive consumption.[45]

During the two wars, however, the question was that with oil becoming an imperative for national security, mustn't it be conserved? This question struck at the heart of the process that made oil what it was. Conservation and saving simply did not go with the ideology and practices of that process. Only a voracious consumption of oil could ensure the survival and growth of its industry, upon which national security now relied. And yet, from the perspective of national security, ensuring future oil supplies was fundamental and demanded conservation. It was clear, however, that conservation was not going to let the industry grow. If it died out or remained small, national security would be threatened too, with limited supplies of and means to sequester oil. For oil to be a useful imperial tool, it needed to have a large magnitude based on conspicuous consumption and profit-oriented corporate organization. In Brian Black's words, "The reliance of the military on petroleum set the tone for humans' twentieth-century commitment to crude".[46] Becoming the lifeblood of militaries, it was then made to saturate the veins of civilian life.

Petroleum in the time of independence

As India's Independence began to loom on the horizon by the mid-1940s, the nationalists started to reflect on the issue of oil as a critical nation-making resource. In a letter to Sardar Vallabhbhai Patel of the INC, industrialist B.M. Birla observed that petroleum consumption would rise dramatically now owing to "rising development standards". He hinted at the need for India to become "self-reliant" with regard to oil.[47] He was perhaps trying to convince the nationalist leadership of the urgency of equating national self-reliance with oil. Building a nation was about making a future with oil. Oil was no longer only about security, but also about advancement and improvement. Petroleum and patriotism got tied together to offer independent India the fruit of development. Post-independence, in an article titled "The Energy Revolution and the Oil Industry in India" published in the *Oil Diary*,[48] K.U. Matthew wrote that for any nation, the greater the per capita oil consumption, the more prosperous it tended to be. He urged the Government of India (GOI) to adopt the American model where focus was on increasing consumption.[49]

Investments in the oil sector were rising with every Five-Year-Plan (FYP) of the GOI. The second FYP (1956–1961) made a Rs. 115 million provision for oil exploration and technical training programmes in exploration, production, refining, and sales.[50] According to the third FYP (1961–1966), the expenditure incurred on oil exploration during the second FYP period was about Rs. 260 million, as opposed to the 115 million earmarked. The third FYP envisaged an expenditure of Rs. 1150 million on further exploration as well as setting up and completing refineries and pipelines.[51] Considering that the GOI spent more than double of what it had earmarked for the oil industry in the second FYP period, it appears that the government was more elastic with its funds when it came to spending on oil.

The demand for fuels and petrochemicals was now far greater than the demand for kerosene. Petrochemicals altered the human relationship with oil by presenting countless inescapable ways in which petroleum could be applied to human life and society. When seemingly non-essential products become outstandingly helpful in easing out daily life, they become essential over time.

Oil shock and mass politics

When the Arab–Israel war began in October 1973, there was a 70 per cent hike in oil prices within a few days. Being import-dependent, India found itself in a precarious energy situation. Domestic production of petrol and kerosene was likely to be slashed by 25 per cent in order to raise their prices and conserve oil for the future. The Petroleum and Chemicals Ministry stated that the domestic output of naphtha needed to increase for fertilizers by cutting the production of petrol and kerosene. Agriculture was expected to be hit strongly as, post Green Revolution, it was highly dependent on petroleum for fertilizers, pesticides, and fuel for tractors, irrigation pumps, and so on.[52]

With the decision of the Ministry to levy extra duties on lubricants, their prices shot up, with transport being worst affected.[53] Prime Minister (PM) Indira Gandhi appealed to the citizens to tone down Diwali celebrations in view of the economic crisis India had been engulfed by.[54]

The Ministry announced a rise in the price of petrol and kerosene.[55] Following this, the Taximen's Union in Bombay demanded a rise in fares. The media reported that transport costs would now soar, raising the prices of all goods ferried by trucks, leading to a general inflation.[56] Taxi drivers complained that there was a drop in their business after fares were raised. The ongoing anti-price-rise movement by women added kerosene to the list of items they were protesting about, and were out on the streets in large numbers. Socialist parties too condemned the move to raise petrol prices.[57]

Domestic budgets were hit as dry-cleaning, groceries, cooking gas, etc., were all affected by the price rise. News reports claimed that people had to switch to coal and wood from cooking gas and kerosene, and buses and bicycles from taxis and cars. There were fears in some cities that all transport would be paralysed. Petrol pumps faced massive jams.[58] Furnace oil was now in such high demand that it was being stolen by organized gangs.[59] In order to curb adulteration of diesel and petrol with kerosene, the GOI pronounced that all kerosene must now be dyed blue.[60] The origin of blue kerosene in India can, therefore, be traced to the Yom Kippur War.

A nationwide strike of petrol dealers was expected to paralyse transport across the country. Madras, Calcutta, Delhi, and Bombay were already facing strikes and protests.[61] With this backdrop of shortage, protests, and inflation, governments had to cut down their own consumption of petroleum products.[62] The PM made a show of austerity by taking a horse carriage to her office one day instead of a car, following which she switched from a limousine to an Ambassador, as the latter consumed less fuel.[63] In the same vein, the West Bengal cabinet decided to stop using limousines for ministers.[64] In some states, ministers were now riding bicycles.[65] Soon after, there was a wholesale shift among all ministers from limousines to Ambassadors. The 1973 oil shock officially marked the end of limousines in India and enshrined the Ambassador as a seal of the government and a symbol of austerity.

After an all-India strike in November 1973 against the economic situation in the country and the government's alleged inaction, the left parties and the Jana Sangh were preparing to move an adjournment motion in the parliament.[66] Likewise, the entire opposition in the Maharashtra Assembly walked out in protest against the state government's failure to supply kerosene for civilian use. Several state assemblies and the parliament witnessed high drama.[67] Oil was now a big enough issue to shake the government, and in addition to security and development, also implied political stability now.

With petrol prices consistently on the rise, some members of the parliament urged for the implementation of "carless Sundays" like in many Western countries.[68] Engineers were devising new ways of running cars. New fuels such as ethanol were being experimented with. The future of cars, it was said, was "alcoholic".[69] Cooking gas was also successfully tested for running cars.[70] The GOI did not take up these ideas and India continued running on limited supplies of petrol. Keeping the oil industry alive was a priority for the government even in times of shortage.

Oil cuts had also hit the pharmaceutical business and healthcare. Petrochemicals were a basic ingredient of modern medicine.[71] Fishing was affected by shortage of fuel for boats, thereby affecting the food people ate.[72] The Baroda Refinery was extracting ingestible protein from oil, which was being used as cattle feed, and this too was impacted by the shock.[73] Petroleum had crept into the daily lives of people and saturated the nation such that it came to a standstill without it. The new uses that were being created for it often had no substitutes.

The web of associations that developed between society and petroleum over a 100 years had made it difficult for society to disengage itself from the complex of oil. This made the demand for oil quite inelastic as it had no easy substitutes anymore, the way it did when it was used merely for illumination. As much as the modern oil industry and governments attempted to make petroleum scientific, predictable, and immune to risk, the GOI had made a classic wildcatter's move by placing its entire stakes on oil, and faced near paralysis without it.

In 1975, the Planning Commission prepared a paper on energy planning in India, according to which, between 1953 and 1974, India's oil consumption rose from 3.5 to 26.3 metric tonnes. In the same period, oil's contribution to overall energy consumption in the country rose from 11.9 to 31.2 per cent. Consumption of non-commercial sources of energy – biomass, dung, wood – steadily came down as the reach of grid-based commercial energy widened and more populations were shifted from energy independence and decentralized renewable sources to dependence on state-provided non-renewable energy. This was essential for industrialization, and consistent efforts were made to achieve this "favourable trend". Non-commercial sources were seen as "wasteful" and primitive.[74]

The oil shock reveals the government's total power over controlling how oil would be used, what activities would be affected, and what would continue. The state exercised enormous control over people's mobility, food, clothing, consumption, working hours, and almost all aspects of life, since most were touched by oil. The fact that people were left in the lurch for several activities but could do nothing about it, is telling of their dependence on an external authority for the provision of the infrastructure, raw materials, and services they needed to execute their basic activities. Oil, thus, became the key with which the state organized the everyday and controlled people's daily lives.

Concluding remarks

Petroleum made a smooth transition from being an *instrument* of the colonial state to becoming a *constituent* of the contemporary state. Colonial concerns about oil, such as railways and defence, remained, but were added to by the politics of the everyday, organized around the idea of citizenship based on rights and duties. By becoming an active fault line that ran between citizens and their government, petroleum provisioning became a vital task of government and an inflammable political ingredient for raucous mass politics in independent India. What this reveals is that oil was not just an instrument of state-making, but could be used inversely by the citizens, as a substance of mass-politics. The government was accountable to the people for the provision of oil, and on failing to do so, could be toppled over.

With this story, I have aimed to challenge the idea that big technology only furthered colonial interests, by bringing out the contradictions and ambivalence imbued in it. Sometimes owing to conflicts within the imperial government about the merits and uses of the technology, sometimes owing to its subversive use at the receiving end that created vulnerabilities for the empire, and sometimes because of confusion, accidents, and problems accompanying new technology, it was not driven by the logic of empire alone. That said, it cannot be denied that oil's seepage into people's private lives allowed the state to seep into them too, and increased its presence in newer arenas. By colonizing the everyday, oil became an apparatus of biopolitics and brought more aspects of life under government. Nevertheless, the state could not ignore the power of the people over this product, nor discipline it fully, as it flowed in and out of legality, with its proliferating black markets and alternate uses that defied state control.

The story of oil captures a double movement in which big and complex technology make possible the everyday technical, political, and social worlds, but simultaneously, the everyday use of that technology is what enables it to become big, and therefore useful for the state. The state and the everyday-life worlds of its subjects/citizens cannot be studied in separation, as they bleed into each other and cohabit the same world. Technically produced objects, such as petroleum products, are an essential part of this puzzle, which make possible and foreclose different kinds of socio-political constellations and modes of power, manifesting the state's and common people's intentions and actions together.

Oil's history in India indicates that the colonial and independent Indian state has existed through developmental functions it has assigned itself. The state's efforts to perpetuate and elaborate itself through these development operations (which are beholden to oil) are met with interruptions and iterations owing to subaltern appropriations of them. In this dialectic between the state and society, although people (wittingly or unwittingly) throw a spanner in the state's operations and extensions, the state has its strategies to co-opt or exterminate informal and ad-hoc networks of the people. While the ability of people to trouble government and escape discipline leaves open several possibilities, the state tends to stay a few steps ahead because of the moral, ideological, and cultural enchantment with development, instituted into popular imagination from the colonial era. Because of oil's importance to development, oil has been turned into not just a tool of government, but also something consumers exerted power over by demanding it, adulterating it, pilfering it, using it for unintended purposes and selling it illegally. Through this process, the state and its subjects/citizens are made and remade, and their relationship reworked. Technology and infrastructure are neither just the context, nor merely the product of this dialectic. They are essential ingredients in making the state, the social, and the individual. The history of petroleum, therefore, is not a mere sequence of events involving technological innovation in pipelines, refineries, and wells, but a set of political calculations and social pressures that ascribed certain political and social possibilities to this substance. Oil, unsurprisingly, is a messy story, entangling high politics of the state with quotidian politics of everyday lives.

Notes

1 Daniel Headrick, *The Tools of Empire*, New York: OUP, 1981; Ian Stone. *Canal Irrigation in British India*, Cambridge: CUP, 1985; Imran Ali. *The Punjab under Imperialism*, Princeton: Princeton University Press. 1988; Michael Adas. *Machines as a Measure of Men*, New Delhi: OUP, 1990; Zaheer Baber. *The Science of Empire*, New York: State University of New York Press, 1996; Richard Drayton. *Nature's Government*, Hyderabad: Orient Longman. 2005; Irfan Habib and Dhruv Raina (eds.), *Social History of Science in Colonial India*. New Delhi: OUP, 2007.
2 David Arnold. *Everyday Technology*, Chicago: University of Chicago Press, 2013.
3 Ritika Prasad. *Tracks of Change*, Cambridge: CUP. 2016.
4 Sarandha Jain, "Building 'Oil' in British India: a Category, an Infrastructure", *Journal of Energy History/Revue d'Histoire de l'Énergie*, n°3, March 2020.
5 Ibid
6 Deepak Kumar. *Science and the Raj*, New Delhi: OUP, 2nd Edition, 2006.
7 Jain, "Building 'Oil' in British India: a Category, an Infrastructure", op. cit.
8 Timothy Mitchell. *Carbon Democracy*, London: Verso, 2011, p. 64.
9 Brian Black, *Crude Reality*, New York: Rowman & Littlefield Publishers, 2012, pp. 1, 7.
10 Timothy Mitchell. *Carbon Democracy*, op. cit., p. 31.
11 Public Works Department (PWD), Railway Stores Branch, Proceeding no. 2679–2683, Part B. (1889, December). New Delhi: National Archives of India (NAI).
12 Revenue and Agriculture Department (RAD), Minerals Branch, Proceeding no. 18–22. (1890, April). NAI.
13 PWD, Civil Works Branch (Coal and Iron), Proceeding no. 13–16, Part A. (1883, October). NAI.
14 Brain Black. *Crude Reality*, op. cit., p. 59.
15 PWD, Civil Works Branch (Coal and Iron), Proceeding no. 13–16, Part A. (1883, October). NAI.
16 PWD, Railway Stores Branch, Proceeding no. 23–40. (1899, September). NAI.
17 Finance and Commerce Department (FCD), Statistics and Commerce Branch, Proceeding no. 263–270, Part A. (1898, March). NAI.
18 FCD, Statistics and Commerce Branch, Proceeding no. 485–486, Part A. (1901, October). NAI.
19 RAD, C.V. Administration Branch, Proceeding no. 8–9, File no. 20, s.no. 3–4, Part B. (1904, August). NAI.
20 Home Department (HD), Judicial Branch, Proceeding no. 63–67, Part A. (1903, September) NAI.
21 Brain Black. *Crude Reality*, op. cit, p. 1.
22 RAD, Geology and Minerals Branch, Proceeding no. 6–18, File no. 108, Part A. (1904, November). NAI.
23 Brain Black. *Crude Reality*, op. cit, p. 59.
24 Ibid, p. 80.
25 Department of Agriculture, Revenue and Commerce (DARC), Stores and Plant Branch, Proceeding no. 1–30, Part A. (1917, March); Proceeding no. 9. (1917, July). NAI.

26 DARC, Petroleum Branch, Proceeding no. 1–55, Part A (1918, September). NAI.
27 DARC, Petroleum Branch, Proceeding no. 1–21, Part A. (1919, April). NAI.
28 DARC, Petroleum Branch, Proceeding no. 1–2, Part A. (1918, July). NAI.
29 Legislative Department (LD), Unofficial Branch, Proceeding no. 187. (1918). NAI.
30 Ibid.
31 David Arnold, *Science, Technology and Medicine in Colonial India*, Cambridge: CUP, 2000, p. 206.
32 LD, Solicitors Branch, File no. 654–S/30. (1930). NAI.
33 Central Board of Revenue, Customs Duties Branch, File no. 369 – Cus II – 33. (1933). NAI.
34 Arnold. *Science, Technology and Medicine in Colonial India*, op. cit., p. 206.
35 Black. *Crude Reality*, p. 136.
36 Political Department, War Branch, File no. 209 – W/39, s.no. 1–59. (1939); Rajputana Agency, Political Branch, File no. P/40 – 8. (1939–41). NAI.
37 External Affairs Department, Frontier Branch, File no. 120-F/42, s.no. 1–67. (1942). NAI.
38 Punjab State Agency, Residency Files, File no. C 8/4-2/46. (1946). NAI.
39 Ibid.
40 Ibid.
41 Ibid.
42 Ibid.
43 Ibid.
44 Ibid.
45 RAD, Geology and Minerals Branch, Proceeding no. 1–4, File no. 136, Part A. (1902, November). NAI.
46 Brain Black. *Crude Reality*, op. cit, p. 134
47 Sardar Vallabhai Patel Papers, Microfilm, Accession no. 62. (n.d.). NAI.
48 Vol. 2, no. 38, 23/09/1962.
49 Ministry of External Affairs (MEA), WANA Section, File no. 6-C (34) WANA/60. (1960). NAI.
50 Planning Commission (PC). *Second Five Year Plan of India*. New Delhi: GOI, 1956.
51 PC. *Third Five Year Plan of India*. GOI, 1961.
52 "Petrol Production to be Slashed". *The Times of India (TOI)*, 19 October 1973.
53 "Price of Lubricants Goes Up". *TOI*, 20 October 1973.
54 "PM for Simple Diwali". *TOI*, 20 October 1973.
55 "Petrol Price Raised by Rs. 1.07 a Litre". *TOI*, 3 November 1973.
56 "Cabbies to Hike Fares". *TOI*, 3 November 1973.
57 "Taxi Fares go up in Bombay". *TOI*, 4 November 1973.
58 "Cabs and Rickshaws go off the Road". *TOI*, 4 November 1973.
59 "Two Gangs of Furnace Oil Thieves Busted". *TOI*, 13 March 1974.
60 "Kerosene may be Dyed Blue". *TOI*, 4 January 1974.
61 "City Unprepared for Dealers' Strike". *TOI*, 14 November 1973.
62 "Drastic Oil Curbs". *TOI*, 4 November 1973.
63 "Indira Sets the Pace". *TOI*, 6 November 1973.
64 "No Use of Limousines". *TOI*, 8 November 1973.
65 "Government Cars face "No Petrol" Prospects". *TOI*, 21 November 1973.
66 "Move in LokSabha Likely". *TOI*, 12 November 1973.
67 "Walkout on Kerosene Fiasco in City". *TOI*, 15 February 1974.
68 "Carless Sunday Urged". *TOI*, 28 November 1973.
69 "IIT Tests show Alcohol Blend can Replace Gasoline". *TOI*, 26 November 1973.
70 "Cheapest Way to Run Car". *TOI*, 6 January 1974.
71 "Oil Cut Will Hit Drug Industry". *TOI*, 14 February 1974.
72 "Kerosene Quota for Fishing Boats". *TOI*, 1 September 1974.
73 "Protein from Crude at Baroda Unit". *TOI*, 11 June 1974.
74 MEA, UI Section, File no. UI/151/151/70. (1970). NAI.

Section III
Environmental issues

9 Designing scientific mining

Evolution and implementation, c. 1860s–1960s[1]

Sahara Ahmed

The greatest developments in the evolution of ecological thought through the nineteenth century were the results of the materialist conceptions of nature, interacting with changing historical conditions. Imbued in these prevailing notions of forest preservation and assuming the sceptre of mastery over nature, the pioneer imperial miners issued their prescriptive observations. By transforming and absorbing myriad categories of information, the colonial rule developed its own genre of knowledge. Experiences of this imperial phase unravel the colonial imperatives to impose control over structures of resources.

This essay endeavours to present vignettes of the mining industry in Bengal and probes into the rationale behind its inception. In the process, it puts forth two arguments: Firstly, the emergence of scientific mining in India, specifically in Bengal was in actuality the organization of knowledge in Europe predicated on the institutionalization of this burgeoning area of study. Secondly, its *modus operandi* was primarily a reserve of governmental proclivities and was often implemented in total disregard of the ecological aspects – precipitating ecological degradation. The fallout of which, involving both human and nature, are in no sense ephemeral.

Environmental implications of mining can be summarized as erosion, sinkholes, loss of biodiversity or the contamination of soil, groundwater and surface water by the chemicals emitted from mining processes. These processes also have an impact on the atmosphere from the emissions of carbon which affect public health and biodiversity. Erosion of exposed hillsides mine dumps and resultant siltation of drainages, creeks and rivers can wreak havoc on the health and sanitation of surrounding areas. Within forests, mining operations disturb the ecosystem and habitats. In farming areas, it disturbs pastures and croplands. In urban environments, mining causes noise pollution, dust pollution and may cause visual impairment.

Environmental pollution in the context of surface mining operations had been identified under the following heads: (i) land pollution, (ii) water

DOI: 10.4324/9781003241980-13

pollution, (iii) air pollution, (iv) ecological pollution. Land pollution pre-cipitated land degradation, soil erosion, water pollution comprised of acid mine drainage, toxic pollutants, hydrological changes; air pollution consisted of dust, gases and noise and ecological pollution comprised of deforestation, wildlife damage, aquatic life damage, aesthetic pollution and sociological changes.[2] Underground mining too has its ill effects on the environment. It has the potential for tunnel collapses and land sub-sidence.[3] It involves large-scale movements of waste rock and vegetation, as in surface mining. Additionally, like most traditional forms of mining, underground mining can release toxic compounds into the air and water. As water takes on harmful concentrations of minerals and heavy metals, it becomes a contaminant. This contaminated water can pollute the region surrounding the mine and beyond. Mercury is commonly used as an amal-gamating agent to facilitate the recovery of some precious ores. Mercury tailings then become a major source of concern and improper disposal can lead to contamination of the atmosphere and neighbouring water bodies. Most underground mining operations increase sedimentation in nearby rivers through their use of hydraulic pumps and suction dredges, blast-ing with hydraulic pumps removes ecologically valuable topsoil contain-ing seed banks, making it difficult for vegetation to recover. Deforestation due to mining leads to the disintegration of biomes and contributes to the effects of erosion.

Coal mining operations can be summed up as strip mining or surface mining and underground mining, but both have their own negative impact on the environment. Strip mining involves the stripping away of earth and rocks to reach the coal underneath. If a mountain happens to be standing in the way of a coal seam within, it will be blasted or levelled changing the landscape and disturbing ecosystems and wildlife habitat. In order to work a mine, trees and other vegetation need to be cleared. This loosens the topsoil which collects as debris in riverbeds inducing floods after a heavy rainfall or causes landslides.

It also leads to contamination of groundwater. The minerals from the disturbed earth can seep into the ground and contaminate waterways with chemicals that are hazardous to our health, for instance, acid mine drainage. Acidic water can flow out of abandoned coal mines. Mining has exposed rocks which contain the sulphur-bearing mineral, Pyrite. This mineral reacts with air and water to form sulphuric acid. When it rains, the diluted acid gets into rivers and streams and can even seep into underground sources of water.

Underground mining allows coal companies to dig for coal deeper into the ground. The problem is that huge amounts of earth and rock are brought to the surface. These mining wastes can become toxic when they are exposed to air and water. Examples of toxins are mercury, arsenic, flu-orine and selenium. The amount of dust generated in mining operations pollutes the atmosphere in the vicinity.

Methane emissions from underground mining are often caught and used as town fuel, chemical feedstock, vehicle fuel – but rarely is everything captured. Methane is less prevalent in the atmosphere as compared with carbon dioxide, but it is 20 times more powerful as a greenhouse gas. Moreover, fires from underground mines can burn for decades. These fires release smoke that comprises carbon dioxide, carbon monoxide, methane, nitrous oxide, sulphur dioxide and other toxic greenhouse gases.

Health hazards caused by coal dust inhalation can cause black lung disease or cardiopulmonary diseases and the probabilities of hypertension, chronic obstructive pulmonary disease and kidney diseases are noticed at a higher rate near coal mines than other places. This has also led to the displacement of large sections of the population.

Attitudes towards scientific mining

The attitudes towards the earth and its investigation underwent considerable transformation in Europe, especially in Britain between the mid-seventeenth and early nineteenth centuries. The labours of the seventeenth and eighteenth centuries provided the basis, both material and conceptual, for the unquestioned flowering of geology in the nineteenth century, contrary to the claims made by early-nineteenth-century geologists.[4] Following the general international trend, great efforts were also made in Europe to put mining on a firm scientific basis.[5] Initially, with regard to earth sciences, the absence of proper training and the lack of proper qualification forced some students to complete their studies at German universities. Till the middle of the nineteenth century, a third alternative existed for obtaining an education in natural sciences in general and geosciences in particular, only in the Faculty of Medicine, where the significant scientific subjects were taught. In Sweden too, chemistry was the main beneficiary of a governmental concern for economic improvement. The concern was first manifested in the 1630s when the government of the day created a powerful board of mines primarily to control the expansion of the copper mine at Falun but also to regulate all aspects of the mining industry. It was not only in Sweden that mining served as a powerful incentive for governmental investment in science and technology but in countries with large mineral deposits, national administration for the control of mining was already common in the seventeenth century.[6]

It was since the eighteenth century that the importance of 'scientific miners' received recognition with the creation of mining academies. A group of scientifically and technologically trained technocrats flourished with the augmentation of these institutions. The primacy given to scientific experiments in comprehending mining attributed to it the status of a science. The most noteworthy being the Freiberg Mining Academy in Saxony particularly with the institution of a Professorship of Mining and Mineralogy, Abraham Werner was also appointed curator of minerals at the Academy. At the end of forty years of service as inspector of mines in the Saxon mining service, Werner also made a significant mark on the Saxon economy.[7]

The mining academies of Central Europe provide compelling evidence of the confidence governments enjoyed in enhancing the returns on the mineral resources on which their economies largely depended.[8] However, the governmental intervention had not only amplified the number of contenders in the arena of mining but governmental proclivities had an overwhelming influence on mining operations. 'Scientific mining' was often relegated to the background giving primacy to the overriding decisions of the governments. An element of vulnerability persisted in the institutions that depended directly on government support, as patronage could soon turn into disfavour riddled by conflicting opinions. In Britain too, governmental support for science tended to be modest. The tribulations of the Berlin and St. Petersburg academies illustrated with brutal clarity that wherever governmental involvement was perceived strongly there lurked the fear of an inhibiting subservience to a politically motivated conception of national interest.[9]

Despite the dangers of unwelcome interference, however, some measure of recognition by the state (allocation of funds) was virtually essential if an academy was to prosper. However, this often stopped short of granting of a name.[10] Government recognition had its own share of pros and cons. Recognition lent an aura of royalty to some institutes while others suffered an inevitable decline due to their dwindling prospects.[11]

The extinction of the indigenous iron industry of the district of Birbhum in West Bengal, India could be attributed to governmental intervention. Scholars, however, differ on the question whether the British policy in this regard was one of 'indifferentism' or a policy deliberately directed towards the destruction of the indigenous industry.[12] However, there was general acquiescence among the proponents of the nationalist school that British industrial might with the connivance of the British Indian Government stamped out the siderurgical industry in India which, judging by the medieval accounts had great possibilities.[13] The accounts indicate that two types of manufacture were in vogue in Birbhum, one by the tribes and the other by the Bengali-speaking people. The latter was concentrated in the north-western part of the district in which ore and timber for preparation of charcoal fuel was easily available.

A recent study on the British attitudes (both official and non-official) towards India's geological resources in India analyses major aspects from 1770, the date of first application by British officers to East India Company for permission to open the mines in India, to 1851, the official date of the founding of the Geological Survey of India.[14] By situating this history of geology firmly within its political, social and economic contexts attributing primacy to the role 'colonial science' certain broad conclusions have been drawn. Firstly, it has been argued that with a few exceptions, the State only reluctantly appropriated the mineral resources of India, and this reluctance was clearly related to the influence of both economics and ideas. Secondly, perceptions of India, both of its environment and society were strongly influenced by geological discourse and lastly, geology in India was not necessarily inferior to metropolitan science.[15] A symbiotic relationship clearly existed between individual geologists and institutions in Europe and India. The methods involved in the gathering and dissemination of geological knowledge about India are described and analysed in terms of 'networks' of exchange and patronage. The role of the Asiatic Society, the requirements to obtain strategic supplies of coal and to be seen as in the role of an 'enlightened' government finally led the Company to establish the Geological Survey of India.[16]

The history of coal mining has an ancient lineage. Large-scale coal mining developed during the Industrial Revolution in England and coal provided the main source of primary energy for industry and transportation in industrial areas from the eighteenth century to the 1950s. Britain developed the main techniques of underground coal mining from the late eighteenth century and progress was further accelerated by improvement in techniques in the nineteenth and twentieth centuries. Oil and its associated fuels began to be used as alternatives since the 1860s. Early coal extraction was small scale, the coal lying either on the surface or very close to it. Typical methods for extraction including drift mining and bell pits were in vogue.

The official history of the GSI begins with the rise of the coal mining industry in India when a coal committee was established by Auckland in 1795 and recommended a geological survey of the coal belt of India.[17] East India Company obtained the services of David Hiram Williams of the Geological Survey of Great Britain. Williams, who surveyed the Ramgarh and Karimpura coalfields, laid the foundations of the GSI in 1854. In its early years, the GSI was housed in the Asiatic Society.[18] The GSI offered great possibilities for industrializing India and even though the efforts of the Survey represented imperialism at its best, its research had compiled a vast record of the mineral wealth of India.[19] P. N. Bose's contributions in this regard deserve special mention. He was one of the early Indians to join the Geological Survey of India as a graded officer. Bose has several firsts to his credit. The first Indian to have graduated from a British University. His early work was on the Siwalik fossils and is credited with the discovery of petroleum in Assam. The establishment of the first soap factory too goes to his credit. The most commendable achievement of his life was the discovery of iron ore deposits in the hills of Gorumahisani in the state of Mayurbhanj which eventually transpired into the legendary Tata Iron and Steel Factory at Jamshedpur following an intimation Bose had sent to J. N. Tata.

The government diluted all efforts at spearheading an industrial utilization drive on grounds that the Survey facilitated the export of minerals to other countries. In turn, these experts from the Survey were transferred to the New Bureau of Mines, an organization with an exclusively administrative function of inspection to assure the implementation of safety rules in the mines. [20]

Inception of mining

The mineral development of India was left largely in the hands of private enterprise which concentrated mainly on rich occurrences of ore and minerals, whose exploitation for export or other use required little risk and practically no capital.[21] The first published reference to the mining of coal in India dates back to the year 1774 when shallow mines were reported to have been developed in the Raniganj field.[22] The first Englishman to be accredited for the discovery of coal in Bengal was probably Mr. Suetonius Grant Heatly, Collector of Chota Nagpur and Palamau. In the same year together with John Summer, he obtained from Warren Hastings, a license empowering them to work the coal mines in 'Pachete and Birbhum'.[23]

The Indian Forest Act of 1878 incorporated the exploitation of mines under the category 'forest produce'.[24] Initially, a clear definition for mines was absent and the term broadly encompassed, quarries, pits and holes of any kind.[25] Records referred to them as quarries, pits and holes of any kind. Fresh pits were opened unabated, old pits were closed and no attempt hitherto had been made to accurately ascertain the number of mines in existence. According to an estimate, there were 146 coal mines approximately at the end of 1895. Coal mines in use were confined to the sub-divisions of Raniganj, Govindpur and Giridih and although there are some in other districts, yet with one or two exceptions they were still insignificant. The report, however, maintained that large coalfields existed which with a better labour supply and railway networks would soon be in operation. With a note of caution, the report expressed its ignorance about the labour requirements of an industry that was still in its infancy.[26]

Another government report indicated some significant observations on the basis of the investigations carried out from Raniganj to Chenacoory and Chanch. The investigations were carried out primarily with the intention of verifying the quality of coal produced and whether it would be suitable for the European system of iron manufacture.[27] The report purported that

> Although inferior to the English and Welsh coals, it is a safe fuel for reduction of ores in blast furnaces with the aid of hot blast. With the 'lange' of the Raneegunge seam, an imperfect coke may be produced, but the great amount of 'wastage' or bad yield in the operation, it would be too costly a fuel for profitable application. It would consequently have to be introduced in its raw or natural condition, in which form, with the blast heated to 600 degrees Fahrenheit, a temperature easily obtained....I am led to the conclusion, which I feel I can state

with much confidence, that this coal would be found very manageable in the blast furnace under the conditions I have already stated, and with a properly constructed furnace, efficient blast power and heating apparatus, a weekly produce from 65 to 70 tons might safely be calculated.[28]

It conclusively maintained that Raniganj and Barool, one containing coal and the other iron ore, respectively, reflected inexhaustible abundance. The working population was also found to be trained and disciplined and available for the minor and less important demands of ironworks besides being connected through railroad in the year 1895.[29] These investigations were further corroborated by other surveys that followed it and guided by those that preceded it.[30]

The requirements of the military department of the East India Company for coal led to the first systematic exploration of coal resources in India from 1808 onwards and henceforth after the inauguration of the first coal mine in 1817 in Raniganj. It was soon joined by a number of competitors in the 1820s.[31] The requirements of the mint and the military department had induced further explorations and the search for higher qualities of coal, paralleled by a growing demand with the introduction of steamboats in the 1840s. In fact, the need for a distribution network had provided the desired impetus for the development of the coal industry with the initiation of another, the railways.[32] It was, however, not only the railways' demand for coal, that induced a change in the development of Indian coal mining but their assurance to continue the unrestricted supply of coal irrespective of the season. Transportation of coal to Calcutta had previously been confined to the rainy season, two or three months of the year, during which river transportation was facilitated by the river Damodar by means of country boats.[33]

In keeping pace with the industrial slumps and booms, triggered off by the international market demand, acquiring mining leases became widespread and the initiation of extensive exploitation of mines began. A group of scientific miners advocated conservation measures to preserve the apparently inexhaustible resources. The war years had impeded such imports and a higher demand was anticipated after the war. A decline in coal production was expected to raise the price of coal.[34] For an interim period, the enhanced rate of coal was evident, but soon the prices began to fall. At this juncture, the mine owners increased production in order to maintain the erstwhile income scale, despite the falling prices.[35] The mine owners, however, resorted to this course of action prior to the onset of the depression. They reduced coal production for a brief period during the depression and repeated this act during the 1930s.

Numerous collieries closed down; however, others in the struggle for survival tried to cope with the declining prices by intensifying production through 'slaughter exploitation'. In this process, often the best quality coal was mined out resulting in overproduction which again depressed prices still further. This development had devastating consequences on the environment. Exploitation and conservation were alien to each other right from the inception of mining operations. The bleak situation had triggered off over-exploitation, leading to subsidence of soil, subterranean fires which often engulfed large areas. These activities caused land degradation, landslides, soil erosion, water pollution, toxic pollutants and hydrological changes as well as air pollution due to dust, noxious gases and noise. These also caused ecological imbalance due to deforestation, which was the reason for the damage to aquatic life as well.

The Coalfields Committee was constituted by the Government of India in 1920 as an initial endeavour.[36] After prolonged discussions with the provincial governments of Bengal, Bihar and Orissa, the Indian Mining Association and the Indian Mining Federation, the Government of India decided as a preliminary measure to engage the services of a recognized authority on 'modern' methods of extraction.[37] Special instructions were given to visit the coalfields and advise on the best means of securing greater economy in the production and consumption of coal and submit a report accordingly to a designated Committee. Mr. Treharnee Rhees, a partner in the well-known firm of Messrs. Forster Brown and Rees of London and Cardiff was assigned the role of principal investigator for the Ranganj and Jharia fields. Rhees reported in 1919:

> ...that in the case of large tracts of coal property the areas let off for working by the landlords have not been so arranged as to conduce to the economical working of the estate as a whole but rather with the object of receiving as much as possible by way of 'salamis' or sums of money paid down as consideration for granting a lease. ...this has led, especially in the case of the smaller takings near the outcrop, to the areas worked being in a number of instances of such small dimensions and fantastic shapes that it is quite impossible to work the coal satisfactorily.[38]

The principal recommendations of Mr. Treharnee Rhees related to the appointment of an inspecting and controlling authority for the supervision of the terms of leases, more efficient methods of coal extraction, including rotation of working and hydraulic stowage, the improvement of the conditions of colliery labour, the more economic use of power and more general employment of electricity, the improvement of methods of coke making, the introduction of coal mixing, the handling and dispatch of coal and more extended employment of screening.

Moreover, the proposers argued that this kind of control was not new to the industry, it had hitherto been exercised under the Land Acquisition Mines Act in respect of coal under railway sidings and branches and they merely wanted to extend it to all the workings.[39] It further stated that some leases already enjoined on the lessee the necessity of allowing free inspection by the officers of the Mines Department and of obeying their 'lawful instructions', while

> others required the lessee to conform to and observe all the provisions of the Indian Mines Act and any rules and regulations made there under and all other Acts of the Supreme or Bengal Council or any other statutory rules for the time being in force so far as they affect the demised premises.[40]

It further harped on the safety provisions incorporated in the Indian Mines Act and the rules framed under it are concerned with the safety of the miner, but the legislation they proposed would be concerned with the safety of the mineral.[41] The two subjects they believed were inextricably linked, and they anticipated no further hurdles in implementing them.

The Coalfields' Committee of 1920 which examined various suggestions made by Mr. Rhees observed that they had found instances in which areas had been let out in a number of minute plots, but the middlemen rather than the superior landlords had been responsible for the worst cases. The major part of the known coal-bearing land in the Raniganj and Jharia fields had already been leased out and any authority controlling unleased areas was unable to remedy existing evils to a considerable extent.

Moreover, the creation of periodic Working Plans for individual collieries was also deemed necessary. For the administration of Mines, The Indian Mines Act, 1923 was passed and it came into force on 1 July 1924.[42] There were 49 sections of this Act, 6 of which were concerned with the appointment, functions and powers of the Inspectorate. The health and safety of workers were provided for in six sections of the Act and provisions were made for hours of work and limitation of employment. The powers of the Central Government to make regulations and the Provincial Governments to make rules were laid down and penalties and procedures were embodied in the Act.[43] The successful implementation of these recommendations, however, veered around the questions it aroused and in placating the myriad sections concerned.

Often decisions were influenced by governmental proclivities and this was reflected in the government's curt response to this. There were sections in government echelons who disagreed with Rhees's recommendations. They felt that the Committee's proposals were a serious departure from the previous conditions which accorded priority to the safety of workers, the colliery interests were at liberty to extract their coal by whatever methods were permitted under their leases, however, wasteful.[44]

Ecological perspectives

All these fields had been worked extensively, 'slaughter exploitation', mining operations had become more and more arduous, 'imposing increasingly onerous duties upon the management, and calling for ever greater knowledge and capacity in those who direct the labour'.[45] Administrative control and better management were advocated by the various committees appointed under a governmental initiative in an endeavour to sustain the industry in the interest of the state. The Report further stated (which was perhaps an expression of its sincerity and emphasized the enormity of the situation) that 'the rapid progress of the mining industry in Jharia had not been accompanied by a corresponding development in such matters as sanitation'.[46]

Sanitation in the coalfield area was a problem which had given infinite trouble during the last few years, prior to 1906. When a plague broke out at Jharia and there was a general scarcity, practically nothing was done; coolies were indifferently housed or left to arrange for themselves here and there, only attempts were made to provide comparatively pure drinking water and conservancy and arrangements were conspicuous by their absence. The outbreak of plague resulted in the formation of a special Sanitary Committee comprising representatives of several leading firms interested and the local *zamindar*, the civil surgeon of the district, the local Mining Officer of the Indian Mining Association and the Sub-Divisional Officer and the Deputy Commissioner as President.[47] The suppression of the plague outbreak and the warding off of any possible recurrence were the primary objects of this Committee and with the objective co-operation of the Raja's Manager, a good deal was done to clear out the Augean stable existing in the shape of the crowded and filthy Jharia bazaar.[48]

With the departure of plague and in the absence of any law or recognized rules which could be enforced, the efforts of the Committee ceased to have any great effect on the smaller collieries and in the adjoining villages or in the midst of the colliery area. The recurring cholera epidemics of 1906 and 1907 stimulated renewed efforts on the part of a few managers, and in the matter of 'cooly lines'.

The good example set was of considerable effect. Little attention was, however, paid to the water-supply question outside a very limited circle, and the result was the disastrous epidemic of 1908, which besides causing enormous mortality, practically brought the working of the mines to a standstill for nearly three months at a time when every extra ton of coal raised meant a record profit. The immediate concern was on more or less elaborate arrangements for improving the water supply of their respective collieries.[49] Another lesson brought home by the 1908 epidemic was the necessity for legislation enabling the enforcement of ordinary sanitary and conservancy rules.

The interests engaged in the coal industry were many and to some extent conflicting. Concerted action was to a certain extent possible where collieries belonged to the Indian Mining Association, but there were many which did not. Moreover, the number of small concerns with very small capital, and of others whose working capital was inadequate owing to the inflated sums paid to promoters when the mines were first opened or taken from their previous owners during the height of the boom, was considerable. A heavy expenditure on water supply and sanitation was unexpected.

Prior to the district being opened up by mining enterprises, the roads were few and poor, and the traffic over them was insignificant. It was pointed out to the late Sir John Woodburn, on the occasion of his visit as Lt. Governor to Jharia in March 1902, that the roads in the district were lamentably deficient.[50] He sanctioned a special grant and passed orders regarding the minimum allotment to be made in future for the maintenance of district roads within the Gobindpur sub-division. This report was suggesting infra-structure development, the basic sine-qua-non for industrial development as understood in the nineteenth and twentieth centuries. Industrial devel-opment could only be ensured by the development of roads, sanitation, availability of potable water for the labourers, electricity and other basic amenities for the workers, requirements which were enlisted at the behest of the local authorities.[51] It further maintained that the proposal for the constitution of a separate administrative district for the coalfield area was laudable but since the coalfield region had expanded considerably such a proposal was no longer feasible. It further justified the prevailing arrange-ment which dovetailed with the development of the railways, had negated impediments in the communication network and improved administrative efficiency. Moreover, it nullified the claim that communication and sanita-tion in the region were not at par with the developments in the coalfield sector. Rather it asserted that the issue of underdevelopment related to the past and the Dhanbad sub-division was definitely a beneficiary as it had received a fair proportion in the total share of grants. This clearly indicated that the initiatives of local enterprises venturing to implement unhindered development were being often thwarted by the Central authority.

Often these dichotomies between Central direction and local initiative inhibited growth and delayed decisions. However, the desire for bound-less profits vouched for the sustenance of this industry. Hence, the report maintained that the issue of sanitation would be accorded foremost impor-tance.[52] The report was optimistic of the stand taken by the Imperial author-ity and indicated that it would execute its plans in accordance with central direction irrespective of local initiatives. But the moment it was put to test, it became manifest at once that nothing of the kind was possible and that an amendment of the Indian Mines Act would be necessary. Besides, it was decided that the expenses incurred would be borne by the collieries, '…one of its provisions was to the effect that the whole cost of the necessary meas-ures should be levied from the collieries and those *zemindary* proprietors who are in receipt of royalties from coal'.[53] This was followed by a consid-erable amount of polemics regarding the issue of non-royalty areas bearing a proportion of the expenses, but since such areas were inconspicuous they were not ultimately incorporated within the ambit of the expenses incurred.

It was also a strange paradox that nature had bequeathed its largest coal reserves in the Damodar valley region (erstwhile Jungle Mahal region). As mentioned earlier, other than coal, several non-coal minerals were associated with the geological formations in the Damodar valley region. Hence on the flip side, industrial growth inexorably signified massive exploitation of mines disturbing the river basin. Besides, the escalating commercial energy consumption with the increase in population and improvement in standards of living mounted pressure on the mines. Jharia coalfield, an isolated basin of the Damodar river, was spread over an area of 450 sq. km primarily in Dhanbad and a small portion in Giridih district. The coalfield had eighteen workable seams (stratum of coal) and forty-six coal horizons.

The Raniganj formation was exposed in the southern Mahuda basin close to the river Damodar and had twelve coal horizons including four workable seams. The seams of the series were thin, medium coking, highly volatile under shallow cover. The Raniganj coalfield was the easternmost coalfield of Damodar river basin. A small portion of this coalfield was also in Dhanbad and Dumka. The coalfield, spread over an area of 1550 sq. km, was developed up to 1050 m in depth and had ten well-developed coal horizons while the Barakar stage was along the northern boundary with seven well-developed coal seams containing good quality coking coal. The deposit of the coalfield had been extracted by underground methods for over 150 years. Extraction of coal affected overlying strata, creating cracks disturbing the water regime in the basin of underground mines.[54] With the productivity of underground mines not keeping pace with rising demand, large opencast mines were a plausible option. It was also contemplated as a viable alternative to underground mining. This would also keep the average cost of coal production low and maintain ecological balance.[55]

A comprehensive health survey was conducted in a village located in *Jharia* coalfield to obtain an integrated vignette of health status existing among the general population of the coalfield.[56] On examination, it was found that 43 per cent of the population was observed to be suffering from some ailments and 70 per cent was in a moderate state of health. Deficiency of food-calorie was as high as 55 per cent among them. Residential rooms were congested and more than five persons shared the same roof in about 50 per cent of the families.

For drinking water, villagers were dependent on a few surface wells. Analysis of the water sample showed that the water was unsuitable for human consumption. It was presumed that due to the prevailing circumstances there was a high incidence of various diseases amongst the villagers.[57] Mining and allied industrial activities caused land degradation, landslides, soil erosion, water pollution due to acid mine drainage, toxic pollutants and hydrological changes as well as air pollution due to dust, noxious gases and noise. These also caused ecological imbalance due to deforestation which is the reason for the damage to wild as well as aquatic life.

The crux of the problem was extensive mining in the watershed basins. The Damodar's catchment area also housed the Jharia coalfields. The region also had substantial deposits of iron ore and bauxite.[58] Minerals, mine rejects and toxic effluents were regularly washed into the river and its tributaries. In certain plants, about 20 per cent of the coal was washed out in the form of slurry, which was deposited in ponds outside the plant. After the slurry settled in the pond, the sediment or *coalfine* was collected manually. Often, the water, which was discharged into the river from the pond after *coalfine* was recovered, carried high amounts of fine coal particles and oil.

The extraction of coal affected the overlying strata, creating cracks. Hence, the water regime in the basin was badly affected by underground mines. With the productivity of underground mines not keeping pace with rising demand, large opencast mines were a popular option.[59] An excessive quantity of sulphates reduces the potability of water. Water pumped out of the mines eventually found its way into the river and added to its pollution load. The other major coal-based polluters were the coke oven plants that heated coal to high temperatures in the absence of oxygen to prepare it for use in blast furnaces and foundries. However, the water discharged after the wash contained oil suspended particles. Often effluents also carried toxic substances such as cyanide.

Studies carried out on pollution loads in the Subarnarekha basin indicated that unplanned and unregulated mining along with the mineral processing industry had serious environmental degradation in different parts of the river basin. Improper mining practices led to uncontrolled dumping of overburden and mine tailings. This exposed earth was easily carried to the river and other water bodies during the monsoons, increasing the suspended solid loads in the water and also silting dams and reservoirs. Topsoil once lost, took years to revitalize. Moreover, minerals and metals when exposed to the surface mingled with surface water contaminated water bodies and eventually harmed consumers. Iron ore was strip-mined in this region. Between Mayurbhanj and Singhbhum districts on the right bank of the river were the richest copper deposits in the country.[60]

Quarrying of building and road materials like granite, basalt, quartzite, dolerite, gravel and even sand created vast stretches of derelict land.[61] Used and abandoned mines and quarries were a source of mineral wastewater and suspended solids. With little or almost no effort to rehabilitate such lands, the wasted countryside testified to the plunder.

Policies and legacies

Governmental inaction indicated a tacit submission to the wasteful methods being practised. Mine owners were complacent that they were at liberty to implement extractive methods which did not jeopardize 'human lives' but since inadequate restriction was imposed on the leases this aspect was also ignored as commercial interests reigned supreme in their minds. As corrective measures were not implemented little or nothing was done to enforce the terms of their leases and no practical restrictions were imposed. Besides, the dichotomy in policy implementation between the Imperial and local authorities was also conspicuous. During the period 1937–1942, the industry witnessed a steadily increasing internal demand, the special discount in rail freight and port terminal charges further helped in increasing the output.[62]

Interestingly, from 1938 onwards, special shipments were being made to China, which were facing a critical coal situation with the westwards advance of Japan. The increase in demand was also accompanied by better prices, and though by 1942, the prices were still at the level of the 1927 prices, they had risen by nearly 75 per cent. During the first three years of the war, there was a considerable increase in industrial activity, there was also some increase in coal production, but there was not enough coal to meet all needs.[63] The inevitable shortage, accentuated to a degree by transport difficulties resulted in an abnormal price increase of coal for the common man.[64]

In the year 1936, the government appointed another committee known as the Coal Mining Committee with Mr. C. B. Burrows as Chairman to enquire into the methods of extracting coal underground and prescribe measures for the safety of workers and to prevent avoidable waste of coal. The Committee observed that those who controlled the industry considered their rights of extraction sacrosanct. The Committee also went into the history of collapses, fires, floods and explosions in explosions in the collieries and suggested various measures to improve the prevailing conditions. On the question of conservation of coal, the Committee felt that measures of conservation should apply to all good quality coals and should be introduced as a matter of public policy in the interest of the community as a whole. They envisaged the implementation of coal conservation in right earnest and prioritized rationalization over nationalization.

The Committee as a whole observed that the coal industry lacked the tradition of development investment, scientific mining and market fluctuations and the demand for coal were met either by slaughter exploitation of mines when the demand was high or by an equally cruel device of exploitation of mine workers when the demand was low.[65] Their observations forcefully state their concern for the industry,

> .... The coal trade in India has been rather like a race in which profit has always come in first, with safety, a poor second, sound methods an 'also ran' (sic) and national welfare a 'dead horse' entered perhaps, but never likely to start (sic.). Neither the Government nor landlords can escape responsibility for allowing this state of affairs to prevail for so long, but this does not alter the facts nor, still less, will it justify further inaction on the part of all concerned.[66]

The years 1942–1945, particularly the first two years triggered a coal famine of unprecedented dimensions. Production declined amounting to over 4 million tons in 1943 over the raisings of the previous year. One of the most important reasons alluded to was that during the depression years of 1931–1936, there was inadequate plant replacement and no replenishments. In the meantime, the war broke out and no replacements were possible. The mines, therefore, were worked with equipment that were inefficient. Besides, even for the coal raised, transportation was insufficient.

Yet another Committee was set up in 1945, the Indian Coalfields' Committee headed by Mr. K. C. Mahindra.[67] The Committee considered that the private ownership of mineral rights in permanently settled areas of Bengal–Bihar had been responsible for a number of harmful consequences and the only solution was the state acquisition of mineral rights. It stated that

> …We have stated earlier that the need for scientific utilization may eventually make the complete regulation of use essential….this private ownership has been attended with many evil which persist even today. There is a wide diversity of royalty rates, salami is rampant and has been responsible, in the main, for a great deal of and vast areas are leased out on a semi-permanent basis and with no prospect of development within measurable time, and without regard to technical considerations; there are besides a number of disabilities from which the private owner suffers in the exercise of his rights.[68]

The appointment of various committees recurrently, indicated not only the commercial importance of this industry but also the concern for sustenance of this industry. But again, little was done to implement the recommendations of this Committee. It also implied that despite the importance attributed to science and technological innovation, it was primarily advocated by a certain section of the colonial officials.

The independent government continued the legacy of appointing committees and organizations to resolve the crisis. In 1951, the Working Party for the Coal Industry, a body comprising of the representatives of the coal industry, labour and government which reported on the question of small collieries, observed that the vesting of mineral rights in private parties had led to fragmentation and presented various difficulties for amalgamation of such fragmented properties.[69] The Working Party, the Indian Mining Federation conceded the need of amalgamation of small collieries but vouched for a voluntary submission. The Indian Colliery Owners Association submitted that mines which were uneconomic to work, due to their size, should be merged with neighbouring collieries.[70]

It was interesting to note the growing exasperation at the governmental nonchalance to act on these repeated recommendations evident even in official circles. The appointment of the Estimates Committee of the Lok Sabha in 1954–1955 expressed the exasperation writ large in the highest offices. The then Coal Commissioner bluntly stated:

> Almost up to 1939, when regulation 77 of the Indian Coal Mines Regulation came into force, the producers of coal had an absolutely free

hand in producing coal from wherever they could and in any manner they thought best. During this period of absolute *laissez faire*, profit was the primary consideration, safe methods were in the picture only here and there and national interests were completely forgotten. The industry and the country are today paying the price for this ruthless and haphazard exploitation of this national wealth during those days. Mining rules had, therefore, to be tightened up. Gradually controls were introduced but even then, the measures of control are only being circumvented and the owners, by themselves have done very little to cooperate in the policy of greater safety in coal mines and the conservation of good quality coal....[71]

At different stages, every committee strongly criticized the existing structure of the industry and indicted it for its unscientific wasteful and unsafe mining methods as well as insensitive treatment of the miner, the producer of coal. Way back in 1937, two eminent members of the Coal Mining Committee advocated nationalization as the panacea for all evils and this was reiterated by latter Government bodies and Committees and also found a very strong advocate in S. Mohan Kumaramangalam.

The whole array of activities led to the takeover of coking coal mines on 16 October 1971. Subsequently, these mines were nationalized on 1 May 1972 and were operated by M/s. Bharat Coking Coal Limited (BCCL). By Coal Mines (Taking over of Management) Ordinance 1973, the non-coking coal mines were also taken over. The mines were nationalized on 1 May 1973 and brought under the management of the Coal Mines Authority Limited (CMAL). Later on, the CMAL and the BCCL were merged and the holding company Coal India Limited (CIL) was formed on 1 November 1975.

Conclusion

The mining industry under the colonial aegis owed its origin primarily to the demands generated by the steamships, the jute and tea industries and within a very short span of time was ascribed the epithet of a 'profitable venture'. Since this industry also catered to requirements of the war industry, in terms of ships and ammunitions, it had a direct bearing on its future prospects. This was evident in the decline in export rates in the 1930s. In an attempt to achieve steady recovery, the mine owners increased production causing environmental degradation. Amidst myriad recommendations, the committees pain stakingly reiterated the importance of scientific extraction procedures and cautioning the mine owners regarding the dangers posed to the dwindling resources and human lives. Governmental inaction was provoked by the prevailing vacillation regarding stringent action and the dichotomies between local and imperial governments. The predominance of central directives was apparent. The legacies of the colonial government were translated into policies of the independent government and hence the perplexing issues persisted unresolved till the 1970s. Instead of ending on a poignant note reflecting on a grim future of rapidly depleting resources due to the misuse and abuse of the same, ending on a pensive note for posterity in the words of the poet kindles a hopeless transient (from his poem, *Shesh Katha – The last word*):

> When I go from hence let this be my parting word,
> That what I have seen is unsurpassable.
> I have tasted of the hidden honey of this lotus
> That expands on the ocean of Light, and thus
> am I blessed – let this be my parting word...[72]

Notes

1 I am indebted to Professor Deepak Kumar for all his encouragement and incisive comments in perceiving and contextualising environmental studies in its historical perspective. However, errors if any are entirely my own.
2 Status of Research on Environmental Pollution in Mines and Mining Areas and Miners Health, Central Mining Research Station, Council of Scientific and Industrial Research, Dhanbad, May, 1986, pp. 10–15.
3 Miranda, M, Blanco-Uribe Q, A, Hernandez, L, Ocha G., J, & Yerena, E, All that Glitters is not Gold: Balancing Conservation and Development in Venezuela's Frontier Forests, 1988, retrieved from http://pdf.wri.org/all_that_glitters_is_not_gold.pdf.
4 Roy Porter, *The Making of Geology*, Cambridge and New York: CUP, 1977, p. 3.
5 Daniella C. Angetter *et al.*, 'Physicians and their Contributions to the Early History of Earth Siences in Austria', in C. J. Duffin, R. I. J. Moody, C. Gardiner Thorpe (ed.) *A History of Geology and Medicine*, London: The Geological Society Publishing House, 2013, p. 446.
6 Ibid., p. 446,
7 Roy Porter, *The Cambridge History of Science*, Volume 4: Eighteenth Century Science, Cambridge: CUP, 2003, p. 10.
8 Ibid, p. 15.
9 Ian Inkster, "Technological and Industrial Change" in Roy Porter (ed.), Cambridge History of Science, op. cit., p. 50.
10 Ivan T. Berend, *An Economic History of Nineteenth Century Europe*, Cambridge: CUP, 2013, p. 50.
11 Karl Gunnar Persson, *An Economic History of Europe, Knowledge, Institutions and Growth*, Cambridge: CUP, 2010, p. 15.
12 Hiteshranjan Sanyal, 'The Indigenous Iron Industry of Birbhum', in *IESHR*, Vol. V, No. 1, 1968, p. 102.
13 Ibid., p. 103.
14 Andrew Grout, *Geology and India, 1770–1851, A Study in the Methods and Motivations of a Colonial Science*, PhD Dissertation (unpublished) School of Oriental and African Studies, London University, 1995, in https://www.academia.edu/6773429/Geology_and_India_1770_-_1851_a_study_in_the_methods_and_motivations_of_a_colonial_science.
15 Ibid., 15.
16 Ibid., p. 15.
17 Ibid., p. 16.

18 Shiv Visvanathan, 'The Rise of Industrial Research', in S. Irfan Habib and Dhruv Raina (eds.) *Social History of Science in Colonial India*, New Delhi: OUP, 2007, p. 295.

19 Ibid., p. 295. Also see Zaheer Baber, 'Science Technology and Colonial Power', in S. Irfan Habib and Dhruv Raina (eds.), op. cit., p. 116.

20 Shiv Visvanathan, 'The Rise of Industrial Research', op. cit., p. 296.

21 Cyril Fox, Annual Address, Year Book of the Royal Asiatic Society of Bengal, vol. 9, 1943, pp. 9–22, cited in Shiv Visvanathan, 'The Rise of Industrial Research', op. cit., p. 296. Also see Roy Porter (ed.) *Cambridge History of Science*, Vol. 4: Eighteenth Century Science, op. cit., p. 185.

22 J. C. K. Peterson, Gazetteers of India, *West Bengal District Gazetteers, Burdwan*, (reprint, 1992), Calcutta, pp. 60–65. Also see Coupland's, District Gazetteer of Manbhum, op. cit., Calcutta, 1910, pp. 50–56.

23 J. C. K. Peterson, op. cit., pp. 50–60.

24 Ibid., p.8

25 *Proceedings of the Dept. of Revenue, Agriculture and Commerce*, Minerals and Geological Survey, No. 31, 27 January, 1874, National Archives of India (NAI), New Delhi, pp. 53–54.

26 Ibid., p. 53.

27 *Proceedings of the Dept. of Revenue, Agriculture and Commerce*, Minerals and Geological Survey of India, No. 18, 9 January, 1874, NAI, p. 53.

28 T. W. H. Hughes, 'Note on the raw materials for iron-smelting in the Raneegunge Field', in *Proceedings of the Dept. of Revenue, Agriculture and Commerce*, Minerals and Geological Survey of India, No. 18, 9 January, 1874, NAI, p. 55.

29 Ibid., p. 55.

30 *Proceedings of the Home Department, Public, 14 May 1858*, Nos. 53–54, The Department of Geolological Survey of India, National Archives of India (NAI) New Delhi, p. 10. See Proceedings of the Home Department, Public, 9 April 1858, No. 169/172, NAI, p. 21; *Proceedings of the Government of India (GOI), Home Branch-Public*, 2 April 1869, NAI, p. 10.

31 Letter from the sub-divisional officer in Raniganj to the Collector of Burdwan district on "the practice amongst colliery owners of granting land to miners", Revenue Proceedings, Miscellaneous Branch, April 1892, B 22, West Bengal State Archives, Kolkata (hereafter WBSA).

32 Ritter C. von Schwarz, op. cit., pp. 13–17.

33 Ibid., p. 17.

34 Dietmar Rothermund, 'The Coalfield', in Dietmar Rothermund and D. C. Wadhwa, (eds.), *Zamindars, Mines and Peasants*, New Delhi: Manohar, 1978, p. 4.

35 A. K. Bagchi, *Private Investment in India, 1900–1939*, Cambridge: CUP, 1972, p. 262.

36 *Coalfield's Committee Report*, National Archives of India (NAI), New Delhi, 1920, p. 20.

37 Ibid., p. 20

38 *Proceedings of the Department of Revenue, Agriculture and Commerce, Minerals and Geological Survey of India*, NAI, February, 1874, p, 53.

39 *The Administration of the Indian Mines Act, 1923*, which came into force on the 1st July, 1924, Vol. VI, No. 3, 22nd August, NAI, 1940, p. 593.

40 Ibid., p. 594.

41 Bye-laws, made under section 32 of the Indian Mines Act are framed by the owner, agent or manager of a mine with the approval of the Chief Inspector of Mines or an Inspector for the control and guidance of the persons acting in the management of, or employed in, the mine, to prevent accidents and to provide for the safety, convenience and discipline of the persons employed in the mine

and interestingly Rules for Coal and Metalliferous Mines are made under Section 30 of the Indian Mines Act by the various Provincial Governments which include sanitary and health provisions, ambulance and first-aid and rescue work, registration of work persons, safety of surface, abandonment of mines, inquiry in the case of accidents, fitness certificates for employment underground of persons who have not completed 17 years of age.

42 W. Kirby, *Administration of the Indian Mines Act Vol. VI No. 3*, 22 August 1940, NAI, p. 20.

43 Indian Coal and Metalliferous Mines and Regulations were formulated under section 29 of The *Indian Mines Act, 1923*, and gazette in the year 1929, NAI, pp. 15–20.

44 Ibid., pp. 21–22.

45 Ibid, pp. 35–37.

46 Ibid, pp. 37–38. Also see H. Coupland, District Gazetteer, Manbhum, Calcutta 1910, pp. 50–60.

47 Ibid, pp. 38–40.

48 H. Coupland, District Gazetteers, Manbhum, op. cit., pp. 50–56.

49 *Report of the Committee on Indian* Mining *Association*, WBSA, 1929, pp. 38–39.

50 *Report of the Committee on Indian Mining Association*, 1 January 1909 to 31 December 1909, Catholic Orphan Press, Kolkata. Excerpts from the Speech of the Chairman, Indian Mining Association, in proposing the health of the place to His Honour, the Lieutenant Governor, Jharia, 30th November, 1909, in *Visit of the Lieutenant Governor of Bengal to Jharia*, pp. 93–95.

51 Extract of the "Englishman" dated, 1 December, 1909, in an Appendix to the Report of the Committee on Indian Mining Association, op. cit., pp. 98–100.

52 Ibid., pp. 98–100.

53 Ibid., p. 100.

54 Anil Agarwal, Sunita Narain and Srabani Sen, *The Citizen's Fifth Report, Part I: National Overview*, New Delhi: Centre for Science & Environment, 2008, p. 25.

55 Certain derivations from my interview with Mr. Arun Kumar Sen, Chairman-cum Managing Director, West Bengal Mineral Development Corporation Limited on 20 March, 2008.

56 Ibid, pp. 30–35.

57 *Report of the Royal Commission on Labour in India, IISCO, Burnpur, Hirapur Works, October, 1929*, Calcutta, 1930, pp. 24.

58 Status of Research on Environmental Pollution in Mines and Mining Areas and Miners Health, Central Mining Research Station, Council of Scientific and Industrial Research, Dhanbad, May, 1986 op. cit., pp. 10–15.

59 Ibid., p. 12.

60 Anil Agarwal, Sunita Narain and Srabani Sen, The Citizen's Fifth Report, Part I: National Overview, op. cit., p. 25.

61 Ibid., p. 25.

62 *Report of the Committee on Indian Mining Association*, WBSA, op. cit., pp. 38–39.

63 Report of the Committee on *Indian* Mining Association, WBSA, op. cit., pp. 38–39.

64 Ibid., p. 39.

65 *Report of the Coal Committee, 1937*, Calcutta: Bengal Secretariat Press, 1938, pp. 20–25.

66 S. Mohun Kumaramangalam, *Coal Industry in India: Nationalisation and Tasks Ahead*, New Delhi: Oxford and IBH Pub, 1973, pp. 20–25.

67 *Report of the Coalfields' Committee, 1945*, Calcutta: Bengal Secretariat Press, 1946, pp. 20–25.

68 Ibid., p. 22.
69 S. Mohun Kumaramangalam, *Coal Industry in India*, op. cit., pp. 20–25.
70 Ibid., p. 22.
71 Ibid., p. 36.
72 Rabindranath Tagore, 'Shesh Katha', from Gitanjali, *A Collection of Prose Translations made by the author from the original Bengali manuscript*, Santiniketan: UBS Publisher in association with Visva Bharati, 2003.

10 On grazing lands and cultivated fodder[1]

Himanshu Upadhyaya

In December 1991, speaking at the inauguration of the National Conference of Composite Feed Producers at Anand, Amrita Patel of the National
Dairy Development Board stated:

> In the coming decades, it is the goal of the NDDB that all livestock be
> removed from degraded gauchar (i.e. grazing land), to convert small
> producer dairy herds to completely stall-fed enterprises.[2]

The audience that Amrita Patel faced was quite different and it is anyone's
guess if this assertive statement would have been received amidst loud
cheers and applaud. To understand pastoral people, we need to pay close
attention to policy discourses on grazing land and how botanical science
looked at grasslands in India during late colonial and post-colonial period.
This chapter aims to revisit some of the representative articulations on grazing starting from the year 1885. Some of these appear in the Proceedings of
the Meeting of Board of Agriculture in India and three Cattle Conferences
as well as in contemporary scientific journal papers.

In 1885, writing his *Memo on Fodder Grasses*, Shortt had noted:

> In India, grass is rarely cultivated for a grazing purposes, nor is it cut
> generally and stacked as hay (if we except cereals and millets that are
> grown for grains and the stalks are stacked). A large number of grasses
> are to be met with in most parts of India, but their rapid growth and
> subsequent dryness from climatic causes unfits them for pasture, but
> the natives from practical experience burn down their pasture-lands
> as soon as the grass becomes coarse and with the first showers of rain,
> young blades shoot up and furnishes excellent pasturage. This is a com-
> mon practice. Most grasses are nutritious and contain much saccharine
> matter in their composition.[3]

The rest part of the text contains an inventory of "a few of the grasses, pulses
and millets whose straw, stalks or haulms, met with on pasture-grounds, are
in general use for feeding cattle"[4].

DOI: 10.4324/9781003241980-14

This view that native peasants didn't pay much attention to the task of cultivating fodder crops was voiced repeatedly in writing of several colonial experts in charge of civil veterinary departments. In his 1909 book, *The Cattle of South India*, W. D. Gunn stated:

> Ordinarily, the natives who lives on a meal of rice, and perhaps a few herbs to season the same with, expects that his cattle in like manner pick up what they can in the way of pasture about the village or its adjacent lands, so that he never troubles himself to grow green food, or perhaps dry fodder for them. The same plant which supplies him with grains feeds his cattle also with its straw. In most towns and villages, cattle are driven out at all seasons to graze abroad, and in the dry season, they more frequently lick the dust only and return home with their stomachs as empty as when they started, to receive perhaps a few handful of straw or rubbish just sufficient to sustain life.[5]

Diminishing grazing land: A constant refrain at Board of Agriculture meetings

If one pays close attention to the proceedings of the Board of Agriculture in India meetings, one can recognize that the issue of curtailment of grazing in reserve forests and diminishing grazing grounds was a theme that finds mention from early on – alas, without much action following those discussions and deliberations. Thus, at the second meeting of the Board of Agriculture in India held at Imperial Agricultural Research Institute, commonly known as the PUSA institute, in 1906, concerns were raised on this theme, but the Board felt that "its full consideration shall be postponed for the want of collection of information from local governments". At this meeting, the then Inspector General, Civil Veterinary Department, Col. Morgan voiced his anguish about the dangers that diminishing grazing grounds might pose for breeding activities in Madras presidency, due to areas that were previously available for grazing getting included in reserve forest categories. He also added that in other parts such as Jhelum colony, the extension of irrigation had led to entire district being brought into arable cultivation, at the expense of pasture lands[6].

At the third meeting of Board held at Cawnpore, in February 1907, when the papers received from local governments were laid on table, it emerged that "in the Punjab canal colonies and in Sindh, the extension of cultivation" was resulting in "a serious contraction of the large breeding herds that supply the work cattle for many districts"[7]. After the short discussion on this subject, where the president briefly recounted the enquiries into the question and instructions issued by local governments, the discussion was sought be summed up by a general opinion that "Grass and fodder reserves have only led to the increase of local cattle up to the limits of supply in good years and subsequent great mortality when scarcity arose". However, there were other officers who countered this opinion from the chair by mentioning that in Ongole and Burma this difficulty didn't exist. In the end, the Board reached a collective opinion that

> the evidence before it is not sufficient to justify any further inquiries on the subject of improved provision of grazing grounds in the large cattle breeding districts, as no further information appears to be available on the subject and it would not lead to any practical results![8]

Summarizing the reports received from district officers in Bengal during the discussion on the subject, Lieutenant-Colonel Raymond stated that

> the encroachment of grazing lands (in Bengal) proceeds steadily owing mainly to equally steady increase in population and to the demands of commerce; the absorption of grazing land is inevitable, but it may be regulated, retarded and controlled by carefully considered legislations. If, however, the population continues to increase, grazing, except on a small scale and on fallow land between crops, is doomed, for it must give way ultimately to land hunger.[9]

Putting his hopes in the ability of the late colonial state to carve out remedies with the help from the Agricultural Societies, Raymond thought:

> They (remedies) will be found in the study by the Agricultural Societies of forage crops and the simplest method of storage of forage.[10]

Summarizing the report submitted by Mr. Oliver, Officiating Superintendent, Civil Veterinary Department, United Provinces, Raymond observed that

> Of 38 districts reported on, in 22 districts, although the cultivated area has increased, no noticeable decrease of grazing grounds has taken place, nor are the cattle reported to have suffered. But in the remaining 16 districts, the reduction in the area of pasturage has told severely on the cattle.[11]

Summarizing the report submitted by Mr. Hewlet, Superintendent, Civil Veterinary Department, Bombay, Raymond stated:

> In Gujarat, there is no flow irrigation and no lands reserved by the Forest Department, other than genuine forest areas – the Godhra forests, for instance. Grazing grounds are, at present, adequate in the Ahmedabad district – as well as in the districts of Broach and Surat. In the Konkan, including Thana, Kolaba and Ratnagiri, there has been no curtailment of grazing grounds, which is simply sufficient. In the Karnataka, Belgaum, Bijapur, Dharwar and Kanara districts, some diminution has taken place in the last 20 years by waste lands – formerly used for grazing – having been brought under cultivation. In the Deccan,

there is but little grazing. The majority of land is under cultivation and affords very little grazing of any description after the middle of November. In parts of Khandesh, there is still ample grazing as also in the vicinity of the Ghats. In Nashik, the grazing area has diminished and is likely to diminish still more in the future owing to keen competition for waste lands for cultivation purposes. In Ahmednagar, Poona, Sholapur and Satara there has been little or no diminution of grazing grounds.[12]

Talking about the Forest Department policy on dealing with the grazing in Bombay presidency, Hewlett had reported on the practice by which "some of the grazing grounds belonging to the Forest department are sold, for grazing purposes, by public auction". However, he felt that "if possible, this system should be discontinued, as it introduces a middleman". Instead, Hewlett proposed that "either the forest department collect grazing fees of 2 to 4 annas from cattle owners, or if this is not practicable, to give the villagers of the villages adjoining the Kurans preference at the auction sale".[13]

In the case of Sindh – especially Sukkur – the introduction of irrigation (and one shall remember that this is from the time before the construction of Sukkur Barrage) had already started a process of bringing land under cultivation that were in past waste land on which any number of camels could graze without restriction. This consequently had led to "deterioration in the stamp of camels bred in Sind", reported the collector of Sukkur (Sind) in a correspondence to Hewlett. There was also a feeling that "Reservation by the Forest Department has not caused diminution of grazing grounds as the Forest areas are available for grazing on payment of a small fee".[14]

In the case of Baluchistan, "in some districts, the diminution of grazing grounds is severely felt by the people, especially in the Quetta district". While in this province, 234 square miles were classified as reserved forest, "69 square miles are available for grazing".[15]

In the case of Ajmer-Merwara (Rajputana), grazing grounds were stated to be sufficient, if the rains are good. The reports suggested that "there has been no increase of forest reserve for the last 20 years, and no encroachment on grazing grounds by cultivation".[16]

The curtailment of grazing grounds in Punjab was so great that it had given rise to correspondence between the Director of Land Records and Agriculture, and the Conservator of Forests. However, reporting on this correspondence, Captain Walker stated that "the curtailment of grazing grounds has been counterbalanced to some slight extent, by the extension of forests". Captain Walker had also recommended "for each group of villages, a certain forest reserve and the grazing rights of these areas to be held by the villages concerned under the usual payment of fees".[17]

In the case of Burma, Raymond concluded from the report submitted by Major Evans that "there is no curtailment of grazing grounds".[18]

At this meeting, the phrase "fodder famine" was discussed by experts repeatedly. Describing the situation in Bengal, Raymond stated that

Fodder famine is rare occurrence in this province. As far as I can see from the Famine Reports of 1873–1874, 1895 and 1897, there was never any great difficulty about fodder, as there was enough to feed the cattle with a reduced allowance.

He was of the opinion that "the proposals for the provision of fodder, in times of famine, as laid down in chapter XV of the Bengal Famine Code appear to be sufficient".[19]

As regards the question of "fodder famine", the views expressed by Oliver in reference to situation in United Province are interesting. His views exhibit a sedentary bias that many colonial officials of his time shared. Oliver considered "occurrence of fodder famine much rarer than an ordinary famine" since its occurrence was expected "not more than once in thirty years". Recounting the experience of his department in resolving the problems posed by fodder famine, Oliver stated that:

Fodder famine are least felt *where the system of stall-feeding is more developed: and all indications point to the conclusion that the best measure to take is the extension of irrigation, which not only minimizes the loss of fodder in a drought, but enables special fodder crops to be grown and – more important still – increases the wealth of the people.*[20]

(emphasis added)

Thus, in the thinking of British officers, stall-feeding was a better mode of animal nutrition compared to free range grazing twice in day from grazing grounds lying at walking distance from villages. They strongly believed that insurance against fodder famine was found in an advice to go for expanding irrigation and through that taking recourse to growing special fodder crops. Animals for whom the question of negative fallout of "curtailment of grazing grounds" and debilitating impacts of "fodder famine" was to be discussed were, in the eyes of British Civil Veterinary experts, almost always those that belonged to a settled cultivator, to whom can be extended irrigation facilities, seeds of special fodder crops or drought-resistant exotic grasses. The extension campaigns that were planned to address the problems posed by "fodder famine" tended to popularize stall feeding amongst peasants. What is also surprising is the fact that such a thinking emanated not from being unaware of transhumant pastoralists and pastoral mobility. Actually, the sentence that precedes the above-quoted opinion in Oliver's note suggests that he didn't seem to advocate an abrupt disruption to pastoral mobility, and he was strongly against interfering with existing arrangements:

As regards utilization of forests, there are *certain well-known regions of migration, where the cattle are transferred seasonally according to the need*, and this should off course not be interfered with.[21]

While Oliver's views are very clearly appreciative of existing arrangements on forest grazing, researchers who have studied the policies on fodder supplies during famine years (see, for example, Saurabh Mishra's *Beastly Encounters of the Raj*) suggest that fodder grasses were baled from forests and those bales were transported to settlements through railways[22]. Oliver also refers to "the Conservator of Forests working out for use in emergencies a table at which *fodder grass can be laid down at each railway stations in the tract liable to fodder famine*" and considered these steps worth emulating for other provinces too[23].

Around the year 1913, the importance of initiating systematic study of fodder grass in forests came to be discussed briefly in Board of Forestry meetings[24]. In a monograph authored by Dr. R. S. Hole, titled "Notes on the Chief Fodder Grasses of Indian Forests", it was reported that:

> A systematic study of forest grasslands is now being undertaken by the Forest Botanist at (Imperial Forest Research Institute) Dehradun with the main objective of determining (i) which forest species are of greatest value and (ii) what are the conditions on which their healthy development and natural distribution chiefly depend. A beginning has recently been made with this work in United Provinces where local forest officers are now preparing maps of their principal grasslands. It is greatly to be desired that similar work should be carried out.[25]

Grazing rights in reserve forests

Thus, discussion on the issue of curtailment of grazing in reserve forests and diminishing grazing grounds was short-lived and didn't lead to any major re-thinking on the part of colonial State. However, the debate on "preservation of grazing in reserve forest areas" re-surfaced at the Board of Agriculture meeting in December 1913, where the sub-committee stated:

> The restrictions of grazing rights now enjoyed – from the point of view of cattle breeding – is to be deprecated, except in such areas as are regarded as already *overstocked*[26]. There still remains in some provinces grazing lands of considerable extent which serve as breeding grounds. There is a danger that these may be encroached upon either for cultivation or as protected forests not open to grazing. The conditions differ so greatly (from province to province) that no specific recommendations can be made. The maintenance, however, of breeding grounds is of such importance that if there is any danger of their being devoted to other purposes, their acquisition on public ground appears *desirable*[27]. If acquired they could be much more efficiently managed than at present, and could be made to support larger herds of cattle.[28]

Thus, grazing lands now came to be discussed through the prism of "efficient management", "whether being of considerable importance for cattle breeding operations or not" and "whether already overstocked or not". Restrictions on grazing in reserve forests, extension of perennial irrigation even to areas known for their excellent cattle breeding traditions and other such measures were the order of the day, given social evolutionary thinking of colonial experts, who considered pastoralism as belonging to the Old World and addressed pastoral people as "vanishing tribes"! Even in cases where grazing lands contributed significantly to cattle breeding operations, the sub-committee had advocated the public acquisition by State, thereby changing the rules of the game.

It is important to reproduce below the discussion on this paragraph in the report of the committee, with a view to understand the emergence of "forest management" discourses and how the same got extended to the question of "grazing in forest":

> As regards increase of grazing area being necessary for the improvement of cattle, Lodge said that the result in some cases had been deterioration, and that restrictions should be imposed so as to keep the numbers of cattle on grazing areas, within limits. He mentioned experiments in selling hay, which had been popularized by offering it for sale and said that by *cutting coarse grass the growth was greatly improved*[29]. *Lantana* destroyed pasture land, which recovered its value after eradication of the pest. He endorsed the principle of establishing an agricultural branch of Forest Department, if the increase of the staff could be obtained...With reference to the question of preservation of grazing grounds, Sampson suggested, in the interest of Madras that *it was undesirable to recommend the exemption of any grazing lands, forest or otherwise, from restrictions, except in recognized tracts.* Smart objected that, in Bombay, there was, he believed, only one recognized tract outside Native States, and the adoption of Sampson's proposal would mean approval of restrictions on grazing almost anywhere in the Presidency. After some discussion in which Chadwick and Col Pease took part and from which it appeared that Sampson's chief objection to unrestricted grazing was based on the evil effects of over-stocking small grazing areas such as are common in Madras, Hailey explained that the intention of the committee had been to avoid restrictions as much as possible.
>
> (emphasis added)

The question of grazing rights in reserve forest came up once again for a brief discussion at the Board of Agriculture meeting in the year 1929, when two general resolutions emphasizing the necessity for conserving and making better use of existing grazing grounds were passed. The subject was again proposed for discussion at the first meeting of Animal Husbandry wing in the year 1933, but unfortunately no discussion on it could take place because it was crowded out due to very heavy programme on the agenda for that meeting.

The same question was revisited at the second meeting of Animal Husbandry Wing of Board of Agriculture and Animal Husbandry in December 1936 at Madras, being on agenda item No. 34!

Before the second meeting of Animal Husbandry wing, there was a preliminary conference convened for the consideration of the subject, "The Better Utilisation of Forest Areas for Grazing" that discussed full reports on this question from all provinces. Those reports had included detailed statistics of incidences of grazing on forest land, percentage of total cattle and livestock populations which utilize forest grazing and percentage of forest land to total areas[30]. When C. G. Trevor report was introduced by Whitehead at the second meeting of Animal Husbandry wing, Arthur Olver was happily surprised to witness the members of committee arriving to the conclusion very closely resembling the opinions which he had formed "that forest grazing was really a minor importance compared to making better use of waste lands"[31]. As if to further buttress Olver's point, Parr pointed out that the C. G. Trevor committee recommendation, although aligned to the terms of reference, "were inadequate since it had concerned itself mainly with forestland which represented only 3 percent of the total grazing area"[32]. Parr proposed that more grassland research and further detailed study of grazing areas needed to be attempted.

Shift to advocacy on cultivated fodder and botanical research on fodder crops

Around the year 1916, if one looks at articles in agriculture and veterinary science journals, one finds that colonial experts and botanists were not devoting more attention to the subject of improving cultivated fodder production in India. In 1916, Sir Albert Howard wrote about introducing *Shaftal* in North West Frontier Provinces and Dr. Harold H Mann published *Fodder Crops in Western India*, which was followed by publication of *Some Wild Fodder Plants of the Bombay Presidency* the very next year. At the 10th session of Indian Science Congress at Lucknow in March 1923, a joint meeting was arranged to discuss the issue of improvement of fodder and forage in India by the presidents of the sections of Agriculture and Botany. Gabrielle Howard who compiled the papers that were read out at this joint meeting provides us with an indication of the background and states:

> Great efforts are being made in all provinces to improve the cattle, but the factor which limits such effort is the amount of the fodder in the country. Improvement in fodder supply must precede or at any rate go hand in hand with the improvement of the breed... Although there is a great field for botanical work in fodder questions, it must not be forgotten that there are limits to what can be accomplished by plant breeding alone. There is not much hope of producing, either by selection, hybridization or introduction, the ideal fodder plant, namely a

crop, which will give a large outturn of nutritious, succulent fodder with the minimum amount of water, which will thrive during that part of the year when all other vegetation ceases to grow and which can be preserved without loss or trouble.

Now, there was more attention to introducing crops like *berseem* (Egyptian clover) in part of India. While the stress on cultivated fodder gained currency, a few experts continued to emphasize the need to improve grazing lands. For example, L. B. Kulkarni stated in his book, *Improvement in Grazing Areas in the Bombay Presidency*:

> Until the last half century, there were in this presidency extensive waste lands such as common grazing grounds, which in good years, at any rate, provided for the village herds a fair quantity of fodder. But owing to the demand for cultivation of food and commercial crops, which tend to increase at the same time the required number of cattle and to reduce the grazing as well as the production of fodder on the cultivated lands, the area available for common grazing by paying a small fee to a private owner has been reduced in almost every part of the Bombay presidency. The very salutary old Mohameddan rule, which provided that there should be one acre of grazing land to every ten acres of cultivated land has largely been forgotten and the areas still available have been and are still being seriously overgrazed with evil results not only for cattle, but also to the grazing land themselves.[33]

Settlement Commissioners like F. G. H. Andersen were prone to see a Kheda peasant as a replica of farmer in England and so grazing came to be spoken in the most condescending terms, such as suggested by following words:

> Grazing is allowed at all seasons and in all-weather without any control, cattle *trample down and spoil* more than half the total crop of grass.[34]

> (emphasis added)

Sir Albert Howard who contributed to significantly on the issue of organic agriculture through his work at Institute of Plant Industries at Indore in the late 1920s and the 1930s and through subsequent writings had voiced a view on India's monsoon grasslands that didn't contain very celebratory description. He stated: "Countries like India, which have no good pastures, ought to pay particular attention to the fodder supply"[35]. In the very next sentence, Howard makes an assertion that "there is little prospect of improving pasturage to any great extent" and thought that "the pressure of population will always present constraints on any extension of the existing areas under grass".

At the 10th session of Indian Science Congress held at Lucknow in March 1923, a joint meeting was organized to discuss the issue of improvement of fodder and forage in India. This was arranged jointly by the presidents of the Agriculture and Botany sections[36]. Gabriele Howard who compiled the papers that were read out at this joint meeting provides us with an indication of the background:

> Great efforts are being made in all provinces to improve the cattle, but the factor which limits such effort is the amount of the fodder in the country. Improvement of fodder supply must precede or at any rate go hand in hand with the improvement of the breed... Although there is a great field for botanical work in fodder questions, it must not be forgotten that there are limits to what can be accomplished by plant breeding alone. There is not much hope of producing, either by selection, hybridization or introduction, the ideal fodder plant, namely a crop, which will give a large outturn of nutritious, succulent fodder with the minimum amount of water, which will thrive during the part of the year when all other vegetation ceases to grow and which can be preserved without loss or trouble.[37]

Speaking at the joint meeting, J. Matson voiced his opinion that the question of fodder deficiency in India was largely understood so far as pertaining to quantity, but there was a need to understand the question from the viewpoint of quality and notably in protein matter specifically. His presentation at the joint meeting also makes a clear impression that the fodder question even on military farms was still at the stage of early experimentations in the 1920s. He urged his botanist colleagues to help identify for irrigated tracts, "a leguminous fodder plant which can grow freely at the time when other crops are making the least demand on the water supply" and candidly admitted that "we have not yet found what we require in this respect". However, as far as non-irrigated tracts were concerned, he thought that Lucerne could do the work based on his experiments. In his speech, Matson had urged his botanist colleagues to devote "the strongest possible emphasis on studying the value of green fodder" since he was of the opinion that in India, "there were no pastures that can be relied upon every year to remedy the errors of previous stall feeding".[38]

Scientific research on Indian grasslands

By the late 1930s, the imperial agricultural and livestock experts were already expressing concerns that the subject of researching Indian grassland was still waiting to be taken up. In his article titled "A short survey of grassland problems in Central Provinces", Dave and Mahetahad expressed a surprise that "the grassland until recently have received so little attention" from Botanists.[39] They recall in this article that "in December 1925, the subject of suppression of spear grass and the improvement of grass areas in Central Provinces was considered at some length by the Provincial Board of Agriculture". They also underlined that "the Botanical section of the Department was entrusted with the task of investigations into the improvement of grasslands in early 1920s". They argued that during the four-year work, "seeds of exotic species, European, Australian, African and American were obtained and experimented with, in order to find out if any of them might prove useful for sowing out our grass areas". Explaining that the results showed that the use of these seeds of exotic species was not found worth recommending in our permanent grasslands, they state that:

> Some of these species turned out to be useful fodder and can successfully be grown during certain periods of the year or in places where irrigation facilities are readily available.[40]

However, for the varying soil and climatic conditions that obtained in large parts of India, it was proposed that the best bet for providing the herbage in permanent grasslands was to rely on the appropriate indigenous species.

The second meeting of Animal Husbandry wing of Board of Agriculture in India held at Madras in December 1936 had discussed C. G. Trevor committee recommendations on the issue of grazing in forestland. There was a feeling sinking deep inside the psyche of experts that grazing land had suffered gradual degeneration and by the mid-1930s, the experiments with foreign grasses and fodder and leguminous crops had reached a fair level of conclusion as to what could be attained by giving more reliance on growing irrigated fodder crops.

The botanical research work that had started to take place on the issue of improving grasslands had started to assign value to the concept of "rotational grazing", "fencing off grassland for some months in a year", "opening up plots of grasslands for grass cutting for some months", etc. In the year 1937, with the publication of the *Grass Farm Manual of Military Farms Department*, the feeding practices at several military dairy farms and grass farms were documented in greater details. This comprehensive document recorded the evolution of methods of fodder production and conservation, dealt with soils and their manuring, talked about cultivation practices on *barani* and irrigated lands, presented detailed account of the selection of adapted grass species and process as varied as haymaking, ensilage, stacking and baling of fodder and other related matters. The question of fodder thus had witnessed considerable building up of scientific and technological knowledge base by the late 1930s. W. Burns who had undertaken some pioneering work on grazing land improvement in Bombay presidency had come to voice his feelings by the late 1930s that even two decades later not many were emulating him and the work on these lines had remained "spasmodic and unorganized"[41]. Being at unease with the growing shift (or probably drift) towards "growing fodder crops and giving up on grass", Harlow, the then Chief Conservator of Forests in Central Provinces put forward his views that clearly sailed against the prevalent current. In proceedings of this meeting, we are told that:

> As regards growing fodder crops and giving up grass, he (i.e. Harlow) stated that the backward tribes in the Central Provinces, amongst whom he had worked, had not enough food for themselves, and therefore it would be impossible to give up land for fodder when there was not enough for human food.[42]

Experiments on foreign grasses in India

On the other hand, Imperial Council of Agricultural Research had been experimenting with foreign grass and cultivated fodder crops with a view to address the animal nutrition question for quite some time. However, by the late 1930s, we can discern a definite tilt toward an agenda of promoting cultivated fodder crops. In 1938, at the second meeting of Animal Husbandry wing of Imperial Council of Agricultural Research, Parr had voiced passionately his concern that the subject of researching Indian grasslands was *still waiting to be taken up* (emphasis added). What was being done at the Imperial Agricultural Research Institute, commonly known as the PUSA institute (Bihar) under the supervision of W. Sayer was experimentation on foreign grasses under Indian conditions. Probably, the tilt towards more focused attention to cultivated fodder crops rather than putting more energy into researching Indian grasslands and native grasses is thanks to this observation by W. Sayer in the concluding part of his paper published in *Herbage Review* (Issue 3) in 1935:

> All our work on foreign grasses has led to the same results. We find that all these imported grasses do well in the monsoon, when we have every fodder grass (of native variety) in abundance. None of them is of any value, when we have no grass, i.e. in the hot weather and in the winter. In *berseem*, we have the best irrigated fodder crop in the world.[43]

Probably, flowing from such an enthusiasm for *berseem* in the thinking of ICAR botanists who designed the research and extension policy, the task of improving Indian grassland was neglected for three decades. This three-decade long neglect took place even after dissenting experts like Parr, Ware and others had voiced passionately the need to undertake more research on grassland. These dissenting botanists had also advocated for researching better methods of management of grazing areas, especially those labelled as "cultivable waste" right from the second meeting of Animal Husbandry wing. This is a telling comment on how the preference for cultivated fodder crops had eclipsed the issue of improvement of grazing areas and grasslands.

By the late 1930s, the question of increasing the production of fodder for livestock had started to attain such a proportion that the feeding practices of Indian peasant were being looked at by colonial agricultural experts with

Malthusian concerns. The question for finding the food for Indian cattle, in their conception was intricately linked to the problem of numbers.

John Russell in his *Review of the Working of Imperial Council of Agricultural Research* opens discussion on the theme "Fodder Crops and Grazing for Livestock" with words that clearly manifest Malthusian thinking:

> Under present conditions the amount of food produced is insufficient for the large number of animals in India and in consequence many of them are inadequately fed. So long as this continues the efforts now being made to improve the quality of the livestock by distributing better bulls cannot attain wide success. The bullock, as the source of power, comes off best, but the cow is not so well fed and in consequence of a low diet gives only poor yields of milk. If it were feasible, the best course would be large reduction in numbers of animals so as to bring the livestock population more into line with supplies of food, but this cannot be done rapidly. Some gradual reduction will no doubt come about by economic pressure as grazing grounds become more closely settled for cultivation, and as the castration of scrub bulls becomes more commonly practiced. Improvements in the farm implements, and particularly in the bullock cart, would reduce the need for so many bullocks in the village. It is commonly stated that the cultivator must have large number of animals in order to obtain sufficient farmyard manure: this is only partially true. The quantity of manure depends on the amount of food eaten, and a given quantity of food produces more manure if fed to one animal than if fed to two. For the present, however, the only practicable solution is to increase the production of animal food.[44]

Russell had pointed out in his report that by the late 1930s, a few provinces in India had taken to growing fodder crops in significant acreage (see Table 10.1). These were Punjab, Bombay, United Provinces and Central Provinces.

Table 10.1 The extent of acreage under fodder crops in provinces (1936–1937)

Name of the province	Total area sown (in million acres)	Area under fodder crops (in million acres)	Area under fodder crops as percentage of total area sown
Punjab	29.8	4.8	16.1
Bombay	34.1	2.6	7.6
United Provinces	43.4	1.4	3.2
Central Provinces	27.5	0.46	1.7

Source: John Russell, *Report on the Work of Imperial Council of Agricultural Research in Applying Science to Crop Production in India*, Delhi: Manager of Publications, Government of India, 1939, p. 43.

A year later, in a note prepared for the second meeting of Animal Husbandry wing of ICAR, held at Madras in December 1937 (1938: 106) it was estimated that "the total area in India under special fodder crops was only about 3 or 4 million acres".[45] At this meeting, the question of "the possibility of crop planning for increased fodder production with reference to leguminous fodder" was debated to a great extent. At this meeting, a sub-committee on animal nutrition presented its report which stressed upon "a need to obtain accurate statistics of the acreage under various fodder crops".[46] For this purpose, the sub-committee on animal nutrition had designed a comprehensive questionnaire with an aim to obtain statistics "showing available fodder crops for the livestock of each province preferably district by district".[47]

Leguminous fodder crops had occupied a higher level of attention in Albert Howard's proposals for developing agriculture in desert like regions. Writing about the introduction of such crops for North West Frontier Province, Albert and Gabrielle Howard wrote in a paper contributed to *The Agriculture Journal of India* that:

> Once leguminous fodder crops like Lucerne, berseem, senji and shaftal become general in North West, the producing power of soil is bound to increase. The work cattle would be better fed and the door will be opened for more intensive cultivation of the land and for the use of heavier implements...Once the Army comes into market for these dried fodders, their extended use is certain. Anyone who has seen the poor feeding of thousands of cattle engaged in moving produce over the main trunk roads in the North West will at once realize how much these fodders would improve the efficiency and reduce the numbers necessary for the work ...The numerous dairies springing up in the large towns are producing milk, inferior both in quality and quantity, to that which would be possible if the albuminoid ratio of the fodder could be improved. For famine reserves, these baled fodders would be of great use. Such produce is easily stored for long periods, is readily transported and the quantity is easily checked by merely counting the bales. It is highly nutritious and therefore would be a useful reinforcement to such material as *bhusa* and dried grass whose function would be the dilution of leguminous hay.[48]

In 1917, *berseem* was introduced at Pusa and first of all, the ICAR got it from Sind[49]. In the annual report on Sukkur Farm, Sindh, for the year 1915–1916, Mr. Mann had reported that "*berseem (Trifolium alexandrium*) fodder commands an excellent market in Sukkur, especially in the hot weather, when other sources of fodder become exhausted". Keatinge had written in a note prepared for the sixth meeting of Board of Agriculture that the question of *berseem* had acquired relevance in

Sindh.[50] Describing the importance of this fodder crop, the Royal Commission on Agriculture (1928) stated,

> If the seed of this crop (*berseem*) can be cheaply grown in quantity, there is at least some ground for the hope that in tracts such as Punjab and Sindh, it may be added greatly to the fertility and wealth of the country.

Sardar Darshan Singh narrated to the participants at the first Cattle Conference (1924: 19) the scheme he had proposed to the Government of Punjab, whereby landholders in Punjab were to be given special concessions in the direction of remission of water rate or land revenue or both, in order to induce them to grow more fodder crops and maintain more cattle on their land[51]. The fodder crops thus recommended was to be catch crops entirely in addition to those previously grown. Parr who was representing the United Provinces at the first Cattle Conference also pointed to the similar concessions being granted by canal authorities for growing Lucerne.[52]

Speaking at the Second Cattle Conference in 1937, W. Sayer (1938: 53–54) recounted what they were doing at the Imperial Agricultural Research Institute, commonly known as the Pusa institute (Bihar) on the issue of cultivated fodder crops and reported that before the year 1935, when they left IARI campus in Bihar and moved to Delhi, some 120 acres were under *berseem*, in rotation with *maize*, and at Delhi campus of IARI, they had some 40 acres of area which was covered by *berseem*.[53]

In the mid-1940s, the issue of animal nutrition had acquired worrying proportions as fodder availability (in terms of roughages as well as concentrates) was considerably lower than requirement. *Memorandum on the Development of Agriculture and Animal Husbandry* (1944) tried to estimate the requirement and availability of roughages and concentrate, but it also contained a few lines on what work needed to be undertaken on the issue of grassland management[54]. Needless to add that by the mid-1940s, the perception of Indian grassland and grazing land then available was coloured by colonial and developmentalist point of view that felt that "since the rainfall being seasonal, grass is available only for a few months in the year". The strategies to attempt neutralizing the seasonal deficiencies flowed from an agrarian and forest management mindset, rather than pastoral one, by advocating "hay and silage making" as well as "reseeding and rotational and controlled grazing". The thinking was in the mode of an intervention through a development project mode and estimated non-recurring cost at Rs. 50 crore and a crore of rupees per year (at 1943–1944 prices) towards maintenance. This mindset had ignored to understand the pastoral and nomadic niches in interacting with grasslands through patterns of mobility.

R. A. Pepperall of the British Milk Marketing Board, who came as an advisor to India in 1945 and penned a report titled *The Dairy Industry of*

India (1948) recommended that for addressing the issue of animal nutrition, village areas should be utilized for the cultivation of green fodder, *rather than for grazing*[55]. The recommendations of the Eighth Meeting of Animal Husbandry wing of Board of Agriculture held at Mysore during 17–22 February 1949 (1951: 28 & 32) indicates that in those early years after India's independence, greater priority was according to the task of bringing more land under cultivation with a view to increasing food production, compared to grazing, and if necessary even the best grazing areas could be alienated from the customary and traditionally established usage and land use classification.[56] Thus, the term "protected areas" was not to be attached to "even the best grazing areas" in expert imagination. Similar view is expressed by Whyte, when writing about "Nomadic and Migratory Grazing", he asked:

> How is the improver of the nomadic way of life and its characteristic types of domestic livestock to approach the problem of the amelioration of semi-arid environment or in some cases the elimination, total or partial, of this type of animal husbandry from the land use system?[57]

To close, let's recall views by Robert Orr Whyte, who worked in India as Grasslands and Fodder Advisor and wrote in 1968, a book titled *Land, Livestock and Human Nutrition in India*. On page 67 of this book, he claims:

> No one can claim that these grass covers are performing any of these roles satisfactorily. Those areas within daily walking distances of the village and along the migratory routes in the plains and mountains are *so hopelessly overloaded* with cattle whose chief role in the conversion of fibrous grass growth into manure or kitchen fuel. Because the grass areas are expected to carry ten to twenty times the correct number of bovines per unit area, the plight of other classes of livestock is very serious.[58]

Similarly, on page 141, he writes:

> Apart from these varied and characteristically Indian uses, the grasslands are of very little economic value. As already stated, *no self-respecting dairy-farmer would consider taking his good cows and buffaloes, in milk or dry, on to the natural pastures for anything other than exercise.*

Very strong words these are on "the economic value of the Indian grasslands". A strange coincidence that these statements found mention in the book published in the year 1968, when Garret Hardin's essay *Tragedy of Commons* also got published. When I first read this sentence around six years back while doing my doctoral research, I was awe-struck with the irony since I grew up in the neighbourhood of a Chaaran settlement, watching over the herd of cows and buffaloes being taken out to graze on a village pasture two times a day (in morning and evening) in the 1980s and I thought that they were not being taken out only for exercise.

Thus, when we engage with the contemporary discussions in official circles, we realize that grazing and grassland were being evaluated from the lens that failed to engage with the complexity of typical Indian uses and the knowledge system shared by pastoral communities. Views that governed much of the policy thinking which exemplified sedentary bias continued to influence and shape the thinking of dairy development planners in India. The continuities started to get probed only around the mid-1990s, when several academicians started to undertake research on pastures and pastoralism in India.

Notes

1 This paper was first presented at Living Lightly Conference, 8–10 December 2016 at India Gandhi National Centre for Arts, New Delhi. The author is thankful to the organizers of academic conference, Ashwnini Chhatre and Vasant Saberwal. Subsequently, this paper was presented at a civil society workshop on Commons at Udaipur, 20–22 January 2017. The second presentation was transcribed and appears on https://agriculturesrc.wordpress.com/2017/07/07/pastoralism/ Last accessed on 3rd November 2019.

2 Cited in Richard P. Cincotta and Ganesh Pangare, 'Population Growth, Agricultural Change and Natural Resource Transition: Pastoralism amidst Agricultural Economy of Gujarat', *Pastoralism in Gujarat and Rajasthan*, 1994. https://www.odi.org/sites/odi.org.uk/files/odi-assets/publications-opinion-files/5405.pdf Last accessed on 3rd November 2019.

3 John Shortt, 'Memorandum on Fodder Grasses', an appendix in *A Manual of Indian Cattle and Sheep*, Madras: Higginbotham, 1885, pp. 161–168.

4 Ibid.

5 W. D. Gunn, *Cattle of Southern India*, Department of Agriculture, Madras, Vol III, Bulletin No 60, Madras: Government Press, 1909.

6 *Proceedings of the Board of Agriculture in India held at* PUSA *on the 15th January1906 and following days*, Calcutta: Office of the Superintendent of Government Printing.

7 For the summary of the information received from various officers on *Curtailment of Grazing Grounds*, see appendix L in the proceedings, pp. 231–233. *Proceedings of the Board of Agriculture in India held at Cawnpore on the 18th February 1907 and following days*, Calcutta: Office of the Superintendent of Government Printing.

8 *Proceedings of the Board of Agriculture in India held at Cawnpore on the 18th February 1907 and following days*, Calcutta: Office of the Superintendent of Government Printing, p. 15.

9 Ibid, p. 231–232.

10 Ibid, p. 232.

11 Ibid, p. 232.

12 Ibid, p. 232.

13 Ibid, p. 233.

14 Ibid, p. 233.

15 Ibid, p. 233

16 Ibid, p. 233.

17 Ibid, p. 233.

18 Ibid, p. 233.

19 Ibid, p. 233. It is very interesting to see that the very first principle in the chapter XV of the Bengal Famine Code, which revolves around concerns for Cattle emphasize that "it is far more effective to bring fodder to cattle that to take

cattle to fodder, and that to keep cattle in a village helps to keep inhabitants of the village together". The very next point is titled as "Growth of Fodder Crops" and advised the colonial State to extend "liberal advances to the cultivators, with a view to construction of temporary wells for the growth of fodder crops, when the fodder famine is imminent". On the matter of grazing, the Famine Code stated, "Areas in which grazing is ordinarily permitted should be thrown open to free grazing, and areas which are ordinarily closed to grazing should be reserved for the supply of grass for the export". The restrictive regime had by then firmly evolved and the Code justified certain exclusions by stating, "*Browsers should not be admitted without payment to any areas which contain forest growth of any importance* and should in no case be admitted to areas ordinarily closed to grazing" (emphasis added). It went on to state, "No grazing should ever be allowed in areas under plantation or regeneration, unless the trees are old enough to be safe from attack".

20 Ibid, p. 235.
21 Ibid, p. 235.
22 See Chapter 4, 'Cattle, Famine and Colonial State' in Saurabh Mishra, *Beastly encounters of the Raj*, Manchester: MUP, 2015.
23 *Proceedings of the Board of Agriculture in India held at Cawnpore on the 18th February 1907 and following days*, op. cit., p. 235.
24 See, for example, Proceedings of the meeting of Board of Forestry held at Dehradun in March 1913.
25 This quote from R. S. Hole's monograph is cited in Robert O. Whyte, *Grasslands and Fodder Resources India* (1961), Indian Council of Agriculture Research on page 12. One also has to understand that this attention to the issue of studying forest grasslands was short-lived in policy discourses in Forestry Research and Forest Management, when one finds Whyte reporting just after citing the above quote that "this laudable scheme appears to have come to nothing".
26 This last italicized phrase at the end of the sentence was added after the discussion on the report of the committee at the meeting.
27 This last word was inserted in the place of "absolutely necessary" when Colonel Pease objected stating that "the general feeling of the committee had not been that the subject was of the importance implied by the wording of the recommendation". He, therefore, suggested substitution of the phrase by the word, "desirable" and the amendment was accepted by the Board.
28 *Proceedings of the Board of Agriculture in India held at Coimbatore in December 1913*, Calcutta: Office of the Superintendent of Government Printing, p. 16.
29 This is what we would like to refer to as the prototype idea that was to get currency and lead to what are presently referred to as "cut and carry" schemes of fodder development. One needs to remember the kind of livestock holder and animals that such schemes could benefit and others whom the very nature of the scheme would exclude. See Purnendu Kavoori *et al.*(eds.), *In the Rainshadow of Green Revolution*, Anand: Foundation for Ecological Security, 2010.
30 Animal Husbandry Wing, *Proceedings of the Second Meeting of the Animal Husbandry Wing of the Board of Agriculture and Animal Husbandry held at Madras from the 14th to the 16th December 1936*, New Delhi: Government of India Press, 1938.
31 Ibid, p. 10.
32 Ibid, p. 10.
33 L. B. Kulkarni, *Improvement of Grazing Areas in the Bombay Presidency*, Bombay Department of Agriculture Bulletin No 112, Poona, 1923.
34 F. G. H Andersen in Bombay Land Revenue Proceedings (India Office Library) Vol 11678, February 1928, p. 59. Cited in Neil Charlesworth, *Peasants and Imperial Rule*, Cambridge: CUP, 2002.
35 Albert Howard, *Crop Production in India*, Humphrey Milford: OUP, 1924.

36 The papers presented at this meeting were compiled and published in a volume edited by the second imperial botanist, Gabriele L. C. Howard, who had also chaired the joint meeting. See Gabriele L. C. Howard, *The Improvement of Fodder and Forage in India*, Bulletin No 150, Agriculture Research Institute, Pusa, Calcutta: Government Printing Press, 1923. In her preface, Gabriele Howard writes, "Two considerations influenced the decision to hold the joint meeting: (1) the great importance of the subject to India and (ii) the growing feeling among the agriculturists that the time has come when the botanists should come forward and assist in solving the pressing problems."

37 Gabriele L. C. Howard, *The Improvement of Fodder and Forage in India*, op. cit.

38 J. Matson, 'Quality Deficiency in the Fodder Supply', in Gabrielle C. Howard, (ed.), ibid.

39 B. B. Dave and D. N. Maheta, 'A short survey of grassland problems in central provinces', *The Agricultural Journal of India*, Vol. XXV, No. 3, 1931, pp. 220–233.

40 Ibid, pp. 220–233.

41 Animal Husbandry Wing, *Proceedings of the Second Meeting of the Animal Husbandry Wing of the Board of Agriculture and Animal Husbandry held at Madras from the 14th to the 16th December 1936*, New Delhi: Government of India Press, 1938. p. 10.

42 Ibid, p. 11.

43 Sayer, W. (1935), *Herbage Review*, No 3.

44 John Russell, *Review of the Working of Imperial Council of Agricultural Research*, New Delhi: Imperial Council of Agricultural Research, 1937, p. 38.

45 Animal Husbandry Wing, *Proceedings of the Second Meeting of the Animal Husbandry Wing of the Board of Agriculture and Animal Husbandry held at Madras from the 14th to the 16th December 1936*, op. cit., p. 106.

46 Ibid, p. 30.

47 Ibid, p. 30.

48 Albert Howardand Gabrielle Howard, 'Leguminous Crops in Desert Agriculture,' *The Agricultural Journal of India*, Vol. XII, Part I, 1917, pp. 27–43.

49 In the year 1915–1916, Imperial Council of Agricultural Research had published a bulletin No 66 that was written by G S Henderson and was titled, *Berseem as a new fodder crop of India*.

50 *Proceedings of the Board of Agriculture in India held at Pusa on the 21st February 1910 and following days*, Calcutta: Superintendent, Government Printing, p. 64.

51 *Proceedings of the First Cattle Conference held at Bangalore on 22nd and 23rd January 1924*, Calcutta: Superintendent, Government Printing, p. 19.

52 Ibid, p. 19.

53 Animal Husbandry Wing, *Proceedings of the Cattle Conference held at Shimla on 25th May 1937*, Simla: Government of India Press, 1940, pp. 53–54.

54 *The Memorandum on the Development of Agriculture and Animal Husbandry*, New Delhi: Imperial Council of Agricultural Research.

55 R. A. Pepperal, *The Dairy Industry of India*, Report on an Investigation with Recommendations, 1945, Delhi: Manager of Government Publication, 1948.

56 *Proceedings of the Eighth Meeting of the Animal Husbandry Wing of the Board of Agriculture and Animal Husbandry in India*, New Delhi: Indian Council on Agricultural Research, pp. 28–32.

57 R. O. Whyte, *Grasslands and Fodder Resources of India*, New Delhi: Indian Council Agricultural Research, 1961, p. 353.

58 R. O. Whyte, *Land, Livestock and Human Nutrition in India*, New York, Washington, London: Frederick A Praeger, 1968.

11 Deforestation, ecological deterioration and scientific forestry in Purulia, 1890s–1960s

Nirmal Kumar Mahato

Introduction

This chapter seeks to explore in what way the scientific forestry was established on the principles of Deitrich Brandis in Purulia.[1] This aspect has not been sufficiently scrutinized in the historical research on forests, forestry and imperialism. Environmental scholarship of south Asia has focused on the relationship between the growth and development of scientific ideas and technologies of natural resource management during the time of modern empires in South Asia. The scientific ideas were very much associated with the forest management and conservation and its effects on the origin of different types of environmentalism.[2] From the early nineteenth century, the commercial travellers, naturalists and surveyors started to document the forested landscape of central and eastern India. In their landscape descriptions, the forests, farms and grassland were described as hybrid landscape. Thus, India's forests were the products of anthropogenic interference and ingrained in agricultural landscapes. Recent scholars tended to describe the Indian forests from 'agroforest' perspective like the scholars of the US, Africa and Southeast Asia.[3]

With a view to implement science and to conserve forest the colonial authority introduced forest management practices and appointed technical bureaucratic experts. Recent studies questioned whether this 'scientific' forestry was at all a scientific policy. These studies also argue that colonial forest policy was shaped and guided by local factors. 'Hence, a complex picture of environmental impacts emerge', as S. Abdul Thaha argues, 'rather than a one way narrative about exploitation, extraction and commanding control over the forests'.[4] There is a debate on the nature of scientific forestry in India. Ravi Kumar argues that considering it as a device of growth and development in India, scientific forestry was introduced in the countryside. He comments on: 'Initially forestry was perceived as science mainly to deal with timber trees and its preview was focused on supply of timber to government departments. But from 1880 onward forestry became a part of the empire discourse of progress'. Colonial science was, thus, regarded as an important agency of progress for the development of colonial society.[5]

DOI: 10.4324/9781003241980-15

Vinita Damodaran argues that the scientific forestry has been seen as 'a mechanistic science where nature, the human body, and animals could be described repaired and control – as could the parts of a machine, by separate human mind acting according to rational laws'.[6] The debate of scientific forestry was embedded within the scientific worldview which is described by Carolyn Merchant as the 'world as dead and inert, manipulability from outside and exploitable for profit ... living animate nature died ... increasingly capital and the market assumed the organic attributes of growth ... nature, women and wage labourers were set on a path as human resources for the modern world system'.[7] In view of this domination over nature, it was inherent in the market economy's use of the both as resources. This domination became a natural trend in colonial Chotanagpur. In the interest of production and profit, the colonial rulers sought to dominate forest, mineral and water resources. That is why, on the one hand, there occurred large-scale deforestation in order to expand agricultural land, and on the other hand, forests were protected in the interest of colonial rulers. Thus, the scientific forestry has recently been described as 'masculine discourse'.[8] This process of domination dangerously threatened the life pattern of the Adivasis (literally 'original inhabitant', an umbrella term designating the indigenous tribal people of India)[9] and as a result, dreadful ecological hazards appeared. The discourse of scientific forestry was totally different from the Adivasi concept of landscape management.[10] Paul Sutter observed that it 'has been concentrated more on the social consequences than the ecological consequences (to the extent that they can be separated) of that change'.[11] Similarly, Mahesh Rangarajan noticed that 'one crucial aspect of historical change often neglected, is the ecological part of the story: when, why and how particular human intervention led to major transformation in the natural world'.[12] Furthermore, the colonial rulers considered the land and natural resources as state property as it primarily satisfied their revenue needs. However, colonial environmental agendas were often marked by internal conflicts because there were no clear-cut policies. In order to maintain ecological balance and continuous supply of timber, conservation measures were taken after the 1860s. But due to Adivasi revolts, the colonial environmental policies could not remain uniform. Moreover, both the colonial government and the Adivasis sought to follow a shared environmental ideology.[13]

Ecological change

In the early part of the nineteenth century, forest was primarily regarded as a 'resource'. The colonial policy of extending cultivable land at the expense of forest resulted in large-scale deforestation. They also exterminated the dangerous predators.[14] H. Coupland mentions the paying of 'rewards ... for the destruction of three tigers and seventy-nine leopards'.[15] Due to the growing demand of the railway system, which required immense quantities of logs of *sal* (*Shorea robusta* Gaertn. f. [Dipterocarpacea]) to make sleepers for the railway, pressure was placed on the forest of Jungle Mahals.[16] By the first decade of twentieth century, Purulia was connected with Asansol, Sini, Chakradharpur, Kharagpur, Gomo, Jharia and Katras. In 1908, narrow gauge rail line of 2'–6' was constructed linking Purulia with Ranchi. Coupland reported: 'this line affords an outlet for the grain and jungle products of the western portion of the district'.[17] Timbers were also required for shipbuilding.[18] The opening of the main line of Bengal Nagpur Railway through Kharagpur and Jhargram (1898) had a profound impact on the forests of the region. The introduction of railways made areas in the interior more accessible. As the forest products could be transported to distance places by the railway, there was a sudden increase in the supply of these products.[19] Pallavi Das rightly noted, 'As railway construction and operation expanded to facilitate increased trade, the railways' timber demand on the forests increased causing deforestation. The railways depended directly on the forests for their sleeper and fuel supply'.[20] Thus, the extension of railway networks was closely associated with the forest management.[21]

Two groups of peoples involved in the process of deforestation. The landowner (*zamindar*) recruited indigenous people on different forms of contract, such as new tillage (*nayabadi*) and *junglebary* (land tenures).[22] The colonial rulers employed European companies to collect wood, such as Midnapur Zamindari Company. Deforestation opened up crop fields for cultivation as well as valuable timber. From 1883 onward, the Midnapur Zamindari Company took on lease of forest land from the *zamindars* and sold the timber for ship-building and the production of railway sleepers.[23]

In the wake of agrarian intervention and woodland destruction, came ecological deterioration. In 1855, Henry Ricketts reported the total absence of trees in Purulia town.[24] In 1863, Major J. Sherwill and Captain Donald McDonald described the landscape as 'hilly, stony and broken', and added: 'The soil is poor'.[25] As Vinita Damodaran argues, 'in the case of Chotanagpur the story of environmental degradation cannot be so easily challenged'.[26] This process can be viewed from different perspectives. Deforestation caused huge amounts of soil to be eroded by rainwater and deposited onto the bed of the river, reducing its depth.[27] The shallowness of the river increased the turbidity of its waters, making them contaminated. This, in turn, affected the health of the hunting and gathering Adivasis, in particular the Savars and Birhors.[28] The colonial authority used *bandhs* (ponds) in disregard of the Adivasi perception of water. They only employed them for irrigation, taking no account of the land–water–vegetation relationship. The agrarian invasion thus accelerated both soil erosion and the filling up of ponds. In 1910, Assistant Settlement Officer Radhakanta Ghosh reported: 'I found that the beds of some *bandhs* have been encroached upon by unscrupulous *pradhan* (village headman) and Bengali settlers who have recently settled in those villages. The beds of those *bandhs* are very fertile and yield rich crop. It is for this reason that the encroachment is made without any regard to the future injuries'.[29] As a result, the ponds silted up quickly and were reduced in size and, accordingly, water-holding capacity.

The clearing of the vegetation surrounding a pond and/or upstream of it accelerated soil erosion. The resulting siltation of the pond[30] started a chain of ecological havoc decrease in water volume, an increase in nutrient concentration, an increase in the productivity of the pond ecosystem and, ultimately, decreasing oxygen levels in the water. This led to a decrease in green plants and their replacement by blue-green algae, which generated toxins and a foul smell, causing the death of the water fauna– a dreadful process known as 'eutrophication'. After the denudation of the soil, evaporation increased, rapidly leading to dryness.[31] This reduced organic matter in the soil, affecting soil texture. The change in the soil microhabitat resulted in a harsh microclimate. The soil regeneration cycle in the area was thus altered and the seed bank jeopardized. Denudation reduced rainfall as well as soil moisture. Temperatures increased and a process of desertification ultimately set in over the whole region. Deforestation combined with monoculture has a devastating impact on tropical environments. In the tropics, a much larger portion of organic matter and available nutrients is contained in the biomass. This organic matter is recycled within the organic structure of the system, partly through the agency of a number of nutrient-conserving biological adaptations, which include mutualistic symbiosis between microorganism and plant remains. With the collapse of this elaborate and well-organized biotic structure, nutrients are rapidly lost to leaching under conditions of high temperatures and heavy rainfall, especially on sites that were poor in nutrients to begin with. E.P. Odum writes: 'For this reason, agricultural strategies of the temperate zones, involving the monoculture of short-lived annual plants, are quite inappropriate for tropical regions'.[32]

Forest management

Ramachandra Guha argues that the scientific forestry in colonial India was developed in response to the revenue and strategic needs of the empire.[33] He notes that 'the large scale destruction of accessible forest in the early years of railway expansion led to the hasty creation of a forest department, set up with the help of German experts in 1864'.[34] Forest service emerged in India with the appointment of Deitrich Brandis (1824–1907) as an Inspector General of Forest of India. He tried to organize forestry in India and Burma on the basis of three principles of German forestry – 'minimum diversity', the 'balance-sheet' and 'sustained yield'. In the concept of 'minimum diversity' Brandis tried to emphasize on producing appropriation to the extent of teak, the predominant commercial woods. It levelled across the inherent diversity of forest. Therefore, the forests were eventually transformed into 'commercially marketable monocultures'. Brandis formulated guidelines for the management of forest and enunciated ground rules on felling. Thus, his approach was to regularize repair, tending and pruning. He advocated diffuse replacement of felled trees by dibbling seed or planting seedling in the gaps, landing and other forest openings. To prevent either under or over-utilization of wood Brandis provided guidelines. In his guidelines, the 'annual yield' would be based on the amount they put on. The useful species gained special attention for conservation and plantation. Thus, the attitude to nature adopted by the Indian forester in the second half of nineteenth century was 'clearly conservationist in character'. Ravi Rajan writes 'by the end of the nineteenth century this utilitarian conservation sentiment became a developmental ideology in its own right'.[35] The Indian forests received the character of German and French forestry though the implementation of the Brandis's three principles.[36]

In his book *Forestry in British India* (1900), Ribentrop categorized the Indian forests into seven forest zones and the forests of South West Bengal belonged to the Central India Deciduous Region.[37] Forests conservancy was started in India in August 1864 and the forests of Bengal were managed by ten divisions which included Bhagalpur, Sonthal Perganas, Patna, Rajshahye, Burdwan, Nuddea (Sunderban and 24 Pergannahs), Cuttack, Chotanagpur, Dacca and Cachar.[38] Among the South West-Bengal districts, Bankura, Birbhum, Burdwan and Midnapur were included in the Burdwan Division and the District Manbhum was integrated with the Chotanagpur division.

In various parts of an ecologically and socially heterogeneous subcontinent, the imposition of the new regime of control had different consequences. Mahesh Rangarajan writes that 'the specific ecological milieu both in terms of forest types and agrarian regimes (land ownership patterns and production system) will build up a better understanding of contrasts between and within different regions'.[39] Thus, this chapter also examines Rangarajan's point in case of the forest management of Manbhum district.

Forests under the *zamindars* and *ghatwals*

With the commercialization of forests,[40] a new form of land use came into existence. Forest could bring larger return per acre than tenanted lands that could be cultivated. The big *zamindars* (landlords) took steps for preserving forests. K. C. Raychoudhury wrote that the *zamindars*, 'particularly in Purulia hold the forests under a peculiar lease and undefined form of tenure, known as the *ghatwali* tenure-originally granted in lieu of watch and ward duties of the *ghats*, i.e., the hilly tracts'. With the opening of new markets, felling went on with an increased tempo and *sal* (*Shorea robusta* Gaertn.f.[Dipterocarpacea]) being easy to regenerate by coppice. This method was followed by all owners. The rotation varied according to the need of forest owner, i.e., for bigger parties the rotation became necessarily smaller. The small holders brought down the rotation to five years or even less and hastened disappearance of forests.[41] Though the *zamindars* maintained the forests as shooting reserves, they ruthlessly destroyed the woods by engaging agents to earn quick profit.[42] In order to protect the original peasantry and their community leadership, a new type of administrative set up like South Western Frontier Agency or Encumbered Estates/ Court of Wards were introduced in many areas of Chotanagpur. But by the twentieth century, a general kind of administration was implemented in these regions of 'zone of anomaly' and the *'zamindars* gained at the expense of headmen'.[43] When the forests were under the managements of Manager, Wards and Encumbered Estates, the forests gradually 'begun to disappear as a result of wanton destruction by tenants'.[44] In 1935, considering the environmental hazards, the government reviewed the role of *zamindars* as forest managers in the British Empire Forestry conference held in South Africa.[45]

Demarcation of protected forests

The government and forest department took a wholesale programme of forest management to combat extensive deforestation.[46] In India, the growth of a forest policy was extraordinarily slow. Throughout the eighteenth century, the whole policy of the government was to extend agriculture and to destroy the forests. In the early nineteenth century demands of forests increased. Colonial government attempted to conserve them. Though a forest conservancy system in some provinces was introduced but it was not raised above the level of revenue administration. Though in 1865, a Forest Act was passed but it took three more decades to implement the terms. 'Social complexity', as Sivaramkrishnan notes, 'was also a factor shaping managerial alternatives.' Here, the social structure was not merely dual character like sovereigns and subalterns, it was complex in nature.[47]

The British Forest Policy of 1894

With the recommendation of Dr. John Augustus Voelcker, a German expert, on Indian agriculture (1893), the first forest policy for British India was implemented. This policy advocated the agriculture expansion in the forest areas. As the forest policy was commercially oriented, it displaced the rights and privileges enjoyed by the forest dwellers.[48] In Manbhum, following the suggestion of Voelker, areas of jungle covering 8.90 and 4.32 square miles were notified as government protected forest in the two temporary settled estates of Matha and Koilapal in 1894. Those were the only government forest sites in the district. All the waste lands and forests in these two estates were originally declared to be 'protected forest'. However, some portions of the forest sufficient for the requirement of the villagers were subsequently assigned to them. The boundaries of these blocks had been demarcated with stone cairns.[49] It is noted that those forests were 'protected' which had not yet been surveyed and it was hoped that it would be reserved for future.[50] Towards the end of December, 1901, a forest officer was deputed by the Deputy Conservator of Forest, Singhbum district to hold the charge of these forests.[51]

With the formation of the province of Bihar and Orissa on 1st April 1912, the Forest Division of Santal Parganas, Palamau, Singhbhum, Chaibasa, Sambalpur, Angul and Puri were taken away from the Lower Province

Name of the District	R.F. in Sq. miles	P.F. area in Sq. miles	Total Sq. miles
Singhbhum		182	
Manbhum	182	14	196
Santal Parganas		292	292

Source: B.N. Prasad, *Bihar Forest Souvenir*, 1965, pp. 18–20.

of Bengal and constituted into Bihar and Orissa Forest Circle and Mr. H. A. Forteath held charge of this new circle with effect from 1st April 1912. At the time, the following districts were under the Chaibasa Division.[52]

Reserve forests

In India, some forest lands became 'reserved' with the emergence Forest Act of 1865. In these forests, the government had the full right of ownership and thus banned all sorts of agriculture. With the Forest Act of 1878, the rules were tightened. In these forests, the peasants who lived in the reserved areas had no right to the forest.[53] Though the 'protected forest' concept came in force in 1894 it took nearby three decades to create a 'reserved' forest in Manbhum district.

The Government of Bihar and Orissa in Council (under the notification No. 7673 III F 67-R and in exercise of the powers conferred by the Sec.19 of the Indian Forest Act of 1878) declared the forests belonging to the Taralal Encumbered Estates situated in the Purulia police station in the district of Manbhum, as 'reserve forest' with effect from 1st September1925.[54] According to the order of Governor-in-Council, J. R. Dain Secretary to Government, in Jaipur Block no. I, the cart track was allowed from Jaipur village towards and Hoyanda onwards as a public right of way. In Block no. II, the cart track from Pirorgoria towards Popo which passed through this block was allowed as a public right of way. In Block no. V, the cart track from Popo to Bali and onwards and towards Hazaribagh was allowed as a public right of way. In Block no. VII, Popo to Perogoria, cart track was allowed as a public right of way.[55]

The government of Bihar and Orissa in Council declared the forest belonging to Kalimati encumbered estates situated in the Bagmundi police station of the district of Manbhum as 'reserve forests' with effect from 1st August 1925. In Kalimati reserved forest, Dolgobinda Singh of Kalimati was allowed to cultivate plot nos. 15, 16, in Kalimati and 14–16 in Perorgoria. Balai Mahato, son of Ratin Mahato of Pirorgoria was allowed to cultivate plot nos. 82–86, 88–92 in Kalimati and 20–22 in Perorgoria. Brajraj Singh, son of Ganga Narain Singh of Kalimati was allowed to cultivate plot nos. 2–13 in Kalimati and 2, 3, 4, 5, 6–14 in Perorgoria. Kujla Bhumij, son of Rasaraj Bhumij of Bagti was allowed to cultivate in plot no. 1340 in Bagti.[56]

Again, the Government of Bihar and Orissa in Council declared the forest belonging to the Mudali encumbered estates situated in Purulia police station in the district of Manbhum as 'reserved forest' with from 1st August 1925. In this forest, the villagers of the Dhanchatani and Lukuichatani were given the right and way routes used by them for coming down to the plains from the top of the hills.[57] In an another notification, the Government of Bihar and Orissa in Council declared the forest belonging to the Barabhum encumbered estates of district of Manbhum as 'reserved forest' with effect from 29th May 1925. The Governor in Council appointed the Sadar Sub-Divisional Officer of Manbhum to discharge within the limits of reserved forests of the Jaipur, Taralal, Kalimati, Mudali and Barabhum Encumbered Estates in the district of Manbum. The power of Forest Officer was vested to him.[58] All the Rangers, Deputy Rangers, Foresters and Forest Guards that might be appointed to the forests of these estates would also discharge the functions which were similar to the officials of the forest department.[59]

The British Government considered the forest from the utilitarian outlook. In 1932, Forest Officer Rai Sahib P. N. Mukherjee reported that there were two reserved forests under the management such as Barabhum Reserve Forest and Taralal Reserve Forest.[60] Revenue orientation became more marked under the state management like Uttarakhand.[61] The income was mainly derived from compensation for Forest Act cesses and sale of minor forest produce. The figures for 1930–1931 included sales of firewood and brushwood in Block no. I. The firewood and brushwood in Block no. II would be sold in auction at Barabazar on 10thApril, 1932. The slight decrease in other income was due to the fact that on account of the decrease in the price of *lac*. Therefore, no settlement of the few *lac* trees could be made.[62] Forest Officer I. A. Habbuck, reported that one reason of the big drop between 1930–1931 and 1931–1932 in Barabhum Reserve Forest was that firewood and brushwood in the coupe under clearance that the year was not actually put up for the auction for till 10th April, 1932. In 1930–1931, a sum of Rs. 1250 was obtained from the settlement of brushwood and firewood in Block no. I. The Forest Officer could not collect revenues from *lac* settlement in the forest in this session, but he collected it in the previous years.[63]

Taralal figures in 1932–1933 show no revenue at all although three foresters were employed on this forest and there was also a Deputy Ranger for other two forests.[64] So the Government decided to sell the Reserve Forest. In Barabhum Reserve Forest, the income was derived from dried *sal* (*Shorea robusta* Gaertn.f. [Dipterocarpacea]) straw, arrears of *lac* rent and *kul* (*Ziziphus mauritiana* Lamk. [Rhamnaceae]) trees for the year 1931–1932. In the year 1932, the miscellaneous receipts so far was Rs. 46 which had been derived from cattle grazing and catching of fish. S.U. Majumdar, Deputy Commissioner reported that 'the former should not have been allowed as it was deadly to regeneration of forest'.[65] Taralal reserve forest was given to a further lease to the proprietor of the Taralal Encumbered

Estate. The forest staff were discharged on the 20th March 1933. Thus, there remained only Barabum Reserve Forests under management. In the year 1932–1933, Rs. 111 was derived as compensation from Forest Act cases and sale of minor forest products. As the price of *lac* had declined very heavily – it hardly covered manufacture cost – there were no applications for *lac* settlement.[66]

Rai Sahib P. N. Mukherjee reported that from Barabhum Reserve Forests, a sum of Rs. 500 had been received during the year 1933–1934 on account of sale of brushwood of Block No. I and Rs. 77 had been received on account of compensation levied on forest offence cases and settlement of minor products such as fishery in a channel and sale of thatching grass.[67] In the year 1935–1936, Rs. 608.6 was derived from the settlement of *palas* (*Butea Monosperma*) trees, straw and sale of *sal* trees. But the expenditure (i.e., Rs. 698.93) was more than the income.[68] In 1935, Rs. 433 was derived from settlement of *sal* trees (Rs. 425.6) and compensation realized from forest offenses cases (Rs. 8).[69] In 1936, an income of Rs 1260.6 was derived from settlement of *sal* trees (Rs. 1192.48), miscellaneous sources (Rs. 33.54) and compensation realized from forest offenses (Rs. 14).[70] For the period of 1937–1939, average revenue of Rs. 725 was derived from the sale of minor produce such as grass, firewood, etc. In the year 1940, the expected revenue was estimated as Rs. 1000.[71]

Several recent studies on forest reservation recognized its territorial basis (Ramachandra Guha 1989, 1990, Mahesh Rangarajan 1992, Ajay Skaria 1999).[72] In his study, Sivaramakrishnan has shown that 'the making of reserved and protected areas displays a regional pattern in Bengal'. He treated southwest Bengal as a 'zone of anomaly to forest reservation'.[73] This section examines to what extent Brandis's three basic principles of forest reservation– 'minimum diversity', the 'balance-sheet' and 'sustained yield' was implemented in Manbhum. First of all, he applied this method in North East Province and Oudh from where it spread to various parts of equatorial tropics.[74]

Minimum diversity

The Barabhum Reserve Forest became a forest of 'commercially marketable monoculture'. In his inspection note written in 1933, Forest Officer Hardayal Singh recorded that 'these forest are good type especially those in Block no. I just near Barabazar. On one side of a channel running in Block no. I, the crop consists only young *sal* (*Shorea robusta*) poles up to 12″–18″, mix with a small percentage of miscellaneous species, i.e., of girth of the log varying from 2″ to 6″'.[75] *Sal* was the most commercially valuable timber. Though of little importance the other species like *nim* (*Azadirachta indica*), *mahanim* (*Melia dubia*), *karanja* (*Pongamia pinnata*), *Arjun* (*Terminalia arjuna*), *piasal* (*Pterospermum marsupium*), etc., were also grown.[76] The Research Officer strongly recommended *sabai* grass planting as there was high demand for fuel.[77]

The forest of Block no. II of Barabazar forest was of much uniform type than that in Block no. I. It contained nothing except small poles and saplings of *sal*, the maximum girth hardly exceeding 12–18 inches. There were other patches containing species like *dhadhaki* (woodfordia), *siuli* (*Nyctanthes*), *sihuri* (Cleistanthus collinus), etc., all growing in a mass and a quite up to the firewood size.[78] Forest Officer P. N. Mukherjee suggested to sow *Simul* (*Bambax malabaricum*), *Tetul* (*Tamaridus indica*) and *Asan* (*Terminalia tomentosa*) trees in this area.[79] Later, many other valuable wood species like Teak, Bamboo, Gamar, Mohua (*Madhuca indica*, Gmelin [Combretaceae]) were introduced.[80]

Balance sheet

The scientific forestry was introduced in India in the form of silviculture plantation. It was introduced in South India in order to generate resources and in this way, the forest department tried to test its viability. 'Silviculture plantations', as Ravi Kumar argues, 'became laboratories of experiments of scientific forestry to test the ability of objective science in manipulating the nature'.[81] The principle of balance sheet was reflected in the silvicultural method followed by the foresters of Barabhum Reserve Forest. After visiting the forest in 1933, Forest Officer Haradayal Singh reported that the young *sal* poles up to 12"–8" girth would be felled provided suitable offers of price were available. The patches were to be vigorously protected from grazing after completion of felling. Removal of dead leaves and *sal* seeds would not be allowed. It would ensure getting coppice regeneration on the cut stumps with enriched surfaced soil to feed them. There were differences of opinion regarding the silvicultural method between Rai Sahib P. N. Mukherjee, Sadar Sub-Divisional Officer, Manbhum and Haradyal Singh, Divisional Forest Officer, Chaibasa Division. The Deputy Commissioner's idea was 'to fill up these patches with *sabai* (*Ischaemum angustifolium*) grass'. Divisional Forest Officer suggested that 'it would be better of teak root and shoot-cutting are planting instead'. Accordingly, teak is one of the first growing species and it is not very much damaged by cattle grazing. Haradayal Sing made an important observation that

> in certain places [it] will improve the crop considerably. But it should be done only by skilled hands and no dishonest contractor should be allowed to enter the area. It is much better to let the crop stand as it is than to ruin it in the name of thinning.[82]

Forester of Barabazar Reserve Forest derived some revenues from cattle grazing. The Divisional Forest Officer Chaibasa Division, who inspected the forest at the end of January 1933, had remarked that a cattle grazing 'is deadly to the regeneration of forest'. He found that the ground was swept of even dead leaves and had suggested that the area should be vigorously protected from grazing. He also mentioned that the Deputy Ranger who was appointed in 1926 had received no training in forest work. So, it would not be possible for him to develop the forest on scientific line.[83]

There was also difference of opinions among the forest officials. While Mr. B. P. Basu, I.F.S., the Forest Research Officer strongly recommended *sabai* grass planting, Haradayal Singh suggested that a few patches in the forest would better be filled up with tree species.[84] *Sabai* was the most valuable grass at that time. It had been brought from Singhbhum district. D. P. Sharma, Deputy Commissioner, Manbhum, writes that its cultivation was extended for the last four years, 1936–1939.[85] In 1940, he reports that most valuable wood species, the Sandal was planted with the assistance of Forest Research Institute, Dehradun and Mysore. Another important wood called *tung* (*Aleurites fordii* [Euphorbiaceae]) which was a cling tree was brought from Mysore in the same year. Its oil was very extensively used for paint and varnish during the time.[86]

Sustained yield

Now we are to examine the main German principle of 'sustained yield' which was adopted by the Indian foresters of Manbhum. The Forest Officials tried to desist themselves from the over utilization of wood. But the district authority tried to extract revenues from forests without taking consideration of over utilization. In 1933, Rai Sahib P. N. Mukherjee, Sadar Sub-Divisional Officer of Manbhum reported: 'the brushwood of block no I of the forest will be fit for the sale next year'. *Sal* tree in Nilmohanpur was numbered but no action was taken for the sale of the large trees due to economic depression. In 1934, Haradayal Singh, Divisional Forest Officer, Chaibasa, reported that the district authority sought to over utilize the forest. In his note, it is found that Block no. I was marked and advertised for sale in spite of the fact that there was absolutely no regeneration on the ground. He felt pity seeing the ruthless felling of all the so-called species by a contractor getting a lease for two years. While the primary aim would be to fill up the blanks but instead, through the above process more blanks were created. Thus, more and more of the surface became exposed and lost more and more of the potential value of the soil.[87]

In the late-nineteenth-century India, with the implementation of forests acts, the foresters emerged as technocrats and important officials in the politics of forest resource utilization. They were interested to implement the forest rules and involved in conflict with the civil administration.[88] The Divisional Forest Officials of were more careful to preserve the forest on Brandis line than the district authorities. The district officials always sought to collect more revenue than the previous year. For this, they were even ready to give permission for grazing. Sivaramakrishnan rightly observes that 'with forest becoming a state resource, control over their disposition, and control over all manner of revenues derived from them, become central to the conflict that arose between foresters and the district administration'.[89] This type of conflict prevailed in the third and fourth decades of twentieth century also.

During the rule of East India Company, demarcation of external boundaries of the empire was the main concern. With the growing knowledge of the landscape and natural resources of the empire, the British Raj sought to concentrate, after 1858, on internal or managerial territorialization. It obviously demanded forest conservancy.[90] As in the Manbhum district, there was 'administrative exceptionalism', and it took three more decades to create protected forests. As most of the forest were under the possession of *zamindars* (landlord) and *ghatwals* (guardian of passes, sometime headman) there were constant conflicts over the jungle rights between them and the *raiyats* (peasants). In spite of that in 1928, the British authority hesitated to dissolve *ghatwali* rights. During the time A. F. A. Hamid, Superintendent of Police, Manbhum wrote

> as this institution is very old one and more or less deeply rooted and as some of the ghatwals are probably under khuntkatti and sthitiban rights, the abolition of the system may led to unrest specially when lands etc. would be snatched from them and resettled.[91]

To meet the demands of World War II (1939–1945), recourse was taken under certain provisions of the Indian Forest Act. Colonial officials started negotiation with the *ghatwals* who held government forest under permanent service of *jagir* and also with the general manager of Wards and Encumbered Estates. As a result, the forest of Bagmundi Estate came under government management with lease of twenty years in 1942. In 1943, the *ghatwali* or private forest of Kaira, Dadha, Bonta, Makuliah, Baroda, Gunda and Ramnagar came similarly under control and management of the Forest department of Bihar through a lease of forty years. The forests of Palkia, Popo and Kenda were taken over by the government in 1944 and the forests of Tundi met the same fate in 1945. In the same year, Barabhum Reserved Forest was transferred from the Civil Department to Forest Department. During this period, demand for timber and poles was increased as timber prices were so high, and the colonial government had given the priority to transportation of war materials. The government started to harvest good timber from remote areas of the forests to meet the war situation.[92]

Forests management in Independent India

In Independent India, commercial industrial zone extracted the chief yield of state forest resources[93] and commercial utilization of forests was enlarged partially due to nationalization of private forests. As Mark Profenburger and Chhatrapati Singh notes, 'Strengthen ties between politicians, business people, and some foresters which were based on remunerative industrial extraction, also drove deforestation, while further eroding the rights of forests communities'.[94] Gadgil and Guha categorized the industry oriented Indian forestry into four phases. The first phase is the traditional 'sustained yield' selection method whereas the second (1960–1965) is the 'programmes of clear-felling ad the plantation of quick growing, chiefly exotic, species'. In the third stage (1975), Farm forestry was introduced and import and captive plantation was implemented in the fourth stage (1985).[95] For the first phase, selection felling was done in the natural forest in order to meet the demands of timber-based industry.

With the passing of the West Bengal Private Forests Act 1948, the scientific forestry was introduced in South West Bengal. According to the provision of this act, all the forest owners submitted their working plans for the area of 431.64 square miles. Cutting cycle was extended up to ten years and preservation of standards was also fixed at the rate of 15–20 acre. It controlled the grazing and fire as well. In the case of experimental afforestation, the North Bengal model such as ploughing the soil, constructing small ridges and sowing the seed was implemented.[96] Due to dry weather, the seedling could not survive so that cattle-proof (6′ ×4′) and contour trenches (18″× 18″ or 24″× 24″, to provide adequate soil moisture) were constructed. Saddling was planted at the interval of 8′× 8′ *thalis*[97] and along with-it box trenches were created for water conservation. Generally, transplants were put in *thalis* (plates) and pregerminated stumps were kept in polythene bags. The species like Eucalyptus, teak (*Tectona grandis* L.f.), *peasal* (*Peterocarpus marsupium, Linn.*), *simul* (*Bombax malabaricum, Syn. Salmaliamalabarica, Linn*) and *sissoo* (*Dalbergia sissoo*) were kept in pot plant and for teak and peasal were raised by stumps. Fruitful results came from planting the species of *gamar* (*Gmelina arborea, Linn.*), *asan* (*T. tomentosa, Linn.*), *bahera* (*Terminalia belerica, Linn.*) and mango (*Mangifera indica, Linn.*). In the cattle proof trench, *gamar* as well as *mahanimb* (*Alilanthus excelsa* Roxb) were generally planted. In order to make plantation success, fire was strictly controlled in the tract.[98] Establishing a nursery at Arabari of Midnapur district afforestation work was started in the centre of laterite zone and different species like *Shorea robusta, Melia azadirachta, Cassia siamea, Tectona grandis, Acacia Arabica, Pongamia glabra, Albizzia* spp. were planted over 395 acres tract. The Private Forest owners neglected the aspects of afforestation of blanks, soil conservation and control of fire and grazing but they concentrated only on felling restriction prescribed by the Working Plan.

In Independent India, Indian Forestry programme was formulated with the passing of Indian National Forest Policy in 1952. As Madhab Gadgil and Ramachandra Guha notes,

> in the National Forest Policy of 1952, the exclusion of local communities from the benefits of forest management is legitimized as being in the 'national interest', namely that the 'country as a whole' is not deprived of a 'national asset' by the mere accident [sic] of a village being situated close to a forest.[99]

By 1957, the Government handled all the difficulties, 10000–15000 acres of forests had been transferred to Hindustan Steel Projects and Durgapur

Group of Industries and Kangsabati canal system (Purulia, Bankura and Midnapur).[100] Thus, at the cost of huge forest big industries were created 'as the "national interest" has virtually been equated with industrial sector...'.[101] After inclusion of Purulia into West Bengal, a new Purulia Division was integrated with the Southern Circle. West Bengal Estates Acquisition Act was also implemented in Purulia since April 1964. With a view to implement soil conservation measures a new Soil Conservation Division was set up with its headquarters at Purulia.[102]

Conclusion

This 'scientific forest policy' adopted by the colonial rulers brought about significant changes in the Adivasi landscape of Manbhum. Like Singhbhum, in the early phase of colonial rule in Purulia, agriculture was pursued at the extent of forest tract. The Forest Department and landlords deprived the Adivasis from their traditional forest rights. However, in some cases, it was allowed to the general people to use some portion for bullock cart, cultivation in some plots, public right of way. The sustainable economy of the region was permanently destabilized and the district became drought prone. Considering the environmental degradations the colonial authorities adopted a broad programme of forest management. But here we observed two contradictory views regarding forest conservation and utilization of forest resources. Forest officials were concerned with forest conservation following Brandis's line while the district officials were interested to extract revenues. As the forest became an important resource for generating revenues, the colonial authorities sought to follow a pragmatic approach for forest conservancy. In 1932–1933, Taralal Reserve Forest was given to lease when no revenue was derived from it. In order to earn revenues, grazing was also allowed in some patches; even revenue was also earned from selling minor forest produce like catching of fishes from small channel. As a result of forest destruction and creation of reserve forests, forest dwellers Adivasis became 'ecological refuges'. Since independence, the Forest Department continued its role as revenue extracting organ but it tried to cut short its felling as well. The forest officials followed the working plan and soil conservation measures but the private forest owners were concerned about quick profit.

Acknowledgement

I express my sincere gratitude to Professor Ranjan Chakrabarti, Former Vice-Chancellor, Vidyasagar University, West Bengal and Professor Such-ibrata Sen, Former Professor of History, Visva-Bharati, Santiniketan, for their sustained help and encouragement. I express my regard to Professor Deepak Kumar, eminent historian of science, who always acts as a perennial source of warm encouragement.

Notes

1 The Manbhum district was bordered to the north by Hazaribagh and Santhal Parganas, to the east by Burdwan, Bankura and Midnapur, to the south by Singhbhum and to the west by Ranchi and Hazaribagh. This Bengal district was emerged as a part of South Western Frontier Agency in 1833 but it was part of Bihar and Orissa during the period of 1912–1956. In 1956, it was divided into two parts, one part became Purulia district of West Bengal and the other part became Dhanbad district of Jharkhand, India.

2 K. Sivaramakrishnan, 'Science, Environment and Empire History: Comparative Perspectives from Forests in Colonial India', *Environment and History*, Vol. 14, No. 1, 2008, p. 42.

3 K. Sivaramakrishnan, 'Transition Zones: Changing Landscapes and Local authorities in Southwest Bengal, 1880s–1920s', in Mahesh Rangarajan and K. Sivaramakrishnan (eds.), *India's Environmental History*, Ranikhet: Permanent Black, 2012, pp. 197–245.

4 Abdul S. Thaha, 'Forest Policy and Ecological Change: The Hyderabad State (1867–1948)', in Deepak Kumar *et al* (eds.), *The British Empire and the Natural World*, New Delhi: OUP, 2011, p. 262.

5 V M Ravi Kumar, 'Colonialism and Green Science: History of Colonial Scientific Forestry in South India, 1820–1920', *IJHS*, Vol. 47, No. 2, 2012, p. 253.

6 Vinita Damodaran, 'Indigenous Forests: Rights, Discourses, and resistance in Chotanagpur, 1860–12002', in K. Sivaramakrishnan and Gunnel Cederöf (eds.), *Ecological Nationalism*, New Delhi: Permanent Black, 2005, p. 118.

7 Carolyn Merchant, *Radical Ecology*, New York and London: Routledge, 1992, pp. 41–60. She observed that the mechanistic worldview is a product of the scientific revolution of the seventeenth century.

8 Vinita Damodaran, 'Gender, Forests and Famine in 19th-Century Chotanagpur', *Indian Journal of Gender Studies*, Vol. 9, No. 2, 2002, pp. 142–144.

9 The word 'Adivasi' means original inhabitant. Recently scholars like Rycroft do not italicize the word in order to normalize its use. D J. Rycroft, 'Looking beyond the Present: The Historical Dynamics of Adivasi (Indigenous and Tribal) Assertions in India', *Journal of Adivasi and Indigenous Studies*, Vol. 1, No. 1, 2014, pp. 1–17.

10 The Adivasi people managed the landscape with their own indigenous knowledge system. Different sacred institutions were created in order to facilitate biological resource management, linked to religious myth and belief system. For details see N. K. Mahato, '*Adivasi* (Indigenous people) *Perception of Landscape*: The Case of Manbhum', *Journal of Adivasi and Indigenous Studies*, Vol. 2, No.1, 2015, p. 1.

11 Paul Sutter, 'What Can US Environmental Historians Learn from Non- US Environmental Historiography', *Environmental History*, Vol. 8, No. 1, 2003. p. 4.

12 Mahesh Rangarajan, *Fencing the Forest*, New Delhi: OUP, 1996, p. 8.

13 Asoka Kumar Sen. 'Collaboration and Conflict: Environmental Legacies and the Ho of Kolhan (1700–1918)', in Deepak Kumar *et al* (eds.), *The British Empire and the Natural World*, op. cit., p. 208.

14 Vinita Damodaran, 'Gender, Forests and Famine in 19th-Century Chotanagpur', op. cit., pp. 142–144.

15 H. Coupland, *Bengal District Gazetteers: Manbhum*, Calcutta: Bengal Secretariat Press, 1911, p. 21.

16 Mark Profenburger, 'The struggle for forest control in Jungle Mahals of West Bengal 1750–1990', in Mark Profenburger and B. MacGean (eds.), *Village Voices, Forest Choices*, New Delhi: OUP, 1996, p. 137.

17 H. Coupland, *Bengal District Gazetteers: Manbhum*, op. cit., p. 185.

18 West Bengal State Archives, Revenue Dept., File No.- 95/7/19, Govt. of Bengal, Forest Branch, May, 1919, Para-7–9.

19 Centenary Commemoration Volume, *West Bengal Forests*, Forest Department, West Bengal, 1964, p. 133.

20 Pallavi Das, 'Hugh Cleghorn and Forest Conservancy in India', *Environment and History*, Vol. 11, No. 1 2005, p. 56.

21 Sebastian Joseph, *Cochin Forests and the British Techno-ecological Imperialism in India*, Delhi: Primus, 2016, pp. 41–42.

22 For *jungleburi*, W.W. Hunter, *A Statistical Account of Bengal*, Vol. 17, Calcutta: Bengal Secretariat Press, 1887, p. 332 and for *nayabadi* see Manbhum District Records (hereafter MDR), Circle Note of Attestation Camp No. II, Barabhum, Session -1904–1910 by Mr. Radhakanta Ghosh, Assistant Settlement Officer, p. 51.

23 MDR, Circle Note of Attestation Camp No. II, Barabhum, Session -1904–1910 by Mr. Radhakanta Ghosh, Assistant Settlement Officer. p. 51.

24 H Ricketts, 'Reports on the Agency Administration', in *Selection from the Records of Bengal Government*, Vol. XX, Calcutta: Bengal Secretariat Press, 1855, pp. 2–3.

25 Note of the Map of *pargana* Pandra, Sherghor, Mahesh and Chatna, Main Circuit No. 5&9, 1862–1863. The survey was conducted by Major J.L. Sherwill and Captain Donald McDonald.

26 Vinita Damodaran, 'Gender, Forests and Famine in 19th-Century Chotanagpur', op. cit, p. 143.

27 H. Coupland, *Bengal District Gazetteers: Manbhum*, op. cit, p. 5

28 Oral History collected from Sri Kalipada Savar, Savar old man, Vill. Sidhatarn, Dist. Purulia, 27th March 2003.

29 MDR, Circle Note of Attestation Camp No. II, Barabhum, Session -1904–1910 by Mr. Radhakanta Ghosh, Assistant Settlement Officer, p. 45.

30 Ibid

31 H. Coupland, *Bengal District Gazetteers: Manbhum*, op. cit, p. 113.

32 E. P. Odum, *Basic Ecology*, Holt-Saunders: College Publishing, Japan, 1983, pp. 211–212.

33 Ramachandra Guha, *The Unquiet Woods*, New Delhi: OUP, 1983. He argues that colonialism 'constituted an ecological watershed in the history of India'. See also Ramachandra Guha and Madhab Gadgil, *This Fissured Land*, New Delhi: OUP, 1992, Chap-V-VIII. However, Richard H. Grove in his work *Green Imperialism*, Cambridge: CUP, 1995, contested this thesis. He argues that the ideological commitment of a section of colonial officials to conservation was much more significant than narrow materialist concerns. Mahesh Rangarajan (*Fencing the Forest*, 1996, p. 7) has commented on this controversy that 'many of the differences between Guha and Grove are due to the difference in the chronological focus of their research. Grove is concerned with the early colonial period; Guha focuses on the late nineteenth century. There is also a marked contrast in their central concerns. Guha examines the broad unity of imperial interests while Grove brings out the divisions among colonial officials'.

34 Ramachandra Guha, *The Unquiet Woods*, op. cit., p. 37.

35 Ravi Rajan, 'Imperial Environmentalism or Environmental Imperialism? European Forestry, Colonial Foresters and the Agendas of forest Management in British India,1800–1900', in Richard H Grove *et al* (eds.), *Nature and the Orient*, New Delhi: OUP,1999, pp. 343–351.

36 Ibid., p. 356.

37 The Seven Forests zones were: (i) Evergreen Forests Zone, (ii) Deciduous Forests Zone, (iii) Dry Forest Zone, (iv) Alpine Forest Zone, (v) Riparian Forest Zone, (vi) Tidal Forest Zone, (vii) Zone without Forests. For details see E. Stebbing, *The Forests of India*, Vol -1, London: John Lane, the Bodley Head Ltd, 1922, pp. 39–45.

38 Ibid, vol -2, p. 375.

39 Mahesh Rangarajan, *Fencing the Forest*, op. cit., pp. 202–203.

40 Ranabir Samaddar, *Memory, Identity, Power*, Hyderabad: Orient Longman, 1998, p. 66.
41 Centenary Commemoration Volume, West Bengal Forests, op. cit., p. 133.
42 West Bengal District Gazetteer, Puruliya. Calcutta, 1985, pp. 53–54.
43 K. Sivaramakrishnan, 'Transition Zones: Changing Landscapes and Local authorities in Southwest Bengal, 1880s–1920s', in Mahesh Rangarajan and K. Sivaramakrishnan (eds.), *India's Environmental History*, op. cit., p. 217.
44 MDR, Notes recorded by D.P. Sharma, Esq. I.C.S. Deputy Commissioner, Manbhum, the 12ᵗʰNovember, 1940 on the working of the Barabum Wards Estate Forests, Para-2.
45 K. Sivaramakrishnan, 'Transition Zones', op. cit., p. 227; Mahato, N.K, 'Environmental Change and Forests Conservancy in South West Bengal, 1890–1964', in Ranjan Chakrabarti (ed.), *Critical Themes in Environmental History of India*, New Delhi: ICHR & Sage, 2020, p. 267.
46 Vinita Damodaran, 'Indigenous Forests: Rights, Discourses, and resistance in Chotanagpur, 1860–12002', in K. Sivaramakrishnan and Gunnel Cederöf (eds.), *Ecological Nationalism*, op. cit., p. 119.
47 K. Sivaramakrishnan, 'Transition Zones', op. cit., p. 199.
48 Sebastian Joseph, *Cochin Forests and the British Techno-ecological Imperialism in India*, op. cit., pp. 49–50.
49 MDR, Land Revenue Administration Report for 1903/1904. Revenue Dept. File No. XV(C), Issue no.135(R), Date of Issue: 27.04.1904, Item No.11.
50 For protected and reserve forests, see John Augustus Voelker, *Report on the Improvement of Indian Agriculture*, London: Eyre & Spottiswoode, 1893, pp. 142–44.
51 MDR, Land Revenue Administration Report for 1903/1904. Revenue Dept. File No. XV(C), Issue no.135(R), Date of Issue: 27.04.1904. Item No. 11.
52 Bihar Forest Souvenir, Forest Department, Bihar, 1965, pp. 18–20; Mahato, N.K, 'Environmental Change and Forests', op. cit., p. 269.
53 David Hardiman, 'Farming in the Forest: The Dangs 1830–1992', in Mark Profenburger and B. MacGean (eds), *Village Voices, Forest Choices*, op. cit., p. 112.
54 MDR, No.259. Forest Dept., Guard File of Inspection Note, By order of the Govt. in Council, J. R. Dain, Secretary to Government, 30ᵗʰJune, 1925.
55 Ibid.
56 Ibid. (under the notification no. 70008 III F 67R and in exercise of the powers conferred by section 19 of the Indian Forest Act, 1878, Act of VII of 1878).
57 Ibid. (under the notification no. 7010 III F- 67 -R and in exercise of the powers conferred by Sec 19 of the Indian Forest Act, 1878, Act of 1878)
58 Ibid
59 MDR, By order, J. R. Dain, the 29ᵗʰ May1925 & 30ᵗʰ June 1925 (under the notification no. 7013 III F- 67 -R and in exercise of the powers conferred by Sec 19 of the Indian Forest Act, 1878, Act of 1878).
60 MDR, Inspection Note of Forest Dept. made by Forest Officer, Rai Sahib P. N. Mukherjee, 31.03.1932.
61 Ramachandra Guha, *The Unquiet Woods*, op. cit., p. 37.
62 MDR, Inspection Note of Forest Dept. made by Forest Officer, Rai Sahib P. N. Mukherjee, 31ˢᵗ March1932.
63 MDR, Inspection Note of Forest Dept. made by Forest Officer, I. A. Habbuck, Para-2.
64 MDR, Inspection Note of Forest Dept. made by Forest Officer, I. A. Habbuck, Para-3.
65 MDR, Inspection Note of Forest Dept. made by Forest Officer, S.N. Mukherjee, February,1933, Para-2–3.

66 MDR, Inspection Note of Forest Dept. made by Forest Officer, Rai Sahib P. N. Mukherjee, 29ᵗʰ March1933 & 30ᵗʰ March1933. Para-2&6.

67 MDR, Inspection Note of Forest Dept. made by Forest Officer, Rai Sahib P. N. Mukherjee, 27ᵗʰ March1934. Para-6.

68 MDR, Inspection Note of Forest Dept. made by Forest Officer, N. L. Bhagat, 17ᵗʰ March1935, Para-6.

69 MDR, Inspection Note of Forest Dept. made by Forest Officer, N. L. Bhagat, 29ᵗʰ March1935, Para-6.

70 MDR, Inspection Note of Forest Dept. made by Forest Officer, S. N. Majumdar, 31st March1936, Para-4.

71 MDR, Inspection Note of Forest Dept. made by Forest Officer, D. P. Sharma, 1940, Paras-2–3.

72 Ramachandra Guha, *The Unquiet Woods*, op. cit; Mahesh Rangarajan, *Fencing the Forest*, op. cit.; Ajay Skaria, *Hybrid Histories*, Delhi: OUP, 1998.

73 K. Sivaramakrishnan, *Modern Forests*, New Delhi: OUP, 1999, pp. 76–78.

74 Rajan, Ravi, op. cit., p. 346; Mahato, N.K, 'Environmental Change and Forests', op. cit., p. 269; Mahato, N.K. *Sorrow Songs of Woods: Adivasi-Nature Relationship in the Anthropocene in Manbhum*, New Delhi: Primus, 2020, pp. 116–118.

75 MDR, D.O. No.151, From Hardyal Singh, Divisional Forest Officer, To, Sahib P. N. Mukherjee, Forest Officer, Manbhum, Date: 26th January,1933, Para-2.

76 MDR, Inspection Note of Forest Dept. made by Forest Officer, S.U. Majumdar, 23ʳᵈ February,1933, Para-5.

77 MDR, D.O. No.2846/22(12), From Hardyal Singh, Divisional Forest Officer, To, Sahib P. N. Mukherjee, Forest Officer, Manbhum, Date: 29ᵗʰ February1934, Para-4.

78 Ibid., Para-6.

79 MDR, Inspection Note of Forest Dept. made by Forest Officer, Rai Sahib P. N. Mukherjee, 26ᵗʰ March1934. Para-2.

80 MDR, Inspection Note of Forest Dept. made by Forest Officer, N. L. Bhagat, 29ᵗʰ March1935, Para-7. See also PDRRC, Inspection Note of Forest Dept. made by Forest Officer, D.P. Sharma, 1940, Para-4.

81 V M Ravi Kumar, 'Colonialism and Green Science: History of Colonial Scientific Forestry in South India, 1820–1920', op. cit., p. 246.

82 MDR, A letter From Hardyal Singh, Divisional Forest Officer, To, Sahib P. N. Mukherjee, Forest Officer, Manbhum, Date: 26th January1933, Para-3.

83 MDR, Inspection Note of Forest Dept. made by Forest Officer, S.U. Majumdar, 23ʳᵈ February, 1933, Paras-7–8.

84 MDR, D.O. No.2846/22(12), From Hardyal Singh, Divisional Forest Officer, To, Sahib P. N. Mukherjee, Forest Officer, Manbhum, Date: 29ᵗʰ February 1934, Para-4.

85 MDR, Inspection Note of Forest Dept. made by Forest Officer, D.P. Sharma, 1940, Para-4.

86 MDR, Inspection Note of Forest Dept. made by Forest Officer, D.P. Sharma, 1940, Para-4; Mahato, N.K, 'Environmental Change and Forests', op. cit., pp. 274–75; Mahato, N. K. *Sorrow Songs of Woods: Adivasi-Nature Relationship in the Anthropocene in Manbhum*, New Delhi: Primus, 2020, pp. 119–20.

87 MDR, D.O. No.2846/22(12), From Hardyal Singh, Divisional Forest Officer, To, Sahib P. N. Mukherjee, Forest Officer, Manbhum, Date: 29ᵗʰ February 1934, Para-2.

88 Ravi Rajan, 'Imperial Environmentalism or Environmental Imperialism? European Forestry, Colonial Foresters and the Agendas of forest Management in British India, 1800–1900', in Richard H Grove *et al* (eds.), *Nature and the Orient*, op. cit., p. 357.

89   K Sivaramakrishnan, 'A Limited Forest Conservancy in South West Bengal, 1864–1912', *Journal of Asian Studies*, Vol. 56, No. 1, 1997, p. 87.

90   Ibid, p. 78.

91   Bihar State Archives, Board of Revenue, File no. S-5 of 1930, Nov., Nos 24–29, Board of Revenue, Bihar and Orissa. E- Enclosed(s) to Progs. no. 24. No. 5983, dated Purulia, the 10th October 1928, From A.F.A Hamid, Esq. I.P.S. To the Deputy Commissioner of Manbhum, Manbhum, Sub: Abolition of Ghatwals, para-11.

92   *Centenary Commemoration Volume, West Bengal Forests*, op. cit., p. 163. Mahato, N.K, 'Environmental Change and Forests', op. cit., pp. 276–277; Mahato, N. K. *Sorrow Songs of Woods: Adivasi-Nature Relationship in the Anthropocene in Manbhum*, New Delhi: Primus, 2020, pp. 120–121.

93   Madhab Gadgil and Ramachandra Guha, *This Fissured Land*, op. cit., 1992, p. 193.

94   Proffenburger, Mark and Chhatrapati Singh, 'Communities and the State: Re-establishing the Balance in the Indian Forest Policy', in Mark Proffenburger and Besty McGean, eds, *Village Voices, Forest Choices*, op. cit. p. 60.

95   Madhab Gadgil and Ramachandra Guha, *This Fissured Land*, op. cit, pp. 186–193.

96   K. C. Roy Choudhury, 'The Forests of the Southern Circle: Its History and Management', in *West Bengal Forests: Centenary Commemoration Volume*, Calcutta, 1966, p. 134.

97   In the case of timber species *thalis* or plates were used and here a mixture containing tank silt, farmyard manure and application of balance nitrogen, phosphorus and potassium (N P K) fertilizer were provided.

98   K. C. Roy Choudhury, 'The Forests of the Southern Circle', op. cit., p. 139

99   Madhab Gadgil and Ramachandra Guha, *This Fissured Land*, op. cit, p.194.

100 K. C. Roy Choudhury, 'The Forests of the Southern Circle', op. cit., pp. 135–136.

101 Madhab Gadgiland Ramachandra Guha, *This Fissured Land*, op. cit, p. 194

102 Ibid., K. C. Roy Choudhury, 'The Forests of the Southern Circle', op. cit., pp. 135–136. Mahato, N.K, 'Environmental Change and Forests', op. cit., pp. 278–279; Mahato, N. K. *Sorrow Songs of Woods: Adivasi-Nature Relationship in the Anthropocene in Manbhum*, New Delhi: Primus, 2020, pp. 121–124.

Section IV
Medical encounters

12 When man meets medicine

Some reflections on *The Death of Ivan Ilyich* and Āyurveda with its epistemological consequences

Jayanta Bhattacharya

Introduction: The case of Ivan Ilyich

In Tolstoy's *The Death of Ivan Ilyich* (1886, translated by Louise and Aylmer Maude), the doctor seemed to imply "if only you put yourself in our hands we will arrange everything -- we know indubitably how it has to be done, *always in the same way for everybody alike*." But he was obsessed with the question if his case was serious or not. The real question was to decide between a floating kidney, chronic catarrh, or appendicitis. We are faced with a number of problems related to medicine, health, body, and disease arising out of reading this classic. If Ivan Ilyich is eager to know of organ localization of his disease, the doctor appears to be omnipotent (and omniscient too) regarding medical decision. Ivan tried to "translate those complicated, obscure, scientific phrases into plain language." The assured authority of the doctor was irrupted by a contradiction drawn from the examination of urine and the symptoms that showed themselves. Finally, "Reviewing the anatomical and physiological details of what in the doctor's opinion was going on inside him", Ivan understood it all. He began to think of the operation that had been suggested to him. To him, "It's not a question of the appendix or kidney, but of life and …death." The observable signs and the patient's symptoms were increasingly matched to findings of pathological science, as in this case and, also, in tandem, with basic tenets of modern medicine. It may be profitable to take into account of German experience during the early nineteenth century. It shows how middle-class doctors imposed their bourgeois ideals of selfhood on suffering peasants. "The person contrasted radically from the self which physicians experienced and expected as a sign of health, namely a modern, secular, and individual self–unitary, self-bounded, internalized, responsible, and cut off from direct divine intervention."[1]

We, the readers of this story, become once again convinced of the fact that the body is a three-dimensional space inside which organize the disease. The *person* of the hapless, wretched, poor fellow Ivan Ilyich transforms into the *pathology* inside the body, with its temporal swings expressed in physiology. Ivan died with all his illness narrative in a domestic setting. He

DOI: 10.4324/9781003241980-17

was living in the era of "The disappearance of the sick-man from medical cosmology, 1770–1870."[2] He was solitary as well as alone. Seen from another perspective, he was not alone as well. American experiences of the mid-nineteenth century make us believe that people like Ivan Ilyich died at home in their beds. "As recently as 1945, most of deaths occurred in the home. By the 1980s, just 17 percent did."[3] Gwande trenchantly comments, "We did little better than Ivan Ilyich's primitive nineteenth-century doctors – worse, actually, given the new forms of physical torture we'd inflicted on our patient. It is enough to make you wonder, who are the primitive ones."[4] Truly speaking, the doctors had a white coat on; they had a hospital gown – both uncertain about how to manage the terminal moments of one's life. Hospitals were reserved for the indigent or those without friends or family and were sites of death rather than cure.[5] Regarding formative years of medical education, as Gwande further brings to our notice, "They are spent in institutions – nursing homes and intensive care units – where regimented, anonymous routines cut us from all the things that matter to us in life."[6]

Coming back to Ilyich, "He wept on account of his helplessness, his terrible loneliness, the cruelty of man, the cruelty of God, and the absence of God." Ilyich seems to be sincerely in search for some metaphors which could fill in the vacuum of his excruciating pain and long drawn illness. Did he also think of a few moral and ethical questions which could redress his suffering? We are not sure.

At this point, to keep in mind, problems may arise when a metaphor expands in a sphere where it is not challenged or complemented by other equally powerful metaphors which are also expanding. In that case, the metaphor in question may go on expanding its application almost indefinitely. As a result of the declining vitality of religious metaphors in Western, or at least European, public discourse, metaphorical ideals such as "healthy behaviour" and "mental health", propounded by doctors and others who are perceived to be "objective" and to have no ideological axe to grind, have expanded to fill the vacuum as it were.[7] Dostoevsky solemnly observed,

> You see, gentlemen, reason is a good thing, that can't be disputed, but reason is only reason and satisfies only man's intellectual faculties, while volition is a manifestation of the whole life…life frequently turns out to be rubbishly, all the same it is life and not merely the extraction of a square root[8]

Since the era of "living anatomy" of Boerhaave by which he tried to unify anatomy and physiology, anatomy or, more precisely, pathological anatomy became the central question of medical metaphors.[9] Simply put, death is a part of human experience, a necessary consequence of life that we all must face. Although death is a rite of passage in which we will all participate – as family members, providers, or eventually, patients – we understand little of what is valued at the end of life. For example, a search of the medical literature published in 1996 identifies 112 papers that contain the keyword "death", a topic that directly affects everyone. In contrast, a similar search reveals more than 1,000 references pertaining to schizophrenia, a disorder with a population prevalence of about 1%.[10] "The exile of the dead" began with the insinuation of commerce into the world of death. So, it is no wonder that atrocity stories are central to people's talk about their encounter with the medical profession. In another poignant observation, "The white man's image of death has spread with medical civilization and has been a major force in cultural colonization."[11] Precisely, a novel image of death has emerged.

In Ilyich's case, nay in the modern world too, the entire cosmos of everyday life seems to be completely filled with metaphors of fabricated "health and youth" of the commodity world or objective scientific metaphors which have destroyed traditional morality and the normal range of predictable moral expectations derived from religion or interpersonal subjective network and bondage. It could not easily confront the moral problems generated by these new social relationships. A vacuum was yet to be filled as perceived in Ilyich's reflection on the absence of God. "To become a patient is to establish a healing relationship with another who articulates society's willingness and capability to help."[12]

We may try to understand how this relationship of health and its concept, unlike Euro-American paradigm of health–commodity–physician, are embodied in physician–healer–social assistance–neighbourliness–community paradigm in Indian medicine.[13]

Gadamer seems to make us aware that modern medicine is so geared to understanding disease and fighting illness that we seldom stop to think about our goal, namely, health. Health manifests itself, as Gadamer notes, by escaping our attention. Hence, the important task of thinking about health requires a patient and disciplined mind, sensitive to subtlety.[14]

At this juncture, it should be noted that in no other journal than the *New England Journal of Medicine* (*NEJM*), the way we see health and medicine in the twenty-first century has been critiqued in clear terms, "Faith in reductionism, which was infused into medicine in the 20th century, has empowered medical research to pursue only isolated problems and to yield targeted, immediately deployable solutions." In this conceptualization, what is missing is "whole-person approach focused on long-term functional status."[15] Another medical philosopher reminds us that within modern Western society, the highly individualistic culture and religious decline linked with medicine's reluctance to relinquish an outmoded form of scientific rationalism can act as reductive influences, stifling conceptual development.[16]

Technology, medicine, and human body

Curiously enough, Ivan Ilyich is archetypal of the quasi-abstracted individual on the canvas of modern medicine, especially in the hospital setting:

The hospital is an intimidating environment for most individuals. Hospitalized patients find themselves surrounded by air jets, buttons, and glaring lights; invaded by tubes and wires; and beset by the numerous members of the health care team – nurses, nurses' aides, physicians' assistants, social workers, technologists, physical therapists, medical students, house officers, attending and consulting physicians, and many others...It is little wonder that patients may lose their sense of reality.[17]

Technologically, every time the stethoscope was (and is) applied to a patient, it reinforced the fact that "the patient possessed an analyzable body with discrete organs and tissues which might harbour a pathological lesion."[18]

Physicians often become the only tenuous link between the patient and the outer world. To accomplish his humane task, the physician does need to incorporate the question of ethics in the realm of medicine and to build a strong personal relationship with the patient. Ruth Richardson reminds us,

> The need for ethical awareness and ethical behaviour applies to *all* dimensions of medicine, not just the practice of clinical medicine. Pathology cannot hold itself somehow distinct from, or immune to, a movement that has been gathering pace since the Nuremberg Code.[19]

Seventeenth-century stalwarts of medicine, like Sydenham and Boerhaave, depended more on clinical history than on technology. Boerhaave instructed,

> Everything pertaining to the case must be listed; nor that least thing neglected which a critical Reader might rightly seek to understand the malady...there must be arrangement according to the surging change of events, and each event must be recorded in its proper place.[20]

Curiously, in 1829, the *Lancet*, expressed concern that the stethoscope could also lead to eavesdropping.

> Auscultation Extraordinary.
> Quoth Rodrick I'll a place contrive
> So dark and safe, no man alive
> Shall to our private meetings grope
> Egad, cries Johnny, that won't do,
> If there's no crack to listen through
> They'll make reports by stethoscope.[21]

Technology is not only constitutive of the models of health and disease. It provides also for their metaphors. Furthermore, with the application of artificial organs such as pacemakers, cochlear implants and advanced limb prosthesis, technology becomes a part of humanity's physical existence, that is, there is a fusion of human being and technology.[22] Before the eighteenth century, medicine was based on the patient's narrative of the symptoms. In addition to this subjective portrait of the illness, the physician observed the patient's appearance and behaviour as well as any signs of disease. During the eighteenth and nineteenth centuries, medical instrumentation enabled and extended the physical examination of patients which made the physician less dependent on subjective narration.[23]

Contrarily, if, rather than as a technological object, society and science view medicine or art of healing in an all-encompassing way, the problems, and, more importantly, the solutions will be understood following this line of thinking. The rise of pathological anatomy and Hospital medicine as the sheet anchor of modern medical knowledge[24] and, subsequently, surgical practice and technological innovations in Europe led to epistemological exclusion of mind and person from the purview of medicine.

In the era of Bedside medicine, cosmological analogies emphasized an image of the body "as a microcosm, a reality *sui generis* subject to its own peculiar laws of growth and decay, comparable to the macrocosm of the physical universe."[25] This particular observation on macrocosm (and, microcosm) of pre-Hospital Medicine period somewhat pertains to Āyurvedic view of medicine of which we shall talk about later on. Notably, in pre-Hospital medicine era, during the medieval period, European medicine was also a logically closed construct. Disease was seen to pass through three stages – first, a rough stage, stadium *cruditus*; second, the increasing stage, stadium *incrementi*; third, a *crisis* leading to a decreasing stadium *decrement*, or to death.[26] There are connotations somewhat similar to it in the important ninth-century Āyurvedic medical as well as nosological text the *Madhavanidāna*.[27]

However, with the advent of Hospital medicine, the sick man became a collection of synchronized organs, each with a specialized function. A dead person began to be conveniently called a cadaver, "as though that made it something different from a person who had died."[28] The body was assumed to be "de-personalised and biography-less" corporeal machine or an "animated corpse." An inhumane attitude of mind has pervaded medical dealings with a too trusting public. The attitude is inhumane because it denies our common humanity. Richardson suspects it may derive from the fact that many doctors learn in the process of becoming doctors to deny aspects of their own humanity.[29] In 1772, Robert Boyle expressed a view of the body that has characterized much of medical practice:

> I think the physician is to look upon the patient's body as an engine that is out of order, but yet so constituted that, by his concurrence with...the parts of the automaton itself, it may be brought to a better state.

In the eighteenth century, when doctors turned to mathematics to produce a Newtonian map of the body, the metaphor of hydraulic pumps was used to express human digestion and blood circulation. Archibald Pitcairne, an eighteenth-century British physician, is exemplary in this regard. In his close attention to Newton's method, ideas of causality, and theory of matter, Pitcairne showed himself more closely acquainted with Newton's thought than many, or most, of his contemporaries. "In his published work he used

Newtonian ideas and arguments, if not wholly Newtonian accounts, to explain such functions as secretion."[30]

This mechanical view of the body as a set of parts that can be manipulated, analysed, and enhanced has reached the ultimate in genetics in which the very object manipulated, assessed, turned into a product, and enhanced, is the DNA itself. "In particular, the science of genetics focuses on the smallest units of the body, studied in isolation."[31] Steven Rose argues, "The core issue is reducibility, which...comes not as second but as first nature to natural scientists."[32] While addressing the core question of Western medicine, Kleinman observes, "The entailments of monotheism foster a single-minded approach to illness and care" that has the "decided advantages of pushing medical ideas to their logical conclusion" and, consequently, "establishing criteria against which orthodoxy and orthopraxy can be certified."[33] Biomedicine differs from Chinese, Indian, or most other systems of medicine by its extreme insistence on materialism as the grounds of knowledge, and by its discomfort with dialectical mode of thoughts.

To speak in a rather harsher tone, through the insistence on the primacy of definitive materialistic dichotomies, biomedicine or Western/modern medicine results in a huge split "between the constructed object of biomedical cure – the dehumanized disease process – and the constructed object of most other healing systems – the all-too-humanly narrated pathos and pain and perplexity of the experience of suffering."[34] Groopman raises the problem at the level of medical education, curricula, and medical thinking. To him, his generation "was never explicitly taught how to think as clinicians. We learned medicine catch-as-catch-can...The trunk of the clinical decision tree is a patient's major symptom or laboratory result, contained within a box."[35] These analyses and observations may be conjoined with the stubborn fact that medicine is, at its core, uncertain science. In such train of thoughts, the body is understood to be inert and subject to modification by external forces. Contrarily, the body can be understood not as subject to external agency, but as "simultaneously an agent in its own world construction. Here the body is not treated as a discrete physicality external to the self. The body is intercommunicative and active."[36] Finally, following Turner, the moot question appears that the body has thus become one of the main battlegrounds on which the struggle to forge a critical perspective, adequate to the changing features of contemporary social, political, and cultural reality is being fought, especially in the medical model where oftentimes the body of the subject is the passive tablet on which disorder is inscribed.[37]

Ivan's story, in its extension, poses before us multiple layers of questions regarding subjectivity, person, metaphors of life, and the body of a patient.

The new horizon of medicine: risk factors

Importantly, technology has replaced and reshaped the traditional meaning of disease, for example, bodily pain (*dolorcorporis*), suspension of joy (*intermissiovoluptatum*), and fear of death (*metus mortis*). The disease has become independent of the subjective experience of the person, and technology has endorsed a new range of disease entities: asymptomatic diseases. Another case in point is *angina pectoris*. It was a descriptive term to be replaced by and transformed into ischemic heart disease – a purely pathophysiological term. The development of molecular biology is also a clear example of this. A great number of new disease entities are based on genetic abnormalities. A variety of genetic tests can detect diseases where the person tested does not feel ill. As a caveat, it may be argued that such descriptions are "not only debatable on its own merits but is also tautology – molecular research leads to more molecular insights than nonmolecular research."[38] It also suggests that our formal systems of medical discourse exclude certain categories of knowledge and speculation.

Quite earlier, Canguilhem insightfully noted,

> A vulgar hierarchy of disease still exists today, based on the extent to which symptoms can – or cannot – be readily localized, hence the Parkinson's disease is more of a disease than thoracic shingles, which is, in turn, more to than boils.[39]

Along with hierarchization of disease, another process of pathologization goes on operating. If the normal does not have the rigidity of a fact of collective constant, but rather "the flexibility of a norm which is transformed in its relation to individual conditions, it is clear that the boundary between the normal and the pathological becomes imprecise."[40]

But in the thirteenth- and fourteenth-century medical practices in Europe, we would come confront some interesting facts. Lanfranchi, the great surgeon of Paris, about the year 1300 is moved to write, "The physicians have abandoned operative procedures to laity, either as some say, because they disdain to operate with their hands, or rather, as I think, because they do not know how to perform operations."[41] Elsewhere, "Occasionally likewise some humor runs down (*reumatizat*) into the chest, spreading over the nerves of the breast or those of the spine between the vertebrae, and sometimes to other places."[42] Medicine was still based on humoral balance within the body and, presumably, harmony between microcosm and macrocosm outside the body. It did very well match with Āyurvedic practice till the late nineteenth century.

Despite the obvious triumph of a medical theory and practice grounded in the hospital, a new kind of medicine, Surveillance medicine, seems to be emerging. It involves a fundamental remapping of the spaces of illness which includes the problematization of normality, the redrawing of the relationship between symptom, sign and illness, and the localization of illness outside the corporeal space of the body. Rose makes us aware of this new characteristic of medicine, "A man may feel entirely well, but if those little squiggles on the paper tell the doctor that he has got trouble, then he must accept that he has now become a patient. That is a powerful persuader."[43] Epidemiology comes to play a big role in this new construction of knowledge about the disease. Moving away from the social origins, clinical epidemiology now focuses on individual behaviour, and the so-called risk factors on the one hand, and depends almost solely on purely statistical ways of communicating its findings. "Epidemiology now reflects the core assumptions of neo-liberalism: it is individualistic, and makes little or no reference to 'social factors', focusing rather on individual risk behaviour."[44]

A symptom or sign for Hospital medicine was produced by the lesion and consequently could be traced through pathological anatomy and used to infer the existence and exact nature of the disease. Surveillance medicine takes up these discrete elements of symptom, sign, and disease and subsumes them under a more general category of "risk factor" that points to, though does not necessarily produce, some future illness. Such inherent contingency is embraced by the novel and pivotal medical concept of *risk*. It is no longer the symptom or sign pointing tantalizingly at the hidden pathological truth of the disease, but the *risk factor* opening up a space of future illness potential. Risks are calculated and assessed in order to rationalize surveillance, and through surveillance risks are conceptualized and standardized into ever more precise calculations and algorithms. "With the 'problematization of the normal' and the rise of 'surveillance medicine', everyone is implicated in the process of eventually 'becoming ill.'"[45] Both individually and collectively, we inhabit tenuous and liminal spaces between illness and health, leading the emergence of the "worried well."

Techniques of surveillance and monitoring are part of medical diagnostics, epidemiological studies, aetiologic research, health care management; they also co-shape individual engagements with illness. In medicine, surveillance data come as digital anatomies for educational purposes and clinical diagnostics that subject the body to imaging techniques, but also as databases of patient collectives that are established in large-scale, at times nationwide, epidemiological studies. The constellations of the body and the medical gaze and their location in space were closely intertwined with specific epistemologies of medical science and practice at different historical periods.[46]

Symptom, sign, investigation, and disease thereby become conflated into an infinite chain of risks. A headache may be a risk factor for high blood pressure (hypertension), but high blood pressure is simply a risk factor for another illness (stroke). Under Hospital medicine, the symptom indicated the underlying lesion in a static relationship. The risk factor, however, has no fixed or necessary relationship with future illness; it simply opens up a space of possibility.

Moreover, the risk factor exists in a mobile relationship with other risks, appearing and disappearing, aggregating and disaggregating, and crossing spaces within and without the corporal body. Pathology in Hospital medicine had been a concrete lesion; in Surveillance medicine illness becomes a point of perpetual becoming. The implication is that *self and community begin to lose their separateness*. The new dimensionality of identity is to be found in the shift from a three-dimensional body as the locus of illness to the four-dimensional space of the time-community.[47]

To note, with the rise of modern public health concept in Europe, there was a division between anatomical space of the body and environmental space. Environmental space was assumed to be the source of incriminating and polluting agents invading anatomical space of the body. Hence, as a logical derivative, this space was to be aggressed, controlled, cleaned, politically corrected, and reconstituted according to the need of the community. Like the body, this space is also repairable.[48] The two spaces are poised against each other. Contrarily, "Ayurveda represents two sciences in one: a biogeography absorbed into therapeutics...a discourse on the world (natural history) is contained within a discourse on man (medicine)."[49] Through its own explanatory logical system, onto an inventory of living creatures, in itself of little importance, "there has been grafted a metalanguage which makes it possible to formulate value judgments."[50] It may be wistfully traced that the question of "value judgment" is conspicuously absent in modern medicine, better not to speak of Surveillance medicine.

Taking up the previous question of the body in Surveillance medicine, this particular perception of extra-corporeal space comes closer to Āyurvedic view of the body and disease causation, where *karma* and extra-corporeal factors may negatively affect a human body. Hence, it may be conjectured that Indian population, especially Hindus, is amenable to the new paradigm of medicine. Though, it must be stressed, at the turn of the twenty-first century, there is nothing like pristine or sacrosanct Āyurvedic medicine or view of life (if it existed at all at any point of time).

Rahul P. Das points to the possibility of the existence of "even several Āyurvedic systems."[51] In modern Āyurvedic practices, the anatomo-clinical method, the positivist representation of an objectified reality, and the dualism of inside and outside are not rejected. Contrarily, according to Langford, "rather subjected to a variety sometimes calculated, sometimes casual maneuvers that subvert, invert, and otherwise play with modern episteme."[52] Despite all these troubling issues, the Āyurvedic assumption of the identity and nature, especially for the vast non-metropolitan population of India is a logical consequence of the leitmotif of Indian world view that "asserts an underlying unity of in the apparent multiformity of creation and strives for a transcendence of dualities, dualities, oppositions and contradictions."[53]

Āyurveda: a journey compliant with modern medicine

The question of ethics (and morality) is ingrained within the cosmos and philosophy of Āyurveda. "Good, evil, happy and unhappy is Life. That (knowledge) in which are declared its nature, and measure, and what is beneficial to it and what injurious, is called the Science of Life"[54] (*Caraka-Saṃhitā*, Sūtrasthāna, 1.40; hereafter CS). Again, "The union of body, senses, mind and soul is called Life. The latter is known again by the names of *Dhāṛ*, *Jīvita*, *Nityaga* and *Anubandha*." (CS, Sū, 1.41). Though, to note, Wujastyk cautions us against the "unwaveringly male gaze" of Āyurveda.[55]

In fact, Āyurveda is not a system of medicine but a dynamic philosophy of life by which one can attain healthy individual and social life so as to perform the functions efficiently and fulfil the social obligations fully, at the end to attain a perfect bliss of liberation.[56] Because Āyurveda constitutes a blend of Vedic metaphysics and traditional pre-modern science "it has earned its high place among the learned and intellectually unique accomplishments of Indian civilization."[57]

Āyurvedic medicine largely relied on its own system of physiology and pathophysiology to explain the nature of humans and disease. Treatment was also decided on this basis, the aim being to restore the normal state. One of the most important features of Āyurveda is the doctrine of three *doṣa*s. The *doṣa*s regulate physiological processes, but they may also initiate pathological processes. The pathophysiological process starts usually in the region of the affected and then spreads to other parts of the body. At this stage of disease development *vāta* plays an important role, as it alone has the capacity to cause movement. If this process continues, the *doṣa* coalesces with other body constituents and leads to a localized disease. In this context, the term *doṣa* is employed in a sense close to its literal meaning (i.e. vitiator); the affected body parts are then summarily termed *dūṣya* (the one to be affected/vitiated). The qualities associated with the *doṣa*s are also important for dietetics and therapy as actions or drugs of the opposite quality treat the respective increase.[58]

Evidently, such a system and view of medicine does not need any precise anatomical knowledge, or any knowledge of organ localization of disease. It does have its own explanatory model premised on (1) a bodily frame with channels through which *doṣa*s, *dhātu*s, and *mala*s flow, (2) surgery without anatomical knowledge but based on *marman*s which served the purpose of regional anatomy, and (3) unaided sensory perception. Without the benefit of knowledge of anatomy, physiology, and organ localization, ecology was an integral part of diagnosis of disease and patient's environment, including flora and fauna, that enabled the doctor to anticipate the course of disease and to take action on it. More interestingly, since even minor surgical procedures like "bloodletting (the fifth of the evacuant therapies) has fallen into disuse, it was removed from the set of *pañcakarman*, and replaced by oily enemas."[59] This is a clear pointer to the question of whole scale rejection of surgical practice (and anatomical too) from the domain of scholastic mainstream Āyurveda. Though, to emphasize at this juncture, nature was always seen to be in harmony with man's living. It becomes apparent that there exists tangible and welcome absence of the idea of mastery and domination over nature.

With the growth and institutional fortification of European medicine in India – the principal "panoptical" site being the Calcutta Medical College – there occurred a violent change in sign systems prevalent in social cosmos. "Among the first results of its foundation was the abolition of institutions in which Vedic and Unani systems of medicine were taught."[60] Modern anatomical knowledge came in the guise of indigenous one – "Once placed in a Sanskrit dress, the European system of anatomy would be accessible all over India for subsequent transfer into Hindi dialects of every province if requisite..."[61]

Importantly, an idea (symbol) is brought into reality indexically, and once there emerges socially in material reality, where its use might spawn a rethinking of the symbol, a new idea, an idea that might change other ideas, change habits, and hence change "actual behavior in the outer world" in a continuous dialectical process.[62] It is through the reconstruction of the indexical parts of a sign system the entire symbolic order can be reconstructed insidiously – without changing the sign-uses of a local cosmos. The indexical parts of these semiotic symbolic assertions, expressed metonymically through actions in the life-world, can be appropriated to reconstitute the existing ones explained mytho-historically. Thus, without changing the prevailing icons of an indigenous society cultural hegemony can be insidiously constructed.[63] During a period of violent change in sign-system (as was the case during colonial medical encounter), defining characteristic of the symbol becomes the overflowing of the signifier by the signified. Within a context of cultural change and during an asymmetric exchange of social forces, some forces spilling out of the domain of social exchange continually escape it. These are floating energies that are not yet fixed or invested in techniques and signs. "The floating signifier is also an explanatory principle for indigenous thought."[64]

Terms and images plucked from the colonial language of medicine and disease began to infiltrate the phraseology of Indian self-expression (or, put otherwise, Indian subjectivity), to become part of the ideological formulation of a new nationalist order.[65] These terms and images, quite explicitly, were primarily moored on superiority of anatomical knowledge, excellence of surgical practices and, at a later period, diagnostic and therapeutic marvels. In Meulenbeld's observation,

> The revival (of Āyurveda) thus led to the construction of a unitary and coherent model of Indian medicine, weaned from inconsistencies and untenable concepts, and, particularly, as free from magical and religious elements as possible. The ancient terms for physiological and pathophysiological processes, nosological entities, etc., were diligently re-interpreted to bring them into line with terms derived from Western medicine. These procedures resulted in the appearance of a type of āyurveda that can best be designated as navyāyurveda or neo-āyurveda[66]

Medicine, during the colonial period, as it can be extrapolated from these analyses, was intricately related with metaphors of higher civilization.[67] There was abundance of terms denoting invasion, war, and killing. Terms like "microbe hunter", "disease killers" were usual choice for microbiologists and physicians of the day.[68]

An example would help us to get at the point. A peasant of Birbhum region of Bengal wrote to a person (to whom he owed some money):

"Wreaking evils on you, I went to town. Hence, I have contacted an alien disease. You must know it…And, on getting cured, I must fully repay the debt whatever I owe to you." This is an account of the "social body" with its own logic, rules, metaphors, and ethical demands. It is also an account of the subaltern. Dwijendranath, the eldest brother of Rabindranath Tagore, provides another differing contemporaneous "elite" account, "Treatment by any means is a wild goose chase! Hence, better not to say anything about *kavirajichikitsa* (Āyurvedic treatment), even the shimmering rays of the nineteenth century knowledge has failed to penetrate its windows." He continues, "Modern medicine starts with *dissecting a corpse*, Āyurveda starts with elaborating on the relationship between *body and mind*." Inspired by "modernity", he innovatively uses the Āyurvedic categories like *vayu, pitta, śleṣmā* (wind, bile, and phlegm) to interpret superiority of western intellect. In his opinion, persons like Danton belong to the category of *pitta* or bile and represent "social dynamics." Finally, he concludes, "By the raging light and scorching heat of English education, orthodoxies are being increasingly banished from metropolis to the fringes of villages."[69]

We can see the emergence of the "medical/anatomical body" as distinct from the "social embedded body" of the former account. Here remains a subtle difference in the conceptualization of the body *as diseased* in modern medicine, and the body *in disease* in popular or Āyurvedic perception that finally transforms into the "diseased body" (a hybrid) in metropolitan Bengali perception. While in the former, the body is embedded in its environs, in the latter, it is the disengaged circumscribed biological body.

Conclusions

Unlike modern medicine, a disease is something that is gradually made manifest. Prodromes (*pūrvarūpa*) develop into full-fledged symptoms (*rūpa*). Secondary affections (*upadrava*) are consequences of the basic morbid process. At the end of this process, recovery takes place or fatal signs (*ariṣṭa*) appear, foreboding death.[70] As these ancient had little to offer effective remedies for any critical condition, prognosis was weighed over diagnosis. It is pertinent to remember what Edelstein observed in Greek context – "while carrying on his business he was not a "scientist" applying theoretical knowledge to the case at hand…these doctors certainly were among the finest examples of unsolicited curiosity and delight in learning."[71] They did not experience (or practice) clinical detachment or medicalization of life or objectification of the body. They were healers. To them health was not viewed from the perspective of absence of disease. They perceived *svasthya* (nearest English equivalent being Health) in the positive sense – to revert to one's original state. Etymologically, *svasthya* connotes *sva+ sthā+ ya* (to get reconstituted to the normal state).[72] They tried for it, often with success, many more times with failure.

One example may be cited here. In the *Suśruta-saṃhitā*, it is categorically mentioned that there are seven hundred sirās by which the body is nourished like garden with water-carriers and like a field by irrigating channels and benefitted with activities such as contraction, extension, etc. Their ramifications are as venation in a leaf, their root is umbilicus wherefrom they spread upwards, downwards and obliquely (Śārīrasthāna, 7.3)[73]

Our attention, as discussed above, should be drawn to the fact of harmonious understanding of man and nature on the one hand, and ecology and physiology on the other. In our aggressive age of biomedicalization and the taming and trimming of nature to the fulfilment of man's inflated prowess and avarice, understanding of this philosophical position can be of some help to assuage as well as share dehumanization of the person of the patient. Tambiah reminds us, "Man's creative freedom consists precisely in his ability to devise cultural perspectives and meaning systems in form and content that cannot be wholly and significantly understood in terms of any objective logic of adaptation."[74] In many ways, Āyurveda represented India subjectivity as well.[75] This particular character of subjectivity – as we come to realize – was reconstituted following anatomical encounter between Āyurveda and modern medicine. A new hybridized genre of Āyurvedic practitioners – Āyurvedic in the garb, but practicing European mode of medicine – began to emerge. Again, beyond clinical detachment or medicalization of life or objectification of the body, there remained another group of Āyurvedic practitioners, to whom health was not viewed from the perspective of absence of disease. They perceived *svāsthya* in a sense, as we come to realize, in a distinctly different sense vis-à-vis modern medicine. At this juncture, we may try to "rediscover the old skills of treating the whole patient rather than just the diseases."[76]

The agony of Ivan Ilyich might be felt from this position of medicine. Here, medicine is not a theory pertaining to "curing machine." Rather, it embraces both the art of healing and the prudence and development grounded on the path-breaking researches of modern medicine. "Ignoring cultural values and social concerns expressed in disputes is not in the best interest of either science or society."[77] Do we, perhaps, need a form of "philosophical spirit" to transcend some of the limitations of our enlightenment heritage to begin the demedicalization of society? If so we must develop courage of a type Kant refers to "the courage to avail yourselves of your own understanding – that is the motto of the enlightenment."[78]

Overcoming all attempts and manoeuvres, the epistemological struggle between modern medicine and empirical knowledge still goes on. The body always exceeds the power frame that attempts to control it. This exceeding is partly possible because of the internal conflicts and contradictions among the various discourses that attempt to control the body.

Notes

1 Quoted in Laura Otis, *Membranes*, Baltimore, London: Johns Hopkins University Press, 1999, p. 175 (n. 1).

2 N. D. Jewson, 'The disappearance of the sick-man from medical cosmology, 1770–1870', *Sociology*, Vol. 10, No. 2, 1976, pp. 225–244. Reprinted in *International Journal of Epidemiology*, Vol. 38, No. 3, 2000, pp. 622–633.

3 Atul Gawande, *Being Mortal*, Hamish Hamilton: Penguin Books, 2014, p. 6.
4 Ibid.
5 Kevin O'Neil, 'Disciplining the Dead', in Gail Weiss and Honi Fern Haber (eds.), *Perspectives of Embodiment*, New York and London: Routledge, 1999, pp. 213–231.
6 Atul Gawande, *Being Mortal*, op. cit., p. 9.
7 Kenneth Boyd, 'Disease, illness, sickness, health, healing and wholeness: exploring some elusive concepts', *Medical Humanities*, Vol. 26, No. 1, 2000, pp.9–17.
8 Feodor Dostoevsky, *Notes from Underground*, Ronald Wilks (trans.), New York: Penguin Books, 2009.
9 Roy Porter, *The Greatest Benefit to Mankind*, New York, London: W. W. Norton &Company, 1998, p. 250.
10 Jayanta Bhattacharya, 'Death, Embodied Approach and Medicine: Problematizing the Normative', in Makarand Pranjape (ed.), *Science and Spirituality in Modern India*, New Delhi: Samvad India Foundation, 2006, pp. 304–323.
11 Ivan Illich, *Limits to Medicine. Medical Nemesis*, London: Penguin Books, 1976, p. 180.
12 Marshall Marinker, "Why make people patients?", *Journal of Medical Ethics*, Vol. 1, No. 2, 1975, pp. 81–84.
13 Jayanta Bhattacharya, "The Body: Epistemological Encounters in Colonial India," in Peter L. Twohig and Vera Kalitzkas (eds.) *Making Sense of Health, Illness and Disease*, Amsterdam and New York: Rodopi, 2004, 31–54.
14 Hans G Gadamer, *The Enigma of Health*, trans. Jason Gaiger and Nicholas Walker, Oxford: Polity Press, 1996.
15 Farshad Fani Marvasti and Randall S. Stafford, 'From Sick Care to Health Care – Reengineering Prevention into the U.S. System', *NEJM*, Vol. 367, No. 10, 2012, pp. 889–891.
16 R. G. Evans, 'Patient centred medicine: reason, emotion, and human spirit? Some philosophical reflections on being with patients', *Journal of Medical Ethics: Medical Humanities*, Vol. 29, 2003, pp. 8–15.
17 *Harrison's Principles of Internal Medicine*, 18[th]edn, Vol. 1, New York: McGraw-Hill, 2008, p. 6.
18 David Armstrong, 'Bodies of Knowledge/Knowledge of Bodies', in Colin Jones and Roy Porter (eds.), *Reassessing Foucault*, London and New York: Routledge, 1994, pp.17–27.
19 Ruth Richardson, 'A potted history of specimen-taking', *Lancet*, Vol. 355, No. 9207, 2000, pp. 935–936.
20 Vincent J. Derbes and Robert Edgar Mitchell, Jr., 'Hermann Boerhaave's (1) Atrocis, nec Descripti Prius, Alorbi Historia (2): The First Translation of the Classic Case Report of Rupture of the Esophagus, with Annotations,' *Bulletin of the Medical Library Association*, Vol. 43, No. 2,1955, pp. 217–240.
21 Quoted in C. Peter W. Warren, 'The history of diagnostic technology for diseases of the lungs', *Canadian Medical Association Journal*, Vol. 161, No. 9,1999,pp. 1161–1163.
22 Bjorn Hofmann, 'The technological inventions of disease,' *Journal of Medical Ethics: Medical Humanities*, Vol. 27, No. 1, 2001, pp. 10–19.
23 Ibid., 14.
24 For a useful discussion on the rise of hospital medicine in India, see Jayanta Bhattacharya, 'The genesis of hospital medicine in India: The Calcutta Medical College (CMC) and the emergence of a new medical cosmology', *IESHR*, Vol. 51, No. 2, 2014, pp. 231–264.
25 N. D. Jewson, 'The disappearance of the sick man,' op. cit., p. 624.
26 H. A. M. J. Ten Have *et al* (eds.), *The Growth of Medical Knowledge*, Doerdrecht and Boston: Kluwer Academic Publishers, 1990, pp. 1–11.

27 G. J. Meulenbeld, *TheMadhavanidāna*, Delhi: MotilalBanarsidass, 2008, pp. 612–613.
28 Katharine Treadway, 'The Code,'*New England Journal of Medicine*.Vol. 357, No. 13, 2007, pp. 1273–1275.
29 Ruth Richardson, 'A necessary inhumanity?,'*MedicalHumanities,Vol.* 26, No. 2, *2000, pp.*104–106.
30 Anita Guerrini, 'Archibald Pitcairne and Newtonian Medicine,'*Medical History*, Vol. 31, No. 1,1987, pp.82–83.
31 Lori Andrews and Dorothy Nelkin, 'Whose body is it anyway? Disputes over body tissue in a biotechnology age',*Lancet*, Vol.351, No. 9095,1998, pp. 53–57.
32 Steven Rose, 'The rise of neurogenetic determinism,' *Nature*, Vol.373, 1996, p. 380–382.
33 Arthur Kleinman, 'What is specific to Western medicine?,' in W. F. Bynum and Roy Porter (eds.), *Companion Encyclopedia of the History of Medicine*, Vol. I, London, New York: Routledge, 1993, p. 17.
34 Ibid, p. 19.
35 Jerome Groopman, *How Doctors Think*, New Delhi: Byword Books Pvt. Ltd., 2011, pp. 4–5.
36 M. L. Lyon and J. M. Barbalet, 'Society's body: emotion and the "somatization" of social theory,' in Thomas J. Csordas (ed.), *Embodiment and experience*, Cambridge: CUP, 1994, p. 48.
37 Terence Turner, "Bodies and anti-bodies: flesh and fetish in contemporary social theory', in *Embodiment and experience*, op. cit., pp. 27–47.
38 Robert A. Aronowitz, *Making Sense of Illness*, Cambridge: CUP, 1998, p. ix.
39 Georges Canguilhem, *On the Normal and the Pathological*, Dordrecht, Boston, London: D. Reidel Publishing Company, 1978, p. 11.
40 Ibid, 105.
41 Henry E. Handerson, *Gilbertus Anglicus*, Cleveland, Ohio: The Cleveland Medical Library Association, 1918, p. 77.
42 Ibid, 44.
43 Geoffrey Rose, 'Sick individuals and sick population', *International Journal of Epidemiology*, Vol. 14, No. 1, 1985, pp. 32–38.
44 Kevin White, *An Introduction to the Sociology of Health and Illness*, London, Thousand Oaks, New Delhi: Sage, 2002, p. 61.
45 Adele E. Clarke, Laura Mamo, Jennifer R. Fishman, Janet K. Shim and Jennifer R. Fosket, 'Biomedicalization: Technoscientific Transformations of Health, Illness, and U. S. Biomedicine', *American Sociological Review*.Vol. 68, No. 2, 2003, p. 172.
46 Susanne Bauer and Jan Eric Olsen, 'Observing the Others, Watching Over Oneself: themes of medical surveillance in society', *Surveillance and Society*,Vol. 6, No. 2, 2009, pp. 116–127.
47 David Armstrong, 'The rise of surveillance medicine', *Sociology of Health &Illness*,Vol. 17, No. 3,1995, pp. 393–404.
48 For insightful discussion on this topic, see David Armstrong, 'Public Health Spaces and the Fabrication of Identity', *Sociology*,Vol. 27, No. 3,1993, pp. 393–410.
49 Francis Zimmermann, *The Jungle and the Aroma of Meats*, Delhi: Motilal Banarsidass, 1999, p. 20.
50 Ibid., p. 130.
51 Rahul P. Das, 'On the Nature and Development of 'Traditional Indian Medicine', *Journal of European Āyurvedic Society*,Vol. 3, 1993, pp. 56–71.
52 Jean Langford, 'Ayurvedic Interiors: Person, Space, and Episteme in Three Medical Practices', *Cultural Anthropology*,Vol. 10, No. 3,1995, p. 360.
53 Sudhir Kakar, *Shamans, Mystics and Doctors*, Delhi: OUP, 1998, p. 330.

54 Translations have been adopted from A. C. Kaviratna and P. Sharma (trans. And eds.), *Caraka-Saṃhitā*, Vols. 1–5, Delhi: Sri Satguru Publication, 2006.
55 Dominik Wujastyk, *The Roots of Āyurveda*.
56 P. V. Sharma, *Essentials of Āyurveda*, Delhi: Motilal Banarsidass, 1998. For a brilliant discussion on this issue seealso, Rahul P. Das, *The Origin of the Life of a Human Being*, Delhi: Motilal Banarsidass, 2004; Dominik Wujastyk, *The Roots of Āyurveda*; G Jan Meulenbeld, 'The Characteristics of a *Doṣa*', *Journal of European Āyurvedic Society*, Vol. 2,1992,pp. 1–5.
57 Horacio Fabrega Jr., *History of Mental Illness in India*, Delhi: Motilal Banarsidass, 2009, p. 336.
58 For a brief, yet insightful, discussion, see Ananda S Chopra, 'Āyurveda', in HelaineSelin (ed.), *Medicine Across Cultures*, New York, Boston: Kluwer Academic Publishers, 2003, pp. 75–83.
59 Francis Zimmermann, 'Terminological Problems in the Process of Editing and Translating Sanskrit Medical Texts', in Paul U. Unschuld (ed.), *Approaches to Traditional Chinese Medical Literature*, Dordrecht and Boston: Kluwer Academic Publishers, 1989, pp. 141–151 (149).
60 'British Medicine in India', *British Medical Journal*, May 25 1907, p. 1245.
61 'Proceedings of the Asiatic Society', *Journal of the Asiatic Society of Bengal*, Vol. 7, No. 2,1838, p. 663.
62 Charles S. Pierce, *Philosophical Writings*, in Justus Buchler (ed.), New York: Dover Publications, 1955, pp. 283–288.
63 Diane P. Mines,'From Homo Hierarchicus to Homo Faber: Breaking Convention through Semiosis', *Irish Journal of Anthropology*, Vol. 2, 1997, pp. 33–44.
64 Jose Gill, *Metamorphosis of the Body*, Minneapolis, London: University of Minnesota Press: 1998, p. 94.5.
65 David Arnold, *Colonizing the Body*, Berkeley: California University Press, 1993, p. 241.
66 G. Jan Meulenbeld, *A History of Indian Medical Literature*, IA, Groningen: Egbert Forsten, 1999, p. 2.
67 Gyan Prakash, *Another Reason*, New Delhi: OUP, 2000.
68 Laura Otis, *Membranes*, op. cit.
69 Jayanta Bhattacharya, 'Anatomical Knowledge and Eat-West Exchange', in Deepak Kumar and Rajsekhar Basu (eds.), *Medical Encounters in British India*, New Delhi: OUP, 2013, p. 52.
70 G Jan Meulenbeld, *The Mādhvanidāna*, p. 612.
71 Ludwig Edelstein, *Ancient Medicine*, Owsei Temkin and Lilian Temkin (eds.), Baltimore and London: The Johns Hopkins University Press, 1994, p. 351.
72 For elaborate discussion, see Minoru Hara, 'A Note on the Sanskrit Word *Svasthya*', *Journal of European Āyurvedic Society*, Vol. 4, 1995, pp. 55–87.
73 *Suśruta-Saṃhitā*, P. V. Sharma (trans.), Vol. II, Varanasi: Chaukhamba Visvabharati, 2005, p. 200.
74 Stanley J. Tambiah, *Magic, Science, Religion, and the Scope of Rationality*, New York: CUP, 1993, p. 153.
75 Sudhir Kakar, *Shamans, Mystics and Doctors*, op. cit.
76 Gordon Horobkin, "Commentary on 'The medicalization of life' and 'Society's expectations of health', *Journal of Medical Ethics*, Vol. 1, 1975, pp. 90–91.
77 Lori Andrew, 'Whose body is it anyway?', op. cit., p. 56.
78 Walter Kaufmann, *Critique of Religion and Philosophy*, New York: Anchor Books, 1961, p. 413.

13 Therapeutic practices of tuberculosis in the Madras Presidency, 1910–1947

B. Eswara Rao

Medical establishments – sanatoria, hospitals, and dispensaries – became a site of new medical knowledge practices and theories while disseminating therapeutic ideas and methods for tuberculosis treatment both in the institutional care and outside. Therapeutics provided respite to patients while expressing different connotations through their discursive practices while alliance with colonial hegemony over existing indigenous medical traditions and people. In turn, various responses and resistance came from indigenous and other medical systems (i.e., Ayurveda, Naturopathy, and Homeopathy) by way of asserting their professional interests of medical traditions, which led to therapeutic modernisation against western medicine. This medical encounter instigated interesting discourses on treatment methods and diet. Last two decades of historical research has focussed on colonial state policy and indigenous medical responses, assertion, modernisation, contestation, and reorganisation in various contexts.[1] Against this background, this essay seeks to address what clinical interactions and multiple levels of medical encounters in the process of the medical practice within and outside institutional space, which led to the emergence of interesting discourses on treatment methods primarily on drug, treatment methods, and diet. This essay also tries to understand how these institutions rationalised western medicine and its related "scientific" and medical ideas over oriental practices in the domain of tuberculosis care.

Western medical therapeutic practices

There was no distinct tuberculosis treatment used for in a sanatorium and in the home. The principles of treatment were "one and the same everywhere."[2] The treatment was based mainly on physical examination, X-ray, and blood examinations. The presence of clinical symptoms such as fever, loss of weight, increased cough, etc., is also associated with an "exudative inflammatory reaction." Temperature was considered one of the best guides in the treatment of tuberculosis. There was professional haziness existed among western medical practitioners over treatment and usage of drugs. A few drugs and injections used in the treatment of tuberculosis by general

DOI: 10.4324/9781003241980-18

practitioners. While highlighting this matter, P.V. Benjamin, Director of the Union Mission Tuberculosis Sanatorium, Madanapalle precisely stated that "Very few of these drugs have any scientific basis. The claims put forward by the unscrupulous advertiser are their only recommendations. Most of these drugs are perfectly useless and some of them are positively harmful."[3] One such drug was sanocrysin, which was prepared with gold. Benjamin found that it is not specific to tuberculosis, but it helps in clearing exudative inflammatory affections of the lungs. Very often, it checks the spread of the disease when ordinary rest treatment alone has failed. It was found that combination of collapse therapy with sanocrysin was "of great use" in mixed types of pulmonary tuberculosis where cavitation or fibrosis and exudative inflammatory processes co-exist.[4]

Both Indian and European western medical practitioners accepted the fact that gold had been used empirically by Ayurveda practitioners in the treatment of tuberculosis from the ancient period in India. Gold was used as a remedy for not only tuberculosis but also for a number of other chronic diseases in the ancient systems of medicine in India. Indigenous medicine prescribed gold for treatment as an oral medicine while western medicine used gold both as an injection and orally. The use of gold in the treatment of tuberculosis started in western medicine from the time of Koch's investigation into the tubercle bacillus. Dr. M. Kesava Pai, Director of the Tuberculosis Institute, Madras, argued that "rational application of gold" in the chemotherapy of tuberculosis was "comparatively [of] recent origin." As a catalytic agent, gold therapy was used under the name *triphal* and *krysolgan* for several years in Europe. Further, Kesava Pai stated that in other countries, for centuries, it was given orally mostly in its metallic insoluble state as a powder or as a confection in combination with other drugs. Modern physicians used it in soluble combination with arsenic orally. The use of metals in an organic solution through an intravenous route was developed in the 1920s.[5] Particularly, European doctor, H. Moellgard, on the therapeutic application of gold-salt did systematic experimental research. He made experiments on guinea pigs and calves. In 1923, the use of sanocrysin as a chemo-therapeutic agent in pulmonary tuberculosis was approved.[6] Ever since Moellgard advocated it, the use varied considerably in the hands of different medical practitioners. Later it was recognised as a successful treatment method of western medicine for tuberculosis. The use of this drug was started in 1926 and continued until 1946 in the Union Mission Sanatorium, Madanapalle [7] in the Madras Presidency. The study by Frimodt-Moller concluded that the treatment gave good results; for instance, 19 out of 27 patients were treated using sanocrysin in a small dose in the Sanatorium, Madanapalle. According to this study, 50% of the cases improved even in stage-III and 84% in stage-I of the disease. A different study by Dr. Kesava Pai revealed that good results were obtained in 31 out of the total 43 patients treated in the tuberculosis Institute, Madras.[8] Dr. Kesava Pai stated that as a remedy for tuberculosis and for a number of other chronic diseases gold had used in many of the ancient systems of medicine in India.[9]

Modern physicians used gold orally in soluble combination with arsenic.[10] In the Tuberculosis Hospital, Madras, treatment began with about 10 cgr and the doses were carried on to about 0.6 gram, but it rarely exceeded this quantity. The total quantity injected into one patient was varied from 1 to 8–50 gram with an average case getting about 5 grams.[11]

There were different medical opinions existed on gold therapy and its effects. A variety of gold came to be used because of change in the preparation. Important attempts were made to reduce toxic effects and increase therapeutic value. Originally, it was believed that gold had a bactericidal action particularly on the tubercle bacilli in the body, but further investigations showed that the positive results obtained were based on the natural defensive forces of the body. It acted like a catalytic agent bringing about the acceleration of the spontaneous healing process. The chemotherapy of tuberculosis was in essence a stimulation therapy.[12] Another view was held by western medical doctors about the negative impact of gold on the body. The use of heavy metals like gold was liable to produce symptoms of metallic poisoning if carelessly given to patients. It was also shown to have a "tendency to produce cumulative effects if given in very large doses or at frequent intervals or to persons who are unable to excrete it."[13] Further, the beneficial effect of gold was in the reduction of the quality of sputum and it helped to change its character. Western medical doctors recognised from Robert Koch's phenomenon that intestinal tuberculosis was a manifestation of an ulcerative type. In such cases, "gold is not likely to do any good." However, doctors acknowledged that gold was well worth trying in the primary type of intestinal tuberculosis and would give satisfactory results as a stimulation of fibrous tissues in the diseased parts.[14] Western medical practitioners argued that gold was not applicable for Indian conditions. The infection of bovine tubercle bacillus in India was rare; therefore, it was believed, intestinal tuberculosis occurs rarely. However, many cases of pulmonary tuberculosis existed. This was one of the reasons for not obtaining good results from gold treatment for pulmonary tuberculosis in India, though the intestinal complication in some degree were associated with the disease of the lungs.[15] Dr. K. Vasudeva Rao, a tuberculosis expert and medical practitioner from Madras argued that one dose of 3 grams was rather too heavy to give good results. He experimented in Madras with 1.5 grams. He found that there was a greater chance of complications occurring from the second course onwards. He concluded, "It is not safe to inject gold in large quantities in order to produce shock. Instead of doing well, it may do more harm." The experiment showed that intra-pleural injections had been discouraging after trying on 10 cases or so, and later, he had given up this type of injection.[16]

There was a disagreement among medical practitioners regarding the dosage of gold to be used in the treatment of tuberculosis patients. Dr. Ukil, All India Institute of Hygiene and Public Health, Calcutta, argued that gold therapy was useful in well-selected cases. Smaller doses would bring improvement and heavy doses would create reactions, including load on the kidneys; and generally, Indian patients tolerated a much smaller dose than European patients. Dr. Ukil was satisfied with the trial of cadmium sulphide.[17] Another participant in this conference argued that doctors should be rational and should study the therapeutic use of other chemotherapeutic drugs such as the sulpharilomide and sulphopyridine group of drugs. Dr. Ukil quoted the evidence that even in America well-known doctors expressed contradicting views and were entirely against gold therapy. Therefore, he warned medical practitioners not to give gold injections indiscriminately and neglect other better and definite methods of treatment like "collapse therapy." Another participant, Dr. Jones said gold treatment had many adverse and irregular results. In America, a large number of the medical practitioners did not accept this. According to Jones "we believe that gold may have some value" which should be classed as a "treatment" comparable to collapse therapy. He warned, "it should not be used instead of collapse therapy."[18] P.V. Benjamin argued that gold had been used in the treatment of pulmonary tuberculosis for 20 years in the United Mission Tuberculosis Sanatorium. Reiterating that even then "it is remarkable there is no agreement as to its value," he admitted that gold was used "too widely and indiscriminately, without scientific control very frequently." His study showed that positive results were observed in 1600 patients at the Sanatorium in Madanapalle, where gold treatment had been given with careful control. He confirmed, "Gold has a place in the treatment of tuberculosis."[19] He further argued that selecting the type of patients for gold treatment was important. This treatment had "definite value" in the case of "exudative cases" and gold could control the disease in the contralateral lung. C. Frimodt-Moller, superintendent of Union Mission Tuberculosis Sanatorium, Madanapalle, urged the medical practitioners to pay attention to the use of other drugs in the treatment of pulmonary tuberculosis. Drugs like carbolic acid, which was discovered by Sir James Roberts, were useful. However, Moller refused to try out carbolic acid treatment until experiments on animals were carried out in one of the recognised research Institutions. According to him, "there is no scientific foundation for this kind of treatment; its effect on lung tissue should therefore be carefully studied in animals before patients are exposed to experimentation."[20] However, it is worth noting that the treatment was used in Tambaram Sanatorium, Madras.[21]

Indigenous medical practitioners argued that gold could be used both in single and in combination with other drugs by correlating the aetiology and pathology of tuberculosis. Ayurvedic practitioners strongly believed that metals possessed immense therapeutic value in the treatment of tuberculosis. They differed with western medicine on the basis that western medicine conceived the treatment of tuberculosis in relation to the bacteria rather than the strengthening of various forces of resistance inherent in the body. Efforts had to be made towards augmenting the forces of resistance in the body to deal with not only the micro-bacterium but also the course of the pathological changes, which occur within the body when infected with the disease. However, disbelief existed among western medical practitioners on the use of gold and its therapeutic value in the treatment of tuberculo-sis.[22] Dr. Dwarakanath, LIM, Madras, argued that the

> result of animal experiments cannot be invoked as proof or as grounds
> for the rejection of the curative value of gold treatment, nor is any sta-
> tistical conclusion possible, the latter also being necessarily useless in
> other methods of specific and non-specific stimulus therapy.[23]

Dr. Dwarakanath conducted a research on patients who were in fairly advanced stage of pulmonary tuberculosis who attended the Government School of Indian Medicine, Madras, from 1938 until 1941. Of the 22 cases, two died and seven patients did not undergo the full course of treatment because they were discharged within a two-week period from the hospital. All comforts were provided and nutritious diet along with gold therapy was given orally. He argued that the majority of cases treated in the medical school even in fairly advanced cases did not show "undesirable reactions." This study proved that the resisting capacity of the "defence mechanism of the body" was important. This depends on the condition of life and on how the patient lives.[24] Moreover, there was no fixed dosage of gold to be used, it was strictly dependent on the condition of individual patients and varied from one to another.[25] He concluded that

> the destruction of the causative organism by chemical means is quite
> out of question and at the present state of our knowledge, the only
> thing that can be done with justifiable hope, is to stimulate the nat-
> ural defence mechanism inherent in the body, to deal, not only with
> the causative micro bacterium but also with the pathological changes
> which occur during the course of the disease, in the organ or organs
> involved.[26]

Ultraviolet treatment was used for advanced pulmonary tuberculosis patients. Generally, in 20–30% of advanced cases, it was found that the intestines or the larynx were affected. Swallowing of the bacilli-containing sputum caused intestinal tuberculosis. A certain percentage of those affected by this disease were treated with ultraviolet treatment. There was a common assumption that the treatment of pulmonary tuberculosis consists simply in the so-called open-air treatment. From what has been said above, it must be evident that nothing can be more wrong than this assumption. There is no doubt that the open air has a great influence in supporting the whole treatment.[27] This treatment was first used for non-pulmonary tuberculosis in 1928 in the Union Mission Sanatorium, Madanapalle.[28]

Before the 1930s, climatic cures, cod-liver oil, and colossal calcium were used in sanatoria and hospitals for the treatment of tuberculosis patients. By the 1930s, these were replaced with artificial pneumothorax[29] and surgi-cal treatment.[30] Majority of cases came for treatment in an advanced stage and were especially difficult to cure with open-air treatment methods. The flow of toxins from the diseased areas even during rest was too high. In such cases, the patients were treated through surgical process. The collapse therapy was applied to make it possible to rest the lungs. Various forms of collapse therapy existed in pulmonary tuberculosis: (1) artificial pneumo-thorax treatment, (2) phrenic nerve, evulsion, and (3) extra pleural thora-coplasty or pneumocystis. The aim of all these methods was to provide rest to lungs. The best results expected from these surgical methods were "when the disease is unilateral; when both lungs are affected their scope is limited."[31]

James Carson in Liverpool first used artificial pneumothorax in 1822 and it originated from the idea of inducing pneumothorax to cure tuberculosis of the lungs. Based on this idea, William Stokes experimented in 1837 and proved that in certain cases of pulmonary tuberculosis the application of artificial pneumothorax was useful. This method of treatment came to be used worldwide from 1925 and it added value due to its success in cases resistant to all other methods of treatment.[32] Dr. T.S. Shetty, ENT Specialist, ENT Hospital, Ramnad, argued in 1937 that

> when I am giving these facts it is not my intention to carry [you] to the domain of a complete cure in cent per cent cases. Many experts who have carried this type of collapse therapy had cures, failures and improvements in each individual cases [that] reacted to the therapy.[33]

From the 1930s, Phrenicotomy and Phrenic Evulsion[34] were extensively used in sanatoria all over the world and it proved very useful as supplement or even substitute to artificial pneumothorax.[35]

T. S. Shetty further argued that the sanatorium methods were much expensive, whereas most victims of this disease were people from the lower and middle classes. He did not agree with prescribing that patients go to the sanatorium:

> how many cases we meet with in our daily practice to whom something could be done and how many such cases can be directed to sanatori-ums and how many sanatoriums cater to the needs of the poor in our country?[36]

Therefore, the new treatment methods were much suitable to Indian con-ditions, he argued. The sanatorium treatment methods were successful in western countries because there people's minds were trained and every lay-man knew the nature of the disease. It was possible to get cases at very early stage in Europe unlike in India where the cases that came for care and treatment were usually in the far advanced state of the disease.[37] S. Krishna Swamy, medical practitioner of western medicine, Madras, stated that arti-ficial pneumothorax was "one of the most valuable and effective weapons against pulmonary tuberculosis." There were also limitations to its applica-tion because in acute cases the disease may spread to the other lung.[38]

According to Dr. M. Kesava Pai, artificial pneumothorax was "one of the best remedial measures known for the cure of the disease." He found in his study that about 72% of the cases in the advanced stages of pulmonary tuberculosis were cured with pneumothorax inflation. Only 6% of the partial pneumothorax cases showed negative results, which led him to conclude that in "the advanced stage of pulmonary tuberculosis artificial pneumothorax is the best method of treatment devised so far for the disease and should be practised wherever possible along with the other lines of treatment." The best results could be obtained with the combination of sanatorium and artificial pneumothorax treatment.[39] The advantages of artificial pneumothorax over earlier treatment methods exhaustive were its effects and efficacy: it gave quick relief even in unfavourable conditions of hospitals in crowded cities and in the homes of the patients.[40] He studied for 14 months ending of 31st July 1928, a total of 84 cases treated with sanocrysin at the TB Hospital, Madras. It was found that in stage-II, 58% of the cases were arrested and every case in this stage benefited, and in 40 % of those in stage-III, the disease was arrested, and 78% benefited. Based on these results, as Kesava Pai declared, there were "appreciable differences in the results obtained from sanocrysin therapy in the two kinds of the sub-acute form of pulmonary tuberculosis, viz., the exudative and fibro-caseous." They matched with those obtained by Frimodt-Moller at Arogyavaram but went contrary to what was usually used by tuberculosis doctors in some parts of Europe.[41]

Another study by Dr. K.S. Sanjivi, M.D. Government Tuberculosis Hospital, Madras, showed that more than 30% of the patients were cured. Initially, patients suitable for artificial pneumothorax were ultimately the candidates for major surgical procedures such as *thoracoscophy* and *pheumolysis, extrapleural pneumothorax* or *thoraeoplasty* but such major surgical work was done only occasionally.[42] For chemotherapy in the 1930s, extensive trials were done with Promin and sulphones. R. Viswanathan, Adviser on Tuberculosis, Government of India, recommended introducing chemotherapy[43] in India after his tour to Europe where he met doctors and researchers on chemotherapy. He argued that streptomycin was useful and showed "definite cures" "definite improvement" for tuberculosis. Therefore, he urged that it be made available to general practitioners and in certain tuberculosis institutions in India. But he recommended its use in small doses until experiments confirmed results.[44] In the 1940s, experiments were done with various drugs such as promin, promizole, sulphone, and streptomycin.[45] From 1949, streptomycin was used in Union Mission Sanatorium, Madanapalle.

Indigenous medical discourse and their assertion

Indigenous treatment methods for tuberculosis existed in India, as the disease was known from the ancient period. Particularly the Vedas mention treatment methods for tuberculosis. The Rig-Veda mentioned the disease and its treatment in the form of herbs. The kinds of herbs used were pale, dusky, tinted, red yellow and black coloured. The earth was considered as the mother of plants, the sun as the father, and the ocean as the root of all life. Surya (sun), Brihaspathi (jupiter), and Varun (God of Water) were believed to impart the medicinal properties to those herbs used for the treatment of tuberculosis (4000 B.C.). But a vivid description of tuberculosis and its treatment began to appear from the time of Charaka and Susrut. Susrut advocated a method of tuberculosis treatment with mostly vegetable products.[46] According to Aswinikumar, tuberculosis originated in Chandralok (Moon world). It was due to Chandra's (moon) excessive sexual indulgence and carelessness with respect to the laws of nature meant for the preservation of health. Brahma was an expert botanist and he deputed physicians to cure the disease of Chandra and others in the Chandralok. The colour of Chandra is coincidentally white, and his son Mercury is deep green in colour. It is inferred that the origin of the existing cause of tuberculosis primarily began in white or fair complexioned people, a condition in which the body had low resisting power.[47] Pigmentation of skin in man was considered as the best means of natural protection from the disease. Therefore, Aswinikumar attributed the origin of the disease to white people whose vital resistance could be lower than that of coloured people or animals. The features such as intelligent and fair complexioned young subjects and those who were having pigeon-shaped chest or barrel-shaped or depressed chest, and yellowish brown or brownish hairs on their back, suffer very easily from tuberculosis than people with other complexion.[48] Ayurveda broadly divided plants into two kinds from a nutritive point of view: nutritive (*algii*) and non-nutritive which was injurious fungi class. The first type has colouration and second type colourless leading a parasitic life over the former. Three important factors were seen as important for germination, growth, and maintenance of plants in life. The growth of plants and colouration are due to red, yellow, and blue rays.[49]

The different colouration of different plants is because of the absorption of different rays derived from different sources in the solar world. These rays act either singly or in combination on plants. For example, white light is the conglomeration of different rays from different sources having different effects on vegetables and animal bodies.[50] All living beings of the animal and vegetable kingdoms in nature depend solely on the interaction and absorption of the different coloured rays. The absorption of light is dependent on the substances, for example, air absorbs blue and partly ultraviolet, water absorbs ultra-red, red, and yellow. All rays can produce heat, light, and chemical actions. These actions vary according to the body on which they fall (either animal or plant). But blue and violet rays are chemically much stronger than green, orange, yellow, and red. According to Ayurveda, the Sun is the life of men, the moon is the thought, mars is the strength, mercury is the speech, Jupiter is the wisdom, Venues is the comfort and Saturn is the trouble.[51] The normal comforts depend on healthy conditions of a system in which there is a regulated supply of all the vitamins found in green plants by the absorption of white rays. According to Ayurveda, patient must eat and wear substances having white and yellow colours.[52] A healthy life depends on good condition of the blood and the nervous system. The normal condition of the nerves is maintained because of proper supply of yellow rays and normal condition of the blood is maintained by normal supply of red rays and sunshine. Therefore, Ayurveda instructed patients to eat and wear red and yellow substances.[53] From the time of Nagarjuna (150 and 200 CE), Ayurveda prescribed a combination of physical hygiene and metallic treatment for tubercular patients, particularly metals like gold and iron in combination with diamond, pearl, and similar substances in addition to vegetables. However, preference had been given to vegetable treatment because tuberculosis belongs to the Fungi class of vegetable kingdom and is colourless.[54]

According to Ayurveda, tuberculosis affects the body when *ojus* (vitality) comes down and then it slowly spreads to different parts of the body. The blood becomes impure owing to infection. To purify the blood, Ayurveda recommends the use of leaches to suck the impure blood from the body. A paste of chandan (sandal), athimathuram (*Glycyrrhiza glabra*), manjistha (*Rubia cordifolia*), and nasasesaram should be applied where the leach sucked to cure the injury. White goat's milk and ghee from such milk were prescribed food for the patient. The affected patients were instructed to be clean inwardly and outwardly and to brush their teeth. The patient's mind had to be free from worries. It was believed that people who live in a place with herds of white goats would not get tuberculosis.[55]

During the early 20th century, a revival movement of indigenous medicine began in the Madras Presidency. It started with the establishment of the Ayurveda College by D. Gopalacharulu in Madras in 1901.[56] The revival movement against western medicine was both defensive and assertive on the professional front. It reconstituted ayurvedic and other indigenous traditions. The movement gained political and social recognition with the Swadesi Movement. It promoted nationalist interests through its platform and gained the support of nationalist leaders. The movement also tried to standardise indigenous medicine and treatment methods according to modern chemistry and pharmacy.[57] It also as Poonam Bala argued that Indian encounters "produced new forms of structures for negotiation of traditional forms of knowledge"[58]

The Government of Madras appointed a committee headed by M. R. Ry Rao Bahadur M.C. Koman in 1918 to investigate the therapeutic properties of indigenous medical drugs, which were considered important by Vaidyans and Hakims. The report of Koman (1921) stated that some drugs were slow in curing and relieving the patient from suffering.[59] Immediately after submission of the report, indigenous medical practitioners' organisations, namely, Dravida Vaidya Mandali and The Madras Ayurveda Sabha made a counter report against Koman's report. M. R. Ry Vaida Visarada, K.G. Natasa Satri, R. Bharata Satrigal, and E.R. Srinivasa Raghava Chariar prepared this counter report. They described the Koman report as "highly inexplicable." The basic philosophical difference between the two medical systems allopathic and ayurvedic came to the fore through their conflict of views and opinions. Koman argued in his report that "I have to select suitable patients at all General Hospitals to whom to fit the medicine." Dravida Vaidya Mandali and Madras Ayurveda Sabha countered by stating that "learned doctors (western medical practitioners) obliged to prescribe patients for medicines instead of medicine for patients." Western medical practitioners should know the uses of medicines "in relation to their particular diseases."[60] For instance, *Talisadi Churnam* drug, as Koman argued, would give a certain amount of relief in pulmonary tuberculosis and in allaying cough, but the action was "very slow."[61]

The Dravida Vaidya Mandali and The Madras Ayurveda Sabha opposed this description of the drug and argued that "learned doctors" (western medical doctor) carefully avoided mentioning at what stage of the disease and at what physical condition of the patient the medicines were given. The effect of medicine/drugs depends on various circumstances particularly the patients' condition, temperament, and finally the stage of the disease. The Dravida Vaidya Mandali described that these descriptions were "good example of 'scientific obscurity' of the allopathic system as they were against the 'empirical obscurity' of Ayurveda."[62] They also opposed misinterpretation of another drug called *Adhatoda Vasika*.[63] According to learned doctors, this drug was commonly used as an expectorant and anti-spasmodic in Asthma, chronic Bronchitis, and consumption. Ayurvedic doctors remarked, "he (Koman) does not know the special use." This drug was a specific remedy in haemorrhage. Its impact was slight in pulmonary tuberculosis, more particularly since the effect was only on *pitta* type of pulmonary tuberculosis. However, the learned doctors did not believe this kind of disease classification. The evidence quoted from Vagbhata that "*Vrisha (vasaka)*" conquers haemorrhage is an excellent example specific for it.[64] The "learned doctors" described the knowledge of many *vaidyans* and *hakims* on drugs as "crude and unintelligent." Countering this criticism, Ayurvedic doctors maintained that "unintelligent [were] not *vaidyans* or *hakeems*" but the learned doctors. The sole fault of "learned doctors" was the absolute disregard of the terminology of the *vaidyans*.[65] Moreover, learned doctors failed to indicate for how long those drugs or medicines were used and in how many cases.[66]

The Ayurvedic drug called *Anacyclus pyrethrum* (pellitory)[67] was used in the treatment of three fairly advanced pulmonary tuberculosis cases and all of them were believed to derive much from it. But learned doctors declared that the improvement noticed in the three cases was entirely due to the special method of treatment. To a great extent, it was the result of the nourishing diet, rest, and open air which they had in the hospital. The ayurvedic drug played only a minor role. Following this, Ayurvedic Medical doctors argued that "the truth is that the learned doctor tries to deceive the laymen by indulging in these meaningless quibbles. If diet and others alone could give the desired result why then resort to medicine at all?," they argued.[68]

Learned doctors did not understand, as Ayurveda doctors did, that "there is no marked distinction between a medicine and diet in therapeutic properties except in the dose administered." Medicine, diet, rest, fresh air, exercise all were considered as "mere aids to nature in its struggle to expel the foreigner (disease) out of the body."[69]

Another familiar method of treatment used in the Madras Presidency was a decoction of *araka* or sealing wax, which reddens the water while it boils. Twelve ounces of this and two beetles were mixed in water after adding sugar and then given to the patient. If this remedy was not efficient, a jam like substance (*lehim*) was made of boiled mutton, several powders, candy sugar, and ordinary sugar. All of these were mixed together and boiled. But some prepared the *lehim* with rabbit instead of mutton. Patients, while undergoing this treatment were not allowed to take fish, coconut, or any sour item in food. The patient had to remain at home and avoid exposure to the sun. This treatment lasted for 14 days and after the treatment, the same number of days had to be spent at home. Prejudices were part of prescribed treatment methods. The prescriptions were also combined with caste and religious prejudice. For example, Brahmins being vegetarians adopted a different method, i.e., sleeping with a living rabbit or placing it on the patients' chest.[70]

Other medical practitioners discourse

Homeopathy originated in Germany and was introduced in India during the 19th century. It was "Indianised" by the end of the 19th century through a combination of emphasis on the mystic concepts of "aura" and "empathetic ethos." Moreover, homoeopathic medical practice was promoted by allopathic medical practitioners in several parts of India. By 1920, homoeopathy became popular in the Madras Presidency. Gary J. Hausman had argued that Homeopathy was transformed by combining with modern and indigenous medicine such as Ayurveda.[71] Dispensaries and hospitals were established in Madras, Bezawada, Conjeevaram, and other parts of the Madras Presidency. It was popular because the state supported it as a "scientific system of medicine" compared to indigenous medical systems and it stood between Allopathy and Ayurveda. Homeopathic medicine was not expensive and hence easily available to the poor, like Ayurveda.[72] Homeopathic practitioners argued that allopathic treatment had failed to provide "quick, gentle and permanent" cure or relief.[73] According to them, allopathic medicine suppressed the patients' disease, leading to various other health problems. When the disease was suppressed, it would remain inside the body and expand rapidly. When the power of the medicine came down or the body became weak the suppressed disease "with thousand heads spread in the body."[74] Homeopathic treatment emphasised the use of high proteins. It was done in two ways namely, control of the factors or causes contributing to the disease; and in the case of the already infected, control of direct infection.[75] The homeopathic system of treatment prescribed about 40 medicines and drugs.[76]

Naturopathy was another therapeutic practice and it became well known during the 1920s. M.K. Gandhi followed Naturopathy and with his influence, many nationalists turned towards Naturopathy with many hospitals and dispensaries also being established.[77] Naturopathy recognised the body's inherent ability to maintain health. Nature as a base for the life force was at the heart of the healing process. The physician was only a facilitator to identify obstacles to health and to enhance the natural healing process. This method emphasised a healthy internal and external environment to maintain good health and recover ill health. Naturopathy treatment was in the form of open-air treatment and sun treatment. Tuberculosis was a general disease with local manifestations, and its cure depended on the soundness of the natural defences of the body. Naturopathic practitioners considered the sun as an immense and inexhaustible source of biotic energy. Heliotherapy could cure surgical tuberculosis in all its forms and at any stage of its development. The principal factors in this treatment were sun and air, the twin gifts of nature.[78] Heliotherapy was considered useful not only in curing tuberculosis but also in its prevention. The main advantage of the sun treatment was the avoidance of surgical interference. A mere exposure to the sun is not heliotherapy, proper dosage of insulation is essential. India is a land of plenty of sunshine. Excessive and uncontrolled exposure to the sun is dangerous as it is the most powerful biotic energy in the universe. The hills were considered the best location for this treatment.[79]

The influence of sun light on the human physiological system starts once the rays are received by the skin, transformed through the pigment into biological energy – transmitted to different organs and parts of the body.[80] Without the intervention of the skin, the strong rays of the sun would kill the organs of the body. The exposure of the body to sun light through the pigment transforms the light energy into biological energy and brings about certain chemical reactions useful in body-defence. The skin is endowed with a rich blood supply as sun light acting on the skin dilates the blood vessels. The parts of the skin habitually exposed to sun light are more actively irrigated by blood than the less-exposed areas. The exposed parts would considerably resist the changes in temperature and infective agents. All practitioners of Naturopathy strongly believed that sun light had immense therapeutic value.[81]

Naturopathy practitioners argued that the skin of humans is universally anaemic and feeble. Civilisation was seen as a source of lavishness and unscientific selection of clothing causing the skin to be unhealthy and inefficient. Unhygienic clothing creates a hot and humid condition in which the skin gets exhausted and becomes weak. Such skin could not fulfil its function and it would lead to weakness and exhaustion.[82] The sun's rays help in curing diseases, particularly tuberculosis by building body resistance and vitality. Primarily, heliotherapy was based on the application of sunlight and exposure to air directly on uncovered skin of patients. The preservation and restoration of the normal functions of limbs and joints; replacement of knife and plaster by simple or therapeutic measures without losing efficiency, permitting only a minimum of interaction with sunlight and air playing on the affected part and the body as a whole finally led to improvement of the general health.[83]

The initial step in heliotherapy consists of very gradual exposure of the patients in bed to direct sunlight and air. The gradual exposure ensures proper development of the pigments in the skin without which the body cannot tolerate sunrays. In effect, dull appetite sharpens, and the patient begins to enjoy food. The patient feels much better and gets restful sleep after sleepless nights and endless agony.[84] Tuberculosis was considered essentially a protracted fight between the seed, bacilli, and the soil, the body of the host, in which the rival forces of the two are delicately balanced. A healthy body is like a rock against the growth of the seed, the bacilli. The symptoms of tuberculosis disease are local and general. Locally, an inflammatory process is setup, which may lead to abscess –formation and necrosis. The general reactions to the disease manifest themselves as disturbances of the various physiological functions of the body.[85]

Diet and nutrition

In contrast to indigenous medicine, western medicine preferred a good diet for treating tuberculosis. K.G. Natasa Sastri, an Ayurvedic practitioner in Madras, criticised western medical instruction of "giving plenty of food in tuberculosis."[86] Pulmonary tuberculosis rests a great deal on the condition of the stomach and the digestive tube. Ayurveda considered reinforcing the bodily resources of vital resistances through self-defence available in the human body. Drugs were used to help maintain the vital organs in a state of comparative health. The *Antiseptic* (an English monthly Medical Journal), in its editorial in 1911 wrote, "we need to fall back upon the conserving measure that establish immunity in the tissues." To attain this end, regular and hygienic living, rest with or without gentle exercise according to the temperature, etc., were recommended.[87] Food having certain properties such as easy digestibility and enough nutritious value were to be given to the patient. In such natural condition, the body would gain weight due to activity and replace the tissues lost due to the disease. Regularity of meals, moderation and exclusion of food articles difficult to digest by the disabled alimentary canal, an increase in those constituents that are easily assimilated and readily accepted by the cells was prescribed.[88]

Homeopathic doctors also considered diet (*pathyam*) an important part of the treatment. The food stuff was of three kinds namely *satvikaaharam* (nutritious and easily digested food), *rajasikaaharam*, *tamasikaaharam*. Indeed, these three classes of food would help body growth and vitality, but each had its own character, which influenced the mental state of the person. Both Ayurveda and Homeopathy considered mental nourishment as important as physical nourishment in the treatment. Some food items were to be avoided because they led to the arousal of sexual erotic feelings (*kamodrekam*) which would lead to the loss of semen. *Satvikaaharam* such as sweet fruits, cow's milk, cow's milk ghee, rice, wheat, vegetables, etc., give good nourishment to the body. *Rajsikaaharam* such as goat/sheep's milk and mutton are useful for the patient, but egg should be avoided because it would cause sexual desire. The peace of the patient's mind depends on what patients eat. *Tamasikaaharam* provokes the mind towards "bad comforts;" therefore, a tuberculosis patient needs to take s*atvikaaharam*. In addition to this, the patient would pray or meditate every day, which would help concentration and mental peace in turn leading to good health. Tuberculosis patients should not talk or argue with others and keep away from their spouse. It was believed that tuberculosis patients may have excessive sexual desire. Thus, both men and women should follow *Brahmacharya* (celibacy). Physical exercise in an open place was necessary. The place where the patient was living and sleeping should be open and well ventilated.[89]

To conclude, both preventive and curative methods were rationalised under a state controlled western medical system. Western medical treatment methods such as sanatorium and climatic cures, cod-liver oil, gold therapy, pneumothorax were commonly used. Artificial pneumothorax came into use only in the 1930s in the Madras Presidency. The sophisticated use of gold therapy was a part of the western medical practice. Some western medical practitioners prescribed this method of treatment with a limitation on the quantity of dosage. However, indigenous medical practitioners rejected the restricted prescription. Overall, western drugs and treatment methods predominately used in institutional care, which eventually tried to claim hegemony and superiority on the basis that indigenous medical treatment methods and drugs had no scientific value and were slow in their effect. This witnessed a counter medical discourse from indigenous medical practitioners. This response coincided with the assertion of indigenous medical practitioners who started a revival movement in the Madras Presidency during the early 20th century. On the other hand, among the western medical professionals, many expressed diverse medical opinions on drugs and the treatment methods of tuberculosis. They differed essentially on issues relating to the efficiency of its treatment method. Scientists, bacteriologists, medical doctors, and health officials failed to provide comprehensive data on the efficiency and length of curative measures, and this created various ambiguities, uncertainties, and apprehensions among the public.

Notes

1 K. N. Pannikar, 'Indigenous Medicine and Cultural Hegemony: A Study of the Revitalisation Movement in Keralam', *SIH*, Vol.8, No.2, 1992, pp. 283–307; Burton Cleetus, 'Western Science, Indigenous medicine and the Princely states: The Case of Ayurvedic Reorganization in Travencore, 1870–1940', in Biswamoy Pati and Mark Harrison (eds.,), *Society, Medicine and Politics in Colonial India*, London: Routledge, 2018; Deepak Kumar and Raj Sekhar Basu (eds.), 'Introduction', *Medical Encounters in British India*, New Delhi: OUP, 2013; Poonam Bala (ed.), *Contesting Colonial Authority*, Lanham: Lexington Books, 2012.
2 P. V. Benjamin, 'The Principles of Treatment of Pulmonary Tuberculosis', *The Indian Medical Journal*, Vol. XXIX, 1935, 17–19, p. 17.
3 Ibid, p. 19.

4 Ibid.
5 M. Kesava Pai, 'On the Efficacy of Sanocrysin Treatment in the Different Types of Pulmonary Tuberculosis, Analysis of 84 cases', *The Madras Medical Journal*, Vol.10, No.10, 1928, p. 172.
6 M. Kesava Pai and P. K. Gunasagaram, 'The Effects of Gold-Salt in Pulmonary Tuberculosis', *The Madras Medical Journal*, Vol.10, No.4, 1928, 53–63, p. 53.
7 Union Mission Tuberculosis Sanatorium-Golden Jubilee (1915–1965) Souvenir, Arogyavaram-Madanapalle, Madras: Diocesan Press, 1965, p. 11.
8 M. Kesava Pai and P. K. Gunasagaram, op. cit., pp. 55–57.
9 Ibid.
10 M. Kesava Pai, 'On the Efficacy of Sanocrysin Treatment in the Different Types of Pulmonary Tuberculosis, Analysis of 84 cases', op. cit., p. 172.
11 Ibid, p. 175.
12 L. R. Dongrey, 'Gold Treatment in Tuberculosis', paper presented in *Second All India Conference of Tuberculosis Association*, November 20–23, 1939, New Delhi, p. 107.
13 Ibid. p. 107.
14 Ibid, pp. 109 and 110.
15 Ibid., p. 110.
16 Ibid, p. 113.
17 Ibid, p. 114.
18 Ibid, p. 115.
19 Ibid.
20 Ibid, p. 118.
21 Govt Order (hereafter G.O.).3546 (Public Health), Dated 6th August 1940, Tamil Nadu Archives (hereafter TNA).
22 C. Dwarakanath, *Gold Therapy in Tuberculosis*, Madras: Nutshell, Gordon and Co., 1943, pp. 10–14.
23 Ibid, p. 19.
24 Ibid, pp. 40–46.
25 Ibid, p. 58.
26 Ibid, p. 59.
27 P. V. Benjamin, 'The Principles of Treatment of Pulmonary Tuberculosis', *The Indian Medical Journal*, Vol. XXIX, 1935, p. 19.
28 *Golden Jubilee Souvenir of Union Mission Tuberculosis Sanatorium, 1915–1965*, Arogyavaram: Andhra Pradesh, 1965, p. 11.
29 Pneumothorax means the introduction of air into the pleural cavity thereby forming a pleural pocket with eventually collapses by putting it into surgical rest. The diseased portion adhere together, the cavity, if any, slowly becomes smaller and smaller. The cavity of the disease is in favourable cases brought under control. See T. S. Shetty, 'Modern Surgical Treatment of Pulmonary Tuberculosis', *Medical Digest*, Vol.5, No.1, April 1937, 1–6.
30 Charles Morehead, *Clinical Researches on Diseases in India*, London: Longman and Roberts, 1860 (second Edition), p. 559 and T. S. Shetty, 'Modern Surgical Treatment of Pulmonary Tuberculosis', *Medical Digest*, Vol.5, No.1, April 1937, 1–6, pp. 1–2.
31 Ibid, pp. 18–19.
32 M. Kesava Pai, 'The Treatment of Pulmonary Tuberculosis by Artificial Pneumothorax', *The Madras Medical Journal*, Vol.8, No.1, July 1925, p. 1.
33 T. S. Shetty, 'Modern Surgical Treatment of Pulmonary Tuberculosis', op. cit., p. 4.
34 "The Phernic nerve being cut off, the hemi diaphragm, which is supplied by the nerve becomes paralysed. The paralysed part becomes immobile and occupies a higher position in the thorax and the lung volume becomes reduced. This leads to a partial collapse of the lower lobe and diminished a respiratory activity especially of the lower portion of the lung. This partial immobilisation produced

by phrenic exairesis has a similar effect like artificial pneumothorax" see Jagdish Chandra Bhattacharjee, 'A Short Note on the Recent Methods of Treatment of Pulmonary Tuberculosis', *Medical Digest*, Vo. 5, No. 1, April 1937, 7–10, p. 8.

35 Jagdish Chandra Bhattacharjee, ibid, p. 8.

36 T. S. Shetty, 'Modern Surgical Treatment of Pulmonary Tuberculosis', op. cit., p. 1.

37 Ibid, p. 2.

38 S. Krishna Swamy, 'Modern Technique in the Treatment of Pulmonary Tuberculosis', *Bulletin of the South Indian Medical Union*, Vol. III, No.5, 1931, p. 15.

39 M. Kesava Pai, 'The Treatment of Pulmonary Tuberculosis', op. cit., p. 52.

40 Ibid., p. 53.

41 M. Kesava Pai, 'On the Efficacy of Sanocrysin Treatment', *The Madras Medical Journal*, Vol.10, 1928, p. 178.

42 G.O. No.3231 (E&PH), Dated 28th July 1940 (TNA).

43 Chemotherapy is the use of chemical substances to treat disease.

44 R. Viswanathan, Recent Developments in Tuberculosis Control in the West and Their Application to India, *Indian Medical Gazette* (hereafter *IMG*), Vol.83, 1948, p. 47.

45 'Chemotherapy in Tuberculosis', *Indian Medical Journal*, Vol.39, No.10, 1945, p. 241.

46 J.N. Misra, 'Treatment of Tuberculosis: Past and Present', *The Medical Digest*, Vol.4, No.12, 1937, 495–501, p. 495.

47 Ibid.

48 Ibid.. 496.

49 Ibid., p. 497.

50 Ibid., pp. 497–548.

51 The solar system classification based on colours which influence all lives in the world. Sun – deep red colour, Moon – white, mars – red, Jupiter – yellow, Venus – White colour, Saturn –black, Mercury – deep green.

52 J.N. Misra, 'Treatment of Tuberculosis', op. cit., p. 499.

53 Ibid.

54 Ibid., p. 500.

55 Ayurvedic System of theory on Consumption and its treatment, in *The Vaidya Kalanidi* (Tamil Journal), Vol.5, No.1, 1917, p. 33.

56 After 1920, several Ayurvedic and other medical systems colleges and schools were established in the Madras Presidency. Among such colleges were the Professional College in Bezawada (1922), Sri Rama Mohana Ayurvedic College in 1923. In response to indigenous medical practitioners the Government of Madras started a government Indigenous Medical College in 1925 in Madras.

57 K.N. Pannikar, 'Indigenous Medicine and Cultural Hegemony', op. cit., pp. 283–307.

58 Poonam Bala (ed.,) *Contesting Colonial Authority*, op. cit., p. vii.

59 Government rejected the aid to three Ayurvedic dispensaries and schools in Madras in 1911. It became controversy and strong resistance came from whole Presidency by demanding to continue the aid. With the pressure of Indian Legislative Council Members, politicians, medical practitioners, and press, the Government appointed Dr. Koman committee to investigate drugs of Ayurvedic and Unani used in the Presidency. G.O.1765 (Local and Municipal), Date 4th November 1919 (TNA); G.O.89 (Medical) Date 13th February 1919 (TNA); G.O.88 (Medical), Date 12th February 1919 (TNA) and G.O.125 (PH), Date 11th February 1921 (TNA).

60 *Report of the Special Committee Appointed by the Joint Board*: The Dravida Vaidya Mandali and the Madras Ayurveda Sabha in Reply to the Report on the Investigation into the Indigenous Drugs by M. R. Ry. Rao Bahadur, M. C. Koman Avergal, LMS, appointed the Government of Madras, Srirangam: Sri Vani Vilas Press, 1921(hereafter Report of the Special Committee of Dravida Vaidya Mandali), pp. 5–6.

61 Ibid, p. 9.
62 Ibid., pp. 10–11.
63 This drug was known in Sanskrit known as *Vasaka*; in Tamil as *Adthodai*; Malayalam as *Adalodam*.
64 Ibid., p. 12.
65 Ibid., pp. 23–24.
66 Ibid, p. 26.
67 It was known as different names in different languages: Sanskrit: akarakarabha, Tamil: Akkilakaram, Malayalam: Akkalakaram.
68 Ibid, p. 84.
69 Ibid, p. 85.
70 Albert Gille, Notes on Some Native Medicines from Southern India, *Man*, Vol.6, 1906, 182–187, p. 183.
71 Gary J. Hausman, 'Making of Medicine Indigenous: Homeopathy in South India', *Social History of Medicine*, Vol.15, No.2, 2002, pp. 303–22, p. 305.
72 Ibid, pp. 307–308.
73 P. Venkata Rao, 'Why Homoeopathy?', *Andhra Homoeopathic Journal*, Vol. IV, No. 4, 1934, pp. 98–106, p. 99.
74 Yeluripati Subramanyam, *'Hindudesa Vaidya Paddatulu'* (Medical Methods in Hindu Country) (in Telugu), *Andhra Homeopathic Journal*, Vol. I, No. 9, 1934, p. 219.
75 Narasimhadevara Kameswara Rao, *Khsya Rogam: Chikitsa* (Tuberculosis and Treatment) (Telugu), Ramachandrapuram, East Godavari: Ananda Trirtha Printings Works, 1942, p. 53.
76 Ibid, p. 74. The prescribed medicine for treatment were *ebrotanum, anacardium, antimonium aarsinicum, argentums album, aarsanicum aayodatum, aaromotalicum, basilinum, baratacarbu, calecariu carba, carbo enimalis, carbo vegitalis, cisctus, ferrumfos, floric acid, hefarsulf, aaodium, calibicrumicum, calicarbu, lesisis, lecopodium, megnicium carbu, megniciamoor, mengnum esaticum, medorisum, milifolium, nitricacid, fanisparis, fasfaric acide, sorinum, ferrojinum, sangineria, sifia, silica, spangia, stanum, sulphur, siflinum, tuja, tuberculinum.*
77 Naturopathy was popular in the Madras Presidency during the 1920s. Dispensaries and hospitals were established at Bezawada, Vetapalem (Guntur district) and other parts of Madras Presidency. People like Kondaparti Veera Bhadra Charulu and Palaparti Narasinga Rao were played major role in dissemination of the naturopathy through journals such as "*Prakrit Patrika*" (Telugu). See Naturopathy in Kondaparti Veera Bhadhra Charyulu, *Just Chikitsa* (Just Treatment) (Telugu), Bezawada: Prakriti Karyalayam, 1924, p. 3.
78 S. N. Sinha, *Tuberculosis and the Sun Treatment*, Calcutta: General Printers and Publishers Ltd., 1941, p. i.
79 Ibid, p. ii.
80 Ibid, p. 2.
81 Ibid, pp. 3–4.
82 Ibid., p. 10.
83 Ibid, pp. 30–31.
84 Ibid.
85 Ibid., pp. 45–46.
86 *Report of the Special Committee of Dravida Vaidya Mandali*, appendix-III, p. VII.
87 Editorial on Diet in Tuberculosis, *The Antiseptic*, Vol. III, No.4 April 1911, pp. 255–257.
88 Ibid, p. 256.
89 Kameswara Rao, Narasimhadevara. *Khsya Rogam: Chikitsa* (Tuberculosis and Treatment) (Telugu), Ramachandrapuram, East Godavari: Ananda Trirtha Printings Works, 1942, pp. 80–82.

14 A case for the social history of homoeopathic hospitals in India

An invitation for its construction and rendition

Dhrub Kumar Singh

'How easy the task if every hospital had its historian! Satisfactory hospital histories are rare'.[1] This is a truism for the remote past as also for contemporary times in the South Asian context. Hospitals as sites for medical education in terms of prognosis and prophylactics still await comprehensive probe and rendering by medical historians. Institutional and social histories of hospitals have hardly been attempted and histories of homoeopathic hospitals and dispensaries are almost rare or absent. This historiographical problem persists and prevails as yet. How and why homoeopathy and homoeopaths engaged with the idea of hospitals and how did the spread of homoeopathy aid the realization and proliferation of homoeopathic dispensaries? From the homoeopathic dispensary records, it is palpably clear that many founders of homoeopathic dispensaries wanted to elevate their dispensary status to that of hospital status. Despite these intriguing but rudimentary facts, historical records in this regard remain unexplored and so appear to be 'notoriously incomplete'.

Homoeopaths with their general rejection of dominant medicine's nosology and owing to their emphasis on individualized and personalized treatment of curing and healing as against the palliating and interventionist efforts of allopaths, promised to restore health and eliminate disease by walking along with the disease and aiding the natural restorative process of the body to health. They proposed to do better than what occupied beds of hospitals could possibly relieve with treatment, and yet they felt and articulated the need to have homoeopathic hospitals like the allopaths. Among homoeopaths, 'there [was] a general recognition that it is better charity to keep man from the need of a hospital bed than to care for him when in one'.[2] And yet in homoeopathic dispensary reports one can palpably and unmistakably discern a wish to upgrade its status into a hospital.

Hospitals in the colonial era and the dearth of homoeopathic hospitals

A fair amount of work has been done on the advent and impact of western medical education within the rubric of east–west[3] encounter in colonial India. We have a clear picture of the founding of native medical

DOI: 10.4324/9781003241980-19

institutions,[4] the precursor of Calcutta Medical College and Hospital.[5] We also have clear documentation and critical appraisals of western medical education[6] in Madras and Bombay Presidencies.[7] The context and contingencies, the ways and aims of medical establishments, including hospitals and dispensaries in nineteenth-century India have come to the fore, because the delineation of public health policies[8] and social history of epidemics[9] have already been attempted. It was because of these varieties of histories pertaining to medical encounter and intervention in the plural medical scenario that many aspects of hospital histories have also been explicated. In other words, in depth studies on medical education, policies on public health, and social histories of various epidemics have invariably provided the context in which many social and cultural aspects of hospitals have come to light. But western medicine did not only mean allopathy. What about those heterodox strands of medical thought and practice which originated in Europe and acquired new role and significance in colonial India? After all, many protagonists of homoeopathy in India were products of the new regime of western medical education of the dominant medicine or allopathy. Many among the first generation of famous 'converted' homoeopaths were themselves alumni of the Calcutta Medical College. Homoeopathy, despite its western origins, has largely been overlooked by historians and homoeopathic hospitals remain even more neglected.

The schism of medical thought and its echo fostered and perpetuated by Hahnemann's rebellion in Europe reached Indian shores through his distant and scattered disciples who gradually paved way for homoeopathy in India. The historical antecedents of homoeopathy in India can be traced to the early part of the nineteenth century. While Dr. John Martin Honigberger[10] became the most recognized name in the early diffusion of homoeopathy in the Indian subcontinent, from the 1840s onwards, European military men were already engaging with homoeopathic literature and experimenting with homoeopathic remedies as an extension of their amateur hobby imbued with philanthropic and humanitarian values. By the 1860s, homoeopathy had found a base in Calcutta where it won local converts who attempted to popularize the pathy from their local dispensaries. The local dispensary was the site for personalized care as several dispensaries were opened for homoeopathic treatment by the closing decades of the nineteenth century. The first homoeopathic hospital, in fact, came up in Calcutta as early as 1851 under the patronage of the Deputy Governor of Bengal, Sir John Hunter Littler, but this was more in the nature of a free dispensary rather than a proper hospital.

Needless to say, doctors and hospitals acquired a new meaning in colonial times. The quest of modernity in India was contextualized and imbued by the constraints of colonialism. Doctors and hospitals played an important role in colonizing the body. In the explication of this wisdom, one has to remember that most doctors were practitioners of the dominant pathy and were custodians of hospitals—which represented the potential of the dominant pathy—both in its surgical and therapeutic aspects. But what about those medical practices which though had their origins in Europe, came to the colony without the trappings of the dominant pathy? In fact, they opposed and stood in contradistinction to the dominant pathy. Did they have hospitals and of what kind? For instance, how did the dispensaries and hospitals of homoeopathy exist as alternatives fostering an alternative medical culture? Who were the people who patronized them, how did these institutions sustain themselves, and what was the attitude of the state towards them? Finally, how did they get accepted among the people and on what grounds?

Hospitals have generally been associated with allopathic medicine and care. In homoeopathic vision cure is associated with personalized care. During the colonial period, homoeopathic institutions received lukewarm response and support from the state and, as such, they could not be visualized on a mega-scale. Surgery was not a predominant part of homoeopathy; therefore, the need for hospitals was not felt so much. One of the founders of the homoeopathic hospital at Banaras in the 1870s perceptively elaborated and explained thus:

> We do not propose treating in this hospital, cases which are strictly surgical. *Surgery is a science in itself*, the efficiency of which in clever hands, no one can dispute. Still there are great many cases that appear surgical, which can be cured under homoeopathic treatment without recourse to surgical aid. Cases, however, that admit of doubt in the first instance, we propose not to take up (emphasis added).[11]

However, homoeopaths tried to empirically observe diseases, especially chronic diseases, therefore, the need for indoor arrangements for observing patients as well as for discerning and understanding the changing panorama of symptomatology became imperative. Hospitals also served the purpose for the enforcement of more stringent precautionary dietary regimen which the homoeopaths prescribed. All these factors if combined definitely point to the need for homoeopathic hospitals. However, the historical fact in the Indian context remains that homoeopathic dispensaries and indoor arrangements were largely sustained by a few locally influential homoeopathic practitioners through collected subscriptions and philanthropy. When homoeopaths began organizing themselves and their healing art under the roof of homoeopathic hospitals and dispensaries, they were moving from their amateurish existence to a more professionally organized existence. This transition of going public through organized hospitals helped them to tide over the blame of quackery levelled against them. Homoeopathic hospitals became the site for the proliferation of this new pathy and ensured its public reception and popularity.

In the eastern part of the subcontinent, even though such dispensaries not only existed but proliferated, they did not operate on such a large scale as state-operated and state-supported hospitals. But it remained a fact that a pathy which sustained itself initially by the classic master–pupil relationship, in its stage of proliferation, required a nursery where new entrants, enthusiasts, and 'converts' could be trained in a more sustainable manner. In doing so, they derived a lot of inspiration from colleges which sprang up in Europe, both in arranging the content of the curriculum and in ways of teaching and imparting skills. An interesting aspect for research which requires probing is how all these efforts were made, sustained, and in the process attained legitimacy within the specific cultural milieu of eastern India? How did the efforts of men like J. H. B. Ironside,[12] Sir Sayyed Ahmad Khan,[13] Rajendra Lal Dutt, Babu Lokenath Moitra, Mahendra Lal Sarkar, P. C. Majumdar, and many others connect and explain the initial proliferation of homoeopathy, homoeopathic dispensaries, hospitals, and colleges? Some were strictly medical men, while others were not. But they came together to accord sustenance to this new and radical pathy.

Once homoeopathy had gained a sound footing, homoeopathic publishing endeavour and book industry started, and though it was largely sustained by groups of likeminded enthusiasts and protagonists, in due course these developments attained a self-propelling economy of their own. Calcutta became the hub of this alternative medical culture. Gradually there were efforts to organize associations and provide them a more professional frame. Evidently, this alternative medical culture has not only endured but has acquired recognition both legal and social and has contributed towards the betterment of human health in one of the most populous regions of India.

The plurality of medical practices in India has always been conceded. However, can the resumé of medical practice and profession in India ever be considered complete without giving serious attention to homoeopathy and homoeopathic hospitals? The best of texts authored by colonial doctors and historical appraisals by medical historians have not paid adequate attention to this glaring omission or lacuna in the medical history of South Asia, which is otherwise an expanding field of research. This lacuna, of course, has a bearing on the medical historiography of the region.[14] My purpose is not to provide a full account of the origin of homoeopathic hospitals and dispensaries but, as a preliminary step, this chapter is an effort to sensitize to their absence in the rendered scholarly historical accounts on various aspects of the institutional medical history of South Asia.

The birth of a new radical pathy

The year 1796 is generally designated as the birth year of homoeopathy, the year when Samuel Hahnemann published 'An Essay on a New Principle for Ascertaining the Curative Powers of Drugs and some Examinations on the previous Principles',[15] where he outlined the three cardinal principles of the new pathy.[16] Hahnemann's rebellion[17] against the dominant and regular mode of medicine and his founding of a new pathy called 'homoeopathy' was based on the maxim *Similia Similibus Curantur*—let likes be treated by likes. In contrast to this dictum, 'regular medicine generally prescribed *allopathically,* [where] treatment was based on principles *other* (from the Greek word 'allos') than symptom similarity. Typically allopathic medicine [tried] to remove or oppose disease causes, and to suppress or palliate symptoms'.[18] The prevailing heroic therapies with their indiscriminate emphasis on purging, drugging, and blood-letting provided the context for Hahnemann's increasing questioning of his own convictions as an allopath and about the crude ways in which medicine was practiced then.

The alternative therapy offered by Hahnemann was based on the vitalistic understanding and appreciation of the body and diseases in contrast to the mechanistic view.[19] He viewed disease as a vitalistic problem affecting and implicating both the physical and mental dispositions and disease symptoms were nothing but manifestations of the deranged vital force which primarily constitute the disease.[20] Since any disease was always the disease of the whole organism, treatment, therefore, needed to be focused on the person as a whole whose individualistic responses to the disease made allopathic 'nosology' redundant. Moreover, since the vital force was responsible for the harmony and equilibrium of the body, the morbid signs and symptoms of the body could be viewed as an expression of its attempt to restore normality. The task of the physician was to assist in the restorative process without impeding the 'homeostatic activity of the human organism'.

During Hahnemann's later medical career an attempt at institutionalizing the new pathy was made with the establishment of the first homoeopathic hospital at Leipzig in 1833.[21] But this effort was short-lived as the hospital was plagued with financial constraints and half-hearted homoeopaths, finally to close down in 1842. Nonetheless, it was outside Germany that homoeopathy was being tried out with fair amount of success during Hahnemann's own lifetime. In Britain, the spread of epidemic cholera incidentally provided the opportune context for the entry of homoeopathy. The devastation wrought by the 'cadaverizing' disease, and the helplessness of dominant medicine instigated the cholera riots of the 1830s in London. Around the same time, in contrast to the varieties of drugs with heavy doses tried out by allopaths, a new breed of physicians with their mild drugs recorded their first success against Asiatic cholera[22] by treating it according to the principles of their system,[23] and they did not embroil themselves too much in the causation controversies of this new malady. In accordance with the fundamental precepts of their 'therapy', homoeopaths remained aloof to conflicting theories that were being propagated about cholera causation and communicability and began using well proven drugs like arsenic, veratrum, ipecac, camphor, and cuprum, as given in their therapeutic law, to deal with the totality of symptoms generated by the disease. The impact of homoeopathic provings of these substances was that their use became 'greatly clarified and each substance or drug was thrown into much sharper focus, its true healing properties being more clearly identified such that they could be employed thereafter with much greater certainty'.[24] The effectiveness of homoeopathy was repeatedly tested in the later outbreaks of cholera epidemic in Britain which continued to offer a fertile testing ground. In 1854, the London Homoeopathic Hospital was adapted to solely cater to victims of the cholera epidemic which broke out that year.[25]

The rise of homoeopathy in India

In the 1830s, as 'Asiatic cholera' took the metropole in its embrace, the 'new school' of medicine, i.e., homoeopathy prevalent in other parts of Europe, found its way to the colony. Some military men practiced it as an amateur hobby from the 1840s. Some Government medical officers stationed at Fort William were known to admire homoeopathy and practiced it. There are evidences that some missionaries also made this 'pathy' a part of their 'do good ethic'. Dr. Mullens of the London Missionary Society was known to distribute homoeopathic medicines to the people of Bhawanipore. Dr. John Martin Honigberger, who despite his use of homoeopathic medicines did not consider himself a homoeopath, though he appreciated Hahnemman's genius and had met the master.

As cholera was a recurrent phenomenon in India, many early homoeopaths encountered it. There are references that this mode of treatment was used by doctors in the General Military Hospital in Bombay, particularly in the treatment of cholera. One of the Judges of Sadr Dewani Adalat, Mr. Ed D'Latour sent homoeopathic medicines for free distribution to inhabitants of Diamond Harbour where cholera was taking its toll. Dr. J. Rutherford Russell, a medical officer at Fort William also practiced homoeopathy. After his retirement, he returned to England to settle as a homoeopathic practitioner and as a homoeopath, he encountered the cholera epidemic of 1848–1849, and even wrote a cholera treatise. Surgeon Samuel Brooking, a retired medical officer under the patronage of the Raja of Tanjore, established a homoeopathic hospital at Tanjore in 1847. In Bengal, in 1850–1851, the Native Homoeopathic Hospital and Free dispensary was established. Rajendra Lal Dutt and a French homoeopath Dr. Tonnere were associated with this hospital although it did not last long.[26]

Clearly, by the 1860s, the Hahnemannian spirit had become well entrenched in choleraic Bengal and doctors like Mahendra Lal Sarkar[27] were creatively engaging with it. Sarkar, an accomplished allopath, who later changed his 'creed' to homoeopathy, was one of the greatest champions of this spirit. His 'conversion' to homoeopathy was not a chance happening, nor was his engagement with it an amateur hobby that he developed.[28] The context of cholera as a recurrent malady and its lack of treatment in allopathy, made Sarkar realize the lacuna in the then prevailing heroic therapies and also allowed him to appreciate homoeopathy in proper perspective. 'Are there non[e] in this country, worthy to take up the mantle of Hahnemann?',[29] Sarkar chided. As one of the best students of the Calcutta Medical College, Sarkar himself was an accomplished allopath but the scientific basis of drug provings made him amenable towards appreciating homoeopathy. For Sarkar, 'provings constitute[d] the soul of therapeutics'.[30]

The logic of the emergence of homoeopathic hospitals in the Indian context

In an era when the dominant pathy, i.e., allopathy was still struggling and grappling to clearly establish the nosology of many 'fevers' and strange behaviour of many prevailing epidemic diseases; at a time when its own pharmacopoeia was still trying to enlarge and refine itself by brushing aside the blame of arbitrariness in the wake of weak and not so clearly defined nosological parameters; in an age when surgery was still to obtain the precise and painless support of anaesthesiology (or science of anaesthesia); in this era of 'uncertainty' when dominant discourse of medicine was still controverting the ill and benign effects of heavy doses of drugs administered on the principle of opposing the disease by their powerful contrary effects to palliate and tame the disease—the side effects and iatrogenic factors of which were starkly manifest—the well-established and constantly being refined parameters of anatomy, physiology, and pathology pushed many adherents and advocates[31] of the dominant medical discourse to appreciate the scientificity[32] of *drug proving*.[33] Even if they did not concede entirely to the '*Similia similibus*' and consequently had a critical view of the alternative ways of understanding body, health, and disease, the scientificity of drug proving drawn out from the elaborate and precise drug pictures along with clear elaboration of disease symptomatology obtained painstakingly by homoeopaths, charmed them towards the new, radical, therapeutic option being advocated by this alternative reform movement of medical ideas and practice. Many allopaths intellectually and pragmatically gradually became amenable towards giving this therapeutic option a fair trial while keeping intact their own reservations regarding the edifice of this new system and the pathy based on it. It is this amenability or the gradual veering of many allopaths towards the new alternative therapeutic option that in homoeopathic literature gets expressed by the term 'conversion' signifying the change of one's faith. It was a shift in conviction created and ushered by the prevailing context of contradictions in dominant medical practice. All those who were 'converted' expressed conviction towards one or more ingredients of the new pathy. It was in this sense that they always tried defending their right as a physician, and it was in this process that all the blame game between allopaths and homoeopaths was enacted.[34] Allopaths outcasted those who veered towards homoeopathy and treated them as corrupted and lesser beings; as physicians fallen in status. The homoeopaths welcomed such 'conversions' as steps towards redemption and their return to the correct mode of physik as true healers or physicians.[35]

Mahendra Lal Sarkar, Rajendra Lal Dutt,[36] Babu Lokenath Moitra,[37] P. C. Majumdar,[38] were the intellectual and pragmatic adherents of homoeopathy in this era of flux and uncertainty[39] in medical ideas and practice. In this gradual amenability of providing a critical but fair trail to many components of homoeopathy by dominant practitioners of medicine lay the root and logic of having homoeopathic dispensaries and hospitals, where not only disease symptomatology but drug proving corresponding to it could be indulged in on a larger scale, consequently refining and updating the then prevalent slim pharmacopeia at hand. Such controlled sites as dispensaries and hospitals also had the added advantage of understanding the merits and demerits of dietary regimens which homoeopaths considered to be of immense value. Moreover, dispensaries and hospitals allowed the possibility of training physicians in this new healing art thereby helping to expand and strengthen the pupilage system of teaching. Hospitals also served as better repertoire of homoeopathic medicine. All these pragmatic considerations paved the way for homoeopaths of classical Hahnemmanian tradition and the not so classical variety to come together on a scientific site, socio-economically represented as homoeopathic dispensaries and hospitals. Sociologically and economically, they represented nodal points in the contact network of homoeopaths and homoeopathic drug companies cutting across regional and national boundaries. The Banaras Homoeopathic Hospital, the Calcutta Homoeopathic Dispensary of Mooktaram Babu Street at Chore Bagan, and several such charitable dispensaries that wanted to expand organizationally into hospitals were examples of such an endeavour.

The *Calcutta Journal of Medicine*, Sarkar's journal which pleaded for reforms in the Indian medical realm by arguing for medical pluralism and scientificity and on that basis supported and asserted the claims of homoeopathy, reported that the Banaras Homoeopathic Hospital and Dispensary was inaugurated by J. H. B. Ironside, a judge of Banaras on 25 September, 1867. Sarkar's editorial comment on the inaugural speech delivered by Ironside on that momentous occasion wished and hoped 'that [this] example of the holy city will soon be followed by Calcutta, [otherwise] so boastful of its enlightenment and liberality'.[40] Sarkar's tinge of sarcasm on the 'enlightenment and liberality' of Calcutta comes from the torrent of vulgar criticism and reprimand which he had faced on the declaration of his inclination towards homoeopathy just a few years ago. His exhortation to emulate this novel and noble example of the holy city of Banaras in establishing a homoeopathic hospital and dispensary at Calcutta reveals that there was no enduring homoeopathic hospital in Calcutta then, though there were individual dispensaries of Rajendra Lal Dutt and many others, including Sarkar's own dispensary. 'Banaras therefore may well be proud to be the first city in India which has had the moral courage to plant the standard of homoeopathy',[41] and for which the citizens and 'their descendants will be [ever] grateful'.[42] Babu Lokenath Moitra, one of the pupils of Rajendra Lal Dutt,[43] was in-charge of the Banaras Homoeopathic Hospital and had been practicing for two and a half years in this city.

> The people had adopted his system without knowing what it was—they had never heard of homoeopathy; he was a perfect stranger here and settled in the city by accident. Thus, homoeopathy established itself in Banaras entirely on its own merit and it is fit therefore that Banaras should be the first city in India *to adopt it practically* (emphasis added).[44]

that is to have a homoeopathic hospital.

J. H. B. Ironside, a man of law and jurisprudence and one of the founding benefactors of the Banaras Homoeopathic Hospital, was himself an amateur homoeopath and was practicing homoeopathy roughly from 1863 to 1864, i.e., four to five years prior to the establishment of the hospital. Like Sarkar, he also had ridiculed homoeopathy before becoming its advocate. Ironside in his speech confessed, 'I am a homoeopath myself and have been so for four or five years. I at one time ridiculed homoeopathy, but then I knew nothing about it; and I can, therefore, understand and make every allowance for people who ridicule it now'.[45] The homoeopathic hospital at Banaras was to serve as the site for the demonstration of the efficacies offered by this mode of healing and cure by taking people into confidence. Again, like Sarkar, Ironside too appealed for a fair trial: 'All we require for homoeopathy is that it should have a place as other systems have, and that it be tolerated as others are'.[46] 'All I ask is, that until this system has been tried and found to have failed, that then, and not till then, should it be rejected'.[47]

Ironside clearly understood and knew the perils of the path. For him, Moitra, and other benefactors the homoeopathic hospital was to be a means to go beyond the role of an individual practitioner by collectively coming together at a creative site and forging links with the patients at large. As homoeopathy arose as a marginalized alternative against the prevalent dominant pathy and without much state support, the urge and pragmatic necessity to validate and legitimize themselves propelled them to establish and carve their own hospitals emulating the structural arrangement and image of the hospitals of the dominant pathy, that is allopathy. It is here that the urge of homoeopaths to come together at a professional site of the hospital and to go beyond outdoor dispensaries to create, elevate, and catapult their dispensary status into indoor homoeopathic hospitals was rooted. Homoeopaths welcomed and celebrated this innovative step forward to get organized and linked to the people at large, but were also aware of the resistance from the prevailing dominant medical practice and its practitioners. As Ironside admitted, 'It is very difficult at all times to introduce anything new into medical use, to get people to believe in it, and adopt it, especially when they have made up their minds as to the absurdity of it'.[48] 'Medical men are specially prone to resist any new doctrine'.[49] 'It is not, therefore, to the members of the medical profession that we must look for any support in a new measure of this sort. The doctors will change their ideas when they see the tide of popular opinion running against them: they must follow their patients—it is to people that we must look for assistance in the introduction of homoeopathy'.[50]

The fact that we have reports of government dispensaries from early 1840s onwards from places like Patna[51] and Cawnpur[52] (Kanpur) authored by government medical men, along with the fact that from the mid-1870s, we start getting independent government reports[53] on government charitable dispensaries particularly in Bengal, clearly establishes the gradual proliferation of dispensaries far and wide after the 1860s. 'By 1842, six dispensaries had been established in Bengal'.[54] In the transition period which followed, there was a shift from pure government dispensaries to government charitable dispensaries along with the proliferation in their numbers. In the 1860s and the 1870s, their number 'increased rapidly, to reach a total of 255 by the end of 1879'.[55] The word 'charitable' hints at public or native participation, definitely of the wealthy and affluent sections of society. The dispensaries varied greatly in the ways they were funded and the services they provided. In the early 1840s, they were mostly initiated and funded by the colonial government. From the 1850s onwards, wealthy natives and zamindars also became involved in setting up dispensaries and supporting and sustaining them to such an extent that, by the 1860s, the government merely supervised them through annual inspections and supplied them with instruments and medicines. Private subscriptions from natives and Europeans provided the required funds.

In contrast, homoeopathic dispensaries evolved, remaining beyond the pale of the government's direct control; their support came solely from private subscriptions, from the 'lovers of the healing art'. The success of homoeopathic dispensaries, in the absence of demonstrable advantages derived from surgery, solely depended upon the practitioner's credibility and reputation for successful treatment or 'miraculous cures'. The fact that homoeopathic dispensaries successfully survived, helped this 'heterodoxy' to make way despite the orthodox opposition of allopathy. As our delineation has demonstrated, from the late 1870s and the early 1880s, homoeopathic dispensaries became a part of the plural medical reality of Bengal.

Homoeopathy and homoeopathist's aim and claim to medical and therapeutic reforms not only asserted their scientificity, but on that basis compelled dominant medicine to concede and engage with plurality to an extent. In this regard, they claimed to be contributing towards making medicine a 'truly liberal profession'. Against the backdrop of the ongoing blame game between allopaths and homoeopaths, D. N. Banerjee, the founder of the Calcutta Homoeopathic Charitable Dispensary[56] appealed:

> Let the relations of physicians be frank, gentlemanly, honest and sincere: envy, malice, backbiting, the mean lie, the equally mean silence, should be eradicated from the relations of gentlemen in any calling, and how much greater the necessity for such reform in a profession upon which is dependent, in great measure, the welfare and prosperity of the world. Let us then with 'progress' for our watchword ever strive to be worthy of our heaven-blessed mission to ameliorate human suffering.[57]

On the scientificity of drug proving and the appeal for reforms based on it, homoeopathy challenged some of the pretensions of the medical profession and desired to change it into a liberal profession. Through invoking categories like 'progress', 'catholicity', and 'plurality'[58] and the amelioration of human suffering, homoeopaths buttressed their own claims for legitimacy.

The Calcutta Homoeopathic Charitable Dispensary was conducted by its founder Babu D. N. Banerjee, in Mooktaram Baboo Street at Chore Bagan.[59] It was run by a managing committee, chaired and headed by a president and assisted by an honorary secretary and had several subscribing members, not only from Calcutta but also from the other presidencies. Many doctors and supporters of this healing art from Europe and America were also subscribing members of this dispensary.

This homoeopathic charitable dispensary was also the site for drug provings. *Ficus Indica* was proved by six provers at this institution and was elaborately dealt with in its annual reports. The managing committee prided upon the fact that *Ficus Indica* was first introduced into the homoeopathic materia medica by them. They conceded that 'although this is an infant institution, it undertakes to discover the knowledge which can contribute to the elevations of pain or of healing disease'[60] by sincere and scientific drug proving and disseminating its results from other provers in other parts of the country and the world.

It was the earnest desire of its founder D. N. Banerjee as well as the president Mr. Kalicharan Banerjee to elevate and catapult this charitable dispensary into a hospital so as to 'secure its permanency and improvement'.[61] All the lovers of the healing art were exhorted to help this institution and all the members were 'kindly disposed to show their best endeavours to'[62] '*change the institution into a hospital*',[63] and also help to 'establish shortly a library in this institution'.[64] The appeals for donations, subscriptions, and funds in this regard, always had an international slant; they were made to the 'lovers of the healing art throughout the globe'.[65] The founder of the Calcutta Charitable Dispensary appealed 'for funds in aid….to the people of [this] country and of Europe and America'.[66] Explaining the wish of the managing committee, he charted the future course of action thus:

> My friends and well wishers, both here and in foreign countries, *it is my earnest desire to establish a hospital*, and firstly, a library in connection with this dispensary. I therefore, pray that the public-spirited donor, my friends and well wishers and lovers of the healing art throughout the globe, will put forward their endeavours to help me in this undertaking (emphasis added).[67]

Help did come from homoeopathic physicians and drug manufacturers. Well-known homoeopathic companies or druggists were profusely acknowledged in this and subsequent annual reports[68] as lending a helping hand in sustaining this institution by their supply of *proven drugs*, thus helping to establish the validity and efficacy of this therapeutic option. The president of the dispensary expressed his gratitude towards Dr. Willmar Schwabe of Germany thus:

> Amongst the European members of this institution I have much pleasure in mentioning the well known name of Dr. Willmar Schwabe of

Germany, who renewed his gift by presenting this year a large collection of medicine and c., and by whose great kindness, the institution, showing sufficient progress, gave greater satisfaction to all the persons interested in that institution than in the past years.[69]

There were many others who from distant lands through subscriptions and knowledge sharing via homoeopathic journals were able to provide moral and material support to this dispensary. Dr. Southerland of America, Boniface Schmitz of Belgium, Oscar Hansen of Denmark and Alexander Villers of Dresden were among the patrons of this dispensary.[70] Among the drug manufacturing companies, Messrs. Mellins Food Co. for India, Ltd., London, Messrs. Burgoyne, Messrs. Bridges, Messrs. Cyriax and Farries, the well-known druggist and Her Majesty's contractors of London, were also acknowledged as according every encouragement to this 'infant institution'.[71] These firms promised to help the 'infant institution by annual subscriptions to make funds for securing the permanency of this charitable dispensary [and] also to promote this art of healing which by [then was] in a flourishing state in this country'.[72] 'Board of directors of Messrs. Mellins Food Co. contributed a sufficient quantity of food for the use of the indigent and helpless patients of this charitable dispensary'.[73] The desire of the managing committee was to usher in a transition from its dispensary status to hospital status for which it appealed for an establishment of 'an endowment fund', apart from gifts of money, books, medicines, and dietary items which helped the dispensary to deliver on daily basis.

We had already mentioned the fact that homoeopathy did not receive state patronage in the magnitude as was the case with dominant pathy which was more aligned and allied to the colonial state. It is not that people associated and organized around homoeopathy and homoeopathic dispensaries and hospitals did not canvass among the higher echelons of state power for support, but their prayers were not heeded to always. The Annual Report of 1888–1889 records the response of the Viceroy to an application that was submitted by Calcutta Homoeopathic Charitable Dispensary by its founder praying for the Viceroy's support to this institution of alternative pathy. Hiding behind protocol, the Viceroy expressed his well-wishing simultaneously remaining non-committal to direct financial support:

In reply His Excellency expressed his wishes that he views with favour the *voluntary creation of such institutions, but, as a rule, His Excellency does not feel call on to give monetary support to local charities*(emphasis added).[74]

The founder and the committee members of the dispensary, fully aware of the fact that their mode of healing was not judged at par with the orthodox and dominant mode of healing by the higher echelons of power generally, tried to mask the frustration caused by the polite but cold response of the Viceroy by the following stoic expression of gratitude:

> It is worthy to know that His Excellency expressed good wishes towards this infant homoeopathic institution, for which expression of kindness I beg to convey the profound venerations of the members of this institution to His Excellency.[75]

The lukewarm response of the state towards homoeopathic hospitals and dispensaries is also attested by the paucity of official sources available in archival repositories. Within the larger rubric of dispensary reports, which are in plenty, homoeopathic dispensary reports appear only sporadically, making it difficult to trace its institutional evolution and trajectory. The mergers, mutations, and their overall growth are yet to be rendered in historical writing. The fact that homoeopathic hospitals and dispensary reports are not found in continuity in the archive also reveals the limitations of the archive for such probings. Moreover, the lack of archival evidences only goes to support the argument that state patronage for homoeopathy was lukewarm. For example, the preceding and latter annual reports of this charitable dispensary, other than for the years 1888–1889 and 1889–1890, are not found in the home department and medical branch of the National Archives of India. This does not preclude the possibility of finding such reports at some other repositories and in some other compilation or under some other head. But the point is that records and sources pertaining to homoeopathic dispensaries and hospitals in India have not received adequate archival attention. In order to piece together the otherwise patchy histories of homoeopathic hospitals so that the evolutionary trajectory of these institutions can be outlined, we need to go beyond the archives so as to tap newer sources. Needless to say, these are yet to be explored and probed. It is precisely for these reasons, that I would like to provoke scholars to probe deeper into the institutional aspects of homoeopathy by hinting at the need to expand on the diversity of sources pertaining to this arena.

Press coverage as a marker of public attention to homoeopathic dispensaries can also be tapped. Press excerpts of the time provide some glimpses into the activities of these dispensaries and hospitals and their public acceptance. For instance, the claims of achievements of this very Calcutta Homoeopathic Charitable Dispensary of Chore Bagan were reported in some of the leading newspapers like the *Statesmen*, *Indian Daily News*, and *Amrita Bazar Patrika*. The *Indian Daily News* reported about the increasing success and accession of desirable members. It also complimented on the proving of the drug *Ficus Indica* by the Dispensary. Similarly, citing from the fourth annual report of the Dispensary, *Hope* of 30 September 1888, brought before the public 'that 1518 patients were treated during the year 1887–88 of whom 66.73 per cent were cured'.[76] It wished 'the Dispensary a long lease of life and increased resources for continuing its useful work'.[77]

All these newspapers reiterated the appeal for funds made by the Dispensary committee.

Through the above narrative, I have tried to pick up the threads of the historical evolution of homoeopathic hospitals in India. Over a century, these threads were woven into the fabric of homoeopathic institutions as they stand today. However, the steps by which this evolution and growth was made possible are yet to be historically assessed and ascertained. It is probable that the Calcutta Homoeopathic Charitable Dispensary of Chore Bagan was a mere upstart and failed to sustain itself due to lack of funds as well as state patronage. Like the Chore Bagan dispensary, there may have been several other attempts elsewhere in the country by homoeopathic doctors and enthusiasts to set up homoeopathic hospitals or upgrade existing homoeopathic dispensaries to hospital status. Some of these efforts may have failed while some may have survived and provided the institutional basis of the present-day homoeopathic institutions. The missing strands or gaps between institutions like the Calcutta Homoeopathic Charitable Dispensary and the homoeopathic hospitals of the present time need to be probed in order to get a more accurate picture of the evolution of these hospitals. What is important to note here is that such efforts existed not only in Calcutta but in cities like Banaras as well, and the above depiction is merely historical evidence of the beginning of such endeavours which went on to constitute the larger fabric of homoeopathic institutions we have in India today. The story of how this alternative pathy has arrived at its present stage and status is yet to be documented, contextualized, and rendered into meaningful historical narrative within the larger rubric of the social history of medicine.

Presence of homoeopathic hospitals; absence of its history

Coming to the present scenario, now that homoeopathy is well entrenched in India[78] and homoeopathic physicians having regular degrees have been recognized legally and accorded more or less equal status to physicians of allopathy,[79] and now that homoeopaths are organized into thriving professional bodies and societies that control and ensure the standard of their profession,[80] there has been a gradual proliferation of homoeopathic colleges which offer regular degree and diploma courses.[81] There are three hundred and seven homoeopathic hospitals in India and we have commensurate number of homoeopathic pharmaceutical companies, exclusive homoeopathic drugstores, and publishing houses specialized in its literature. Calcutta possesses the honour of having the most prestigious homoeopathic college and hospital—the National Homoeopathic Medical College and Hospital, Calcutta. Today, many homoeopaths are trusted and enjoy the confidence of their patients.

Apart from homoeopaths attached to government establishments, there are many streams of homoeopaths catering to different clientele. With the burgeoning of the middle class, a particular category of homoeopaths have adapted to advertising and marketing of their skills against chronic as well as life-style diseases targeting a particular class in society. In contrast, there are homoeopaths having modest dispensaries of their own. Some homoeopaths are also attached to charitable dispensaries of religious establishments like gurdwaras (Sikh worship places), affirming the ethics of charity and cure. Then there are those homoeopaths who practice for a nominal fee, but cater to a large geographical area and clientele of the semi-urban or rural hinterland. Their medicines are sought by the rural folk, not only for their family members, but also for the upkeep of their cattle and pets.

Homoeopathy and homoeopathic hospitals have endured the test of time and have become an integral part of today's plural medical reality of South Asia in general and India in particular. But, histories of homoeopathic hospitals encompassing their socio-economic and cultural contexts still await the deployment of the 'historian's craft'. If homoeopathic hospitals have existed and endured for the last one and a half centuries, then why are there no histories of homoeopathic hospitals? This cursory essay is more a chiding and an invitation to address this lacuna.

Notes

1 S. S. Goldwater, 'Concerning Hospital Origins', in Arthur C. Bachmeyer and Gerhard Hartman(eds.), *The Hospital in Modern Society*, New York: The Commonwealth Fund, 1943, p. 1.
2 Andrew R. Warner, 'The Possibilities of Future Development in the Service rendered by a Hospital to a Community', in Arthur C. Bachmeyer and Gerhard Hartman (eds.), op. cit., p. 15.
3 Deepak Kumar, 'Unequal Contenders, Unequal Ground: Medical Encounters in British India, 1820–1920', in A. Cunningham and B. Andrews(eds.), *Western Medicine as Contested Knowledge*, Manchester: MUP, 1997, pp. 172–190.
4 Zhaleh Khaleeli, 'Harmony or Hegemony? The Rise and Fall of the Native Medical Institution, Calcutta: c.1822–1835', *South Asia Research*, Vol. 21, No. 1, 2001, pp. 77–104.
5 Mel Gorman, 'Introduction of Western Science into Colonial India: Role of Calcutta Medical College', *Proceedings of the American Philosophical Society*, Vol. 132, No. 3, 1988, pp. 276–298.
6 Larger presidency medical colleges had hospitals attached to them.
7 Anil Kumar, *Medicine and the Raj: British Medical Policy in India, 1835–1911*, New Delhi: Sage, 1988. For Bombay Presidency see Mridula Ramanna, *Western Medicine and Public Health in Colonial Bombay 1845–1895*, Hyderabad: Orient Longman, 2002. For Madras Presidency, see D. V. S. Reddy, *The Beginnings of Modern Medicine in Madras*, Calcutta: Thacker Spink and Co., 1947.
8 Various studies on colonial public health policies in the Indian subcontinent are available. See, for instance, Mark Harrison, *Public Health in British India: Anglo-Indian Preventive Medicine 1859–1914*, Cambridge: CUP, 1994; Mark Harrison, *Climates and Constitutions: Health, Race, Environment and British Imperialism in India 1600–1850*, New Delhi: OUP, 1999; Kabita Ray, *History of Public Health: Colonial Bengal, 1921–1947*, Calcutta: K. P. Bagchi & Co., 1998; Sandeep Sinha, *Public Health Policy and the Indian Public: Bengal, 1850–1920*, Calcutta: Vision Publications, 1998; Poonam Bala, *Medicine and Medical Policies in India: Social and Historical Perspectives*, Lanham: Lexington Books, 2007.

9 Many studies on various epidemic diseases in colonial India have been made. See, for instance, David Arnold, *Colonizing the Body: State Medicine and Epidemic Disease in Nineteenth-Century India*, Delhi: OUP, 1993; I. J. Catanach, 'Plague and the Indian Village, 1896–1914', in Peter Robb(ed.), *Rural India: Land, Power, and Society under British Rule*, London: Curzon Press, 1983, pp. 216–43; Ira Klein, 'Plague, Policy and Popular Unrest in British India', *Modern Asian Studies*, Vol. 22, No. 4, 1988, pp. 723–55; Arabinda Samanta, *Malarial Fever in Colonial Bengal,1820–1939: Social History of an Epidemic*, Kolkata: Firma KLM Pvt. Ltd., 2002; Jane Buckingham, *Leprosy in Colonial South India: Medicine and Confinement*, Basingstoke: Palgrave Macmillan, 2002; Sanjoy Bhattacharya, *Expunging Variola: The Control and Eradication of Smallpox in India, 1947–1977*, New Delhi: Orient Longman, 2006; Sanjoy Bhattacharya, Mark Harrison, and Michael Worboys, *Fractured States: Smallpox, Public Health and Vaccination Policy in British India, 1800–1947*, New Delhi: Orient Longman, 2005.

10 Dr. John Martin Honigberger is considered the 'proverbial' introducer of homoeopathy in India. He had travelled far and wide before coming to Punjab in 1829 where he impressed Maharaja Ranjit Singh and remained at his court. In 1834, during his sojourn back to Europe, he had met Samuel Hahnemann and was considerably impressed by him. It is said that he bought huge amounts of homoeopathic medicines from Hahnemann's pharmacist. But as his book reveals, he did not entirely subscribe to Hahnemannian principles. Neither was he in favour of the mania of system making in the realm of medicine. Honigberger claimed that his path was a *medium system* approach, i.e., a smooth and middle course. He utilized many homoeopathic drugs with advantage in his treatment. See John Martin Honigberger, *Materia Medica* (in 2 vols), Vol. 1, 1st pub. 1852, rpt; Delhi: Low Price Publications, 1995.

11 *Calcutta Journal of Medicine*, Vol. 1,No.1, Jan. 1868, pp. 28–29.

12 J. H. B. Ironside was a judge of Banaras as well as a patron and practitioner of homoeopathy. He supported the establishment of Banaras Homoeopathic Hospital and Dispensary.

13 Sir Sayyed Ahmad Khan is remembered as the founder of the Anglo-Mohameddan Oriental College which subsequently became the Aligarh Muslim University. He was also the founder of the Aligarh Scientific Society which aimed at disseminating a scientific spirit by translating European books on science. But few know that Sayyed Ahmad Khan was also secretary to the committee which ran the Banaras Homoeopathic Hospital and Dispensary. See the exchange of letter between him and J. H. B. Ironside in *Calcutta Journal of Medicine*, Vol. 1, No. 11, 1868, pp. 46–70.

14 Official histories of the medical profession in India as late as 1923 did not pay any attention to homoeopathic physicians. One of the important commentators of the medical professional in India, Sir Patrick Hehir of the Indian Medical Service dismisses this entire arena of medical activity in a few cursory lines. See Patrick Hehir, *The Medical Profession in India*, London: Henry Frowde and Hodder & Stoughton, 1923.

15 See Peter Morrell, *Hahnemann and Homoeopathy*, New Delhi: B. Jain Publishers, 2003, p. 56. After this essay, Hahnemann had recorded the findings of his further proving which were first compiled in his *Fragmenta de viribus medica mentorum positivis* (1805) and later in his *Materia Medica Pura* (1811). Thus, 1796 is regarded as the birth year of homoeopathy.

16 To revise the prevailing polypharmacy, Hahnemann proposed three points: (i) that the scientific mode of ascertaining drug action upon human being is by experimenting them upon a healthy individual, (ii) that the healing properties of drug correspond to its disease-producing properties upon the healthy human organism, and (iii) that as a necessary consequence of the above two

propositions the drug must be administered in such a dose that will not produce too great an aggravation of the exciting or natural disease. These principles were to constitute the basis of the new pathy that he founded.

17 The consideration of Hahnemann's revolt is important because the Hahnemannian questions were once again raised by M. L. Sarkar on the eve of his appreciation of homoeopathy. Both Hahnemann and Sarkar were trained doctors of the regular dominant medicine. Both had to grapple with cholera, the former in the first half of the nineteenth century had an indirect encounter with it; the latter in the second half of the same century had a direct encounter with the malady.

18 Phillip A. Nicholls, *Homoeopathy and the Medical Profession*, London: Croom Helm, 1988, p. 3.

19 The regular dominant medicine followed the Cartesian–Mechanistic conception of the living organism, which perceived the human body as a complex piece of machinery.

20 For Hahnemann, it was not conceivable, nor provable '[...] by any experience in the world, that, after the removal of all the symptoms of the disease and of the entire collection of the perceptible phenomena, there should or could remain anything else besides health, or that the morbid alteration in the interior could remain uneradicated [...]'. Samuel Hahnemann, *Organon of Medicine*, translated by William Boericke, New Delhi: B. Jain Publishers (Indian Reprint), 6th edition (First published in 1843), 1979, p. 97.

21 Dr. Franz Hartmann played an instrumental role in the establishment of Leipzig Homoeopathic Hospital and he also worked there as the chief physician. In 1833–1834, he was also the co-editor of the yearbook of the Homoeopathic Hospital at Leipzig. See *Pioneers of Homoeopathy*, New Delhi: B. Jain Publishers, 2003, pp. 143–144.

22 Nicholls, *Homoeopathy and the Medical Profession*, op. cit., p. 11.

23 Thomas Lindsley Bradford, *The Life and Letters of Dr. Samuel Hahnemann*, (1895), Calcutta: Roy Publishing House, 1970 (First Indian edition), p. 256.

24 Morrell, *Hahnemann and Homoeopathy*, op. cit., p. 34.

25 Dr. Maclouchlin, a medical inspector who was surveying the results of the treatment of cholera in different hospitals in London found the London Homoeopathic Hospital's record so outstanding, that it was eventually concealed by the Medical Council in its Blue Book. It was only when a Member of Parliament who also happened to be a patient of homoeopathy demanded that the results of the London Homoeopathic Hospital be revealed, that a second Blue Book with separate figures from the non-homoeopathic hospitals was prepared and produced before the Parliament along with a letter by the medical inspector Dr. Maclouchlin, an orthodox medical man. The London Homoeopathic Hospital revealed a death rate of 16.4 % against the 51.8 % shown by other hospitals. Dr. Maclouchlin noted in his official report that he had witnessed even extreme cases of cholera revive under homoeopathic treatment and testified in the conclusion of his letter that: 'If it should please the Lord to visit me with Cholera I would wish to fall into the hands of a homoeopathic physician'. See Harald C. Gaier, *Thorsons Encyclopaedic Dictionary of Homoeopathy*, London: Thorsons, 1991, pp. 164–165.

26 For more examples and anecdotal accounts of this nature, see Surinder M. Bhardwaj, 'Homoeopathy in India', in Giri Raj Gupta (ed.), *Main Currents of Indian Sociology*, Vol. IV. New Delhi: Vikas, 1981, pp. 31–54. Also see the chapter 'Rise and Development of Homoeopathy in India's Past History', in Sharat Chandra Ghose, *Life of Dr. Mahendra Lal Sircar*, Calcutta: Hahnemann Publications, 1935, pp. 27–83; and Gary J. Hausman, 'Making Medicine Indigenous: Homoeopathy in South India', *Social History of Medicine*, Vol. 15, No. 2, 2002, pp. 303–322.

27 Dr. Mahendra Lal Sarkar was one of the earliest and most famous 'converts' to homoeopathy in nineteenth-century India. Sarkar was born on 22nd September 1833, graduated in medicine from Calcutta Medical College and obtained the degree of Doctor of Medicine (M.D.) in 1863, and declared his faith in homoeopathy in 1867. He lived as an accomplished and much sought-after homoeopath all his life. He died on 23rd February 1904. He founded an institution called Indian Association for the Cultivation of Science (1876) for the promotion of scientific spirit and basic research in science among natives in India. Sarkar's journal, the *Calcutta Journal of Medicine* which he single-handedly edited for more than twenty years played an instrumental role not only in the dissemination of homoeopathy but also in animating the medical debates of those times.

28 His 'change of creed' was very much a professional decision—a decision taken as a 'physician awakened to a sense of awful responsibility of his calling'. Mahendra Lal Sarkar, *A Sketch of the Treatment of Cholera*, 2nd edn (1st edn. Pub. in 1870). Printed by P. Sircar, Calcutta, Anglo-Sanskrit Press, 1904, pp. iii to v. However, Sarkar was also influenced by the death of his mother who fell a victim of cholera when she was only thirty-two years of age and when Sarkar himself was barely four! See Ghose, *Life of Dr. Mahendra Lal Sircar*, op. cit., p. 2. Cholera epidemics rekindled the memory of his mother and challenged him as a doctor.

29 *Calcutta Journal of Medicine*, Jan 1868, op. cit., p. 25.

30 Ibid, p. 25.

31 This tension is well expressed in Sarkar's book *A Sketch of Cholera* in the context of epidemic diseases.

32 See Dhrub Kumar Singh, 'Choleraic Times and Mahendra Lal Sarkar: The Quest of Homoeopathy as "Cultivation of Science" in Nineteenth Century India', *Medizin, Gesellschaft und Geschichte*, No. 24, yearbook of the Institut für Geschichte der Medizin der Robert Bosch Stiftung, Germany, 2005, pp. 207–242. Engagement with homoeopathy was construed as an engagement at a higher level with science. Homoeopathy was seen by its practitioners as more scientific. Assessing and conceding to the plurality of medicine on a scientific basis, so as to derive benefit from the plural tradition of medicine, was Sarkar's objective.

33 Sarkar along with many physicians regarded this as Hahnemann's singular contribution against the prevailing arbitrariness of administering medicine in heavy doses which itself had become a malady.

34 To understand this in the British context, see Nicholls, *Homoeopathy and the Medical Profession*, op. cit.

35 The entire *Calcutta Journal of Medicine* not only articulates the intellectual and pragmatic necessity of choosing the right therapeutic option but also defends the right of a physician to do so. Similar views regarding the status of physicians, the context of Sarkar's 'conversion' and his right to choose his therapeutic option, the possibility of dialogue between allopaths and homoeopaths, and the consequent bearing on the profession as a whole also got articulated in nineteenth-century Bengali journals like *Chikitsa Samelini*.

Sarkar's eclectic outlook was not without reason and rationale. Neither was he the first to tread such a path. Dr. John Martin Honigberger, the 'proverbial' introducer of homoeopathy way back in the early years of 1850s defended his right as a researcher-physician. He also defended his pragmatic right and choice by deriding homoeopathy as advocated by Hahnemann, supporting his position as a '*medium* between two extremes; a system grounded on experiments, which [he] advisedly adopted, and which success impelled [him] to pursue' (p. vi). He 'regard[ed] the *two* medical systems, *Alloeopathia* and *Homoeopathia* as two opposite poles' (p. xi). 'Mine is a smooth and middle course', he averred and categorically stated that 'I am not under the influence of the mania of *system making*' (p. vi). His, as he claims, was a *medium system* approach. All three—cholera, homoeopathy, and Hahnemann—find a place in Honigberger's account.

Sarkar too seriously engaged himself with all three. Can one surmise that Sarkar had read Honigberger's account? In all probability, yes. See Honigberger, *Materia Medica*, Vol. 1. op. cit.

36 Rajendra Lal Dutt was one of the initial propagators of homoeopathy in Bengal who provided it a firm foundation. It was he who brought homoeopathy into high esteem, both among the elite and the laity of Bengal. It was he who chided Sarkar to have an insight into the working of remedies based on homoeopathic principles.

37 Babu Lokenath Moitra was the physician in-charge of Banaras Homoeopathic Hospital. He had also been a pupil of Rajendra Lal Dutt.

38 Like Sarkar, P. C. Majumdar was also a product of Calcutta Medical College. He received his LMS degree from Calcutta Medical College in 1878. He was induced and encouraged by his father-in-law Dr. Bihari Lal Bhaduri, who himself was an alumnus of Calcutta Medical College, to appreciate homoeopathy. Majumdar's father-in-law served as Government Medical Officer in parts of Orissa and Bihar but retained his close contacts with Dr. Salzer and Rajendra Lal Dutt. Dr. Bhaduri himself relinquished allopathy to become a homoeopathic physician. Majumdar earned immense experience and insight working as an assistant to Dr. Salzer. Later on, he established himself as the foremost homoeopathic practitioner. He in collaboration with Dr. D. N. Roy, were instrumental in establishing the first Homoeopathic teaching institution, the Calcutta Homoeopathic Medical School which later became a college in 1881. Majumdar also edited the *Indian Homoeopathic Review*, the oldest homoeopathic journal in India after Sarkar's *Calcutta Journal of Medicine*.

39 Sarkar's uncertainly speech 'On the supposed uncertainty in medical science and on the relationship between disease and the Remedial Agents' was the turning point of his 'conversion' to homoeopathy leading to the consequent episode of his being thrown out of the Bengal chapter of the British Medical Association.

40 *Calcutta Journal of Medicine*, Jan 1868, op. cit., p. 26.

41 Ibid, p. 27.

42 Ibid, p. 27.

43 This is the same Rajendra Lal Dutt who had chided Sarkar to witness the truth of homoeopathy in his perturbing years of uncertainty and was instrumental in Sarkar's 'conversion'.

44 *Calcutta Journal of Medicine*, Jan 1868, op. cit., pp. 27–28.

45 See the inaugural speech of J. H. B. Ironside delivered on 25 September, 1867, on the occasion of the inauguration of the Banaras Homoeopathic Hospital and Dispensary and as reprinted by Sarkar in his journal, *Calcutta Journal of Medicine*, Jan 1868, op. cit.., p. 27.

46 Ibid, p. 26.

47 Ibid, p. 26.

48 Ibid, p. 27.

49 Ibid, p. 27.

50 Ibid, p. 27.

51 S. Davis, *Report of Government Dispensary, Patna. Prepared with Inputs sent by Ram Eshur Awushtee, In charge Government Dispensary January 1840*, 1840.

52 J. McRae, *Report on Cawnpur Dispensary, 1st February 1841*, 1841.

53 In the intervening period, dispensary reports were clubbed together or were part of the general reports on lunatic asylums, vaccination, and dispensaries.

54 *General Report on Lunatic Asylums, Vaccination, and Dispensaries in the Bengal Presidency, 1868*, p. 50,V/24/664, OIOC in Christian Hochmuth, 'Patterns of Medical Culture in Colonial Bengal, 1835–1880', *Bulletin of History of Medicine*, Vol. 80, No. 1, 2006, p. 57 (39–72).

55 *Dispensaries*, 1880, V/24/744, OIOC, p. 3 in Christian Hochmuth, Ibid, p. 57.

56 Information on the Calcutta Homoeopathic Dispensary has been gleaned from the two reports found in the National Archives of India (hereafter NAI). Though

this dispensary was founded around 1883, reports prior to 1888–1889 and following 1889–1890 are not available.

57 NAI, Home Department, Medical Branch B, November 1889, Pro. Nos. 5 to 7. 'Annual Report of the Calcutta Homoeopathic Charitable Dispensary for the year 1888–89', p. II.

58 Plurality can be applied with reservation to some of the hardcore and strict Hahnemmanians. But many like J. H. B. Ironside from the realm of homoeopathy and Bhola Nath Bose from the realm outside it appealed to the spirit of plurality. They advocated the toleration and scientific appreciation of alternate ways of looking at the body, health, and disease, as long as such alternatives could be proved to be bringing benefit to mankind.

59 The annual report of the year 1888–1889 informs us of its sixth year of existence, hence this dispensary probably came into being around 1882–1883.

60 'Annual Report of the Calcutta Homoeopathic Charitable Dispensary for the year 1888–89', op. cit., p. 1.

61 Ibid, p. I.

62 Ibid, p. I.

63 Ibid, p. I; emphasis added.

64 Ibid, p. 5.

65 Ibid, p. II.

66 Ibid, p. II.

67 Ibid, p. II.

68 NAI, Home Department, Medical Branch B, December 1890, Pro. Nos. 97 to 99. 'Annual Report of the Calcutta Homoeopathic Charitable Dispensary for the year 1889–90'.

69 'Annual Report of the Calcutta Homoeopathic Charitable Dispensary for the year 1888–89', ibid, p. 3.

70 Ibid, p. I.

71 Ibid, p. I.

72 Ibid, p. 3.

73 Ibid, p. 3.

74 Ibid, p. 4.

75 Ibid, p. 4.

76 Ibid, p. 13.

77 Ibid, p. 13.

78 There are over 200,000 homoeopathic practitioners in India. See Eswara Das, *History and Status of Homoeopathy around the World*, New Delhi: Jain Publishers, 2005, p. 7.

79 In India, homoeopathy is integrated into the services of Primary Health Care (PHC). Consequently, we find the offer of homoeopathic treatment in government dispensaries coming under the rubric of the Central Government Health Scheme (CGHS). Ute Schumann, 'Dimensions of Health Planning in the Homoeopathic Medical System', in R. N. Pati (ed.), *Health, Environment and Development*, New Delhi: Ashish Publishing House, 1992, p. 107.

80 The Central Homoeopathy Council Bill was passed by the parliament in 1973. In 1975, the Central Council of Homoeopathy was constituted by the Government of India and, in 1982, homoeopathic practitioners (professional conduct, etiquette, and code of ethics) regulations were framed. Das, *History and Status of Homoeopathy around the World*, op. cit., p. 111.

81 There are around 180 Homoeopathic Medical Colleges in India, 31 of them conducting postgraduate courses in different specialties. Thirty-two colleges from among the total are maintained by the state. All these colleges after the publication of standardized undergraduate education regulations in 1983–84 are affiliated with universities and are conducting a uniform standardized five and a half years degree course. See the official website of AYUSH, Ministry of Health and Family Welfare, Government of India, and Das, Ibid, p. 7.

15 Saviour sisters

Services of the Delhi female medical missionaries in late colonial India

Ch. Radha Gayathri

Every society expects women to go through the process of childbirth and society considers as its duty to find ways and means to minimize the risk entailed in such a process. There was no reliable statistics, yet it was estimated in 1936 that the number of women dying every year during childbirth was nearly one and half lakh and the inclusion of the deaths due to abortion would swell the number more. As per the WHO statistics in 2017, the Maternal Mortality rate in India is 145 per 1 lakh live births which translates to an average of about 35,000 deaths in a year.[1] It is far away from the sustainable development goal of 70 by 2030.

The present chapter tries to analyse the contributions and experiences of female medical missionaries through their narratives who worked with Delhi Female Medical Mission, i.e., St. Stephen's Hospital at Delhi. It was started by efforts of missionary women who had no formal medical training and St. Stephen's hospital gradually grew into huge medical institution, which is surviving even to this day.

> A female medical mission may be defined to be the practice of medicine by a lady for the purpose not merely of curing but of Christianizing her patients…. This is a key which may be said to fit every lock. She would find an entrance where the educational missionary would find it closed. She would soften bigotry, remove prejudice, dispel ignorance, drive away gloom, and unobtrusively but effectually deposit the all-pervading leaven of the gospel in numberless hearts and homes…[2]

Dr. Elmslie, a Presbyterian medical missionary from Kashmir thus pleaded, before the London Zenana committee in 1871, for the need of female medical missionaries for India. In a way, these words sum up the perceived role of women medical missionaries in India. Christian missionaries along with colonial administrators, Indian social reformers, and the nationalist played a very important role in the reform movement aiming at the general betterment in the condition of women in India during the late eighteenth and nineteenth centuries. They were the pioneers who have strived for the introduction of western health care system for women in India. Large number of

DOI: 10.4324/9781003241980-20

unmarried women came as missionaries to India from 1880 onwards and by 1900, two-third of missionaries in India were women. Female missionary doctors, be it through the church or mission hospitals, Dufferin Fund, or the state sponsored hospitals, have contributed immensely for the promotion of western medical systems for maternal health care and medical education in India. However, these women missionaries appear as mere 'adjuncts' even in the published records and histories of various missions, finding mention usually in the end as 'women's work'.[3]

Research on this area was done in last few years by some scholars like Rosemary Fitzgerald, Kumari Jayawardena, Ruth Brouwer, Leslie Flemming, Jharna Gourly, etc.[4] They showed that modern missionaries were not merely religious agents but were also cultural and social agents. While discussing about female medical missionaries, many issues and questions emerge like the motive behind the female medical missions, relation between the church and the colonial rulers, the attitude of missionaries towards Indians women and health systems, the response of the Indians, especially Indian Christians, success of missionaries at evangelization, Indianization of church, and church institutions, the depictions of the native *dais*, etc. The role of Indian Christians and their contribution in the promotion of church in the colonial context is another issue which has not been explored much. In earlier missionary records, Indian Christians exist only as loyal and nameless employees. This becomes more important from 1920 onwards as Indianization of churches has started by then.

Missionary roles gain more significance when set in the colonial context. Jeffery Cox tried to further analyse the dynamics of exchange between the missionaries and the indigenous people in terms of exchange, even though it was a skewed and asymmetrical exchange. Thus, he says:

> Missionaries and Indian Christians were in many respects engaged in a common enterprise, creating something new that was neither European nor Indian but simultaneously indigenous, foreign, and hybrid. Furthermore, missionaries in India were important to Indians, and in some circumstances were, and remain, respected and admired figures in the Indian Christian community and in the collective memory of educational and medical institutions that they helped to build. Missionaries also found themselves in a zone of 'transculturation'.[5]

Women missionaries, single and professional, were another distinct feature of modern missionary movement. From the early nineteenth century onwards, these single, independent women missionaries and nuns began travelling to Asia and Africa unlike their predecessors who were mere adjuncts to their male relations. They came in response to a belief that they had been 'called' to this work and they started women's groups, girls' schools, orphanages, and convents. Through these women's institutions, they were able to enter the homes of married or secluded women to teach languages and other skills and spread the gospel. They were touring for 2–3 years and did not stay permanently in field.

Missionary enterprise in India

The missionary activity had two main components, i.e., the church or the ecclesiastical establishment and the missionary societies. The Anglican chaplains for the Europeans in the East India Company's civil and military services constituted the ecclesiastical part of the Church of England. The Church Missionary Society (CMS), the Society for the Propagation of the Gospel in Foreign Parts (SPG), the Church of England Zenana Missionary Society (CEZMS), the Zenana Bible and Medical Mission, the Society of St. Hilda, and the Cambridge Mission to Delhi (which was closely related to the SPG) were different missionary societies associated with the Church of England. Local committees of Europeans, who were eager to spread the Christian faith among the Indian population, generally invited and supported financially missionary societies into the cities of United Provinces and the Punjab.[6] Various societies associated with different denominations were established in North India during the eighteenth and nineteenth centuries. By the turn of the nineteenth century, missionaries were well spread in major towns and cities of North India.

Missionary activities

Integral to the process of evangelization was the social work like spread of education, medicine, running of orphanages, old age homes, etc. Western progress and enlightenment were provided as ideological and institutional alternatives. The missionaries also realized quite early that the massive task before them was not purely religious. Though conversion to Christianity was the main objective of the women medical missionaries, they also strived to bring about some social changes, which they believed would follow from religious change.[7] It is in this context that modern missionaries sought to reform the conditions of women in India and thereby promoted female education and health care systems. As a result, from the early years of British rule in India, Christianity began to assume an additional role in challenging prevalent social customs and family structures that affected the status of women. Colonial administrators and missionaries discovered common interests of social reform or imposition of western values and this convergence of interests often looked like an identity of interests to outsiders.

Women's work for women

Women played a crucial role in almost all areas of Protestant and Catholic missionary endeavour. Although women had done missionary work for years, they had done so in subordinate capacity as assistants to officially church-designated missionaries. Women's presence in the field was not only an indicator of peaceful intentions and thus ensured of friendly reception by the local people but also served as role model of female behaviour. 'In effect, missionary wives were not only 'married to the job' but they were often married for the job'.[8] From the 1860s onwards, there was a change of attitude towards the role of women in the missions and there was a realization that the women missionaries were needed to approach women who are in seclusion. 'Objections to women missionaries receded and were replaced by increasingly insistent pleas for "virtuous" and "valiant" women – "the more highly cultivated and refined the better"– to dedicate their lives to missionary service overseas'.[9] These women missionaries not only sought the conversion of heathen women but also strived to bring some relief to these secluded women.

By the beginning of the nineteenth century, almost all the missionary societies started forming the women's branches and women missionaries were commissioned to carry forward 'women's work for women' started by the female relatives of the male missionaries. One of the earliest Ladies Auxiliaries to be established was of the SPG and it sent women as missionaries to be attached to SPG missions in various places in India. In 1890, the CMS laws were revised to cover cases of female candidates received by the Ladies Committee. Similarly, London Missionary Society also saw the emergence of women's auxiliary in the nineteenth century. 'Women's work' in the LMS was progressively transformed from a 'labour of love' conducted by missionary wives to a professional employment carried out by single women missionaries.[10] By 1915, there were 333 women engaged in education, medical work, and evangelism. 'Not violence, not emotion, but logic, led the SPG to pronounce in 1902 that as women's work was a vital part of the missionary effort of the church, it should be organised and maintained by the society'.[11]

This sudden emergence of single women coming forward to work in distant lands is better understood in the background of the feminist movement, especially 'Quaker movement' witnessed in west during the nineteenth century.[12] During this period, a network of Christian women emerged proclaiming global sisterhood and they ventured out of their homes into the male world of work. They travelled alone to far off countries in the name of the 'noble cause' to serve God and improve the condition of women. These middle-class single women were mostly professionally trained, and they entered paid employment in large numbers. 'Rather than a trope of domestication, it is a trope of emancipation, which organizes the representations of the 'mission of sisterhood' within the missionary texts'.[13]

One of the outcomes of the bitter struggle for higher education in the west and of the feminist movement of the early twentieth century was the emergence of women professional doctors. Some of these early women doctors like Clara Swain, Elizabeth Beilby, Edith Pechey, Fanny Butler, etc., worked in India and these doctors had the additional burden of social reform besides the day-to-day work in the hospitals and evangelization. Some scholars felt that undeniably the missionary women were philanthropic yet there was an issue of 'opportunity'. The perception that foreign medical women were needed in India played an important role in opening medical schools to British women.

Women in the North India Missions were generally engaged in four spheres of activity, i.e., education, medicine, zenana visiting, and itineration. A second sphere of activity was the provision of Western medical care to women and girls. Medical attention, it was also felt, would be an important means for missionaries to get access to females secluded in the zenanas, many of whom were reluctant to consult male physicians. Zenana missionaries, seeing the miseries of some of the women and children, would act as go-betweens, reporting to a doctor and bringing medicines.

In the years before medical mission, untrained women missionaries also tried to introduce new methods of tackling health problems. Using the home nursing skills, most missionaries also ran 'veranda dispensaries' and dispensed simple medicines on their evangelical tours through the rural areas.[14] This was not very satisfactory; so some missionaries studied what medical books they could lay hands on, learned the use of simple remedies, and did what they could to relieve sickness among zenana women. On their first furlough itself, some of them sought opportunities to increase their knowledge. For example, Miss Rose Greenfield of the society for Female Education in the East came to teach in zenanas but seeing the need for medical aid, she attended clinics conducted by her brother in the guise of a nurse.[15]. During this period when women had no access to medical education in Europe and the United States, some institutions in the West gave short courses in medicine and midwifery to missionaries.[16] The medical work of missions started as secondary activity but later on became the central purpose of medical missions.

Western women physicians made space for their work in indigenous society with great efforts as local women mostly consulted practitioners of traditional medicine like *hakims* and *vaidyas*. It was felt that these women carried the Gospel of God into these dark areas.

> Called to the inner most recesses of harem and zenana to take pity on mother or child, the woman Missionary doctor came 'as the first streak of God's pure sunlight which permeates those polluted prisons; as the lowly yet true herald of that Sun of Righteousness risen with healing on his wings'.[17]

Contrary to the norms of medical practice in Western societies, women physicians in India had to reach out and make themselves available to those who were sick, wherever they were. House visits were important confidence-building measures, before patient could be expected to visit the dispensary. These medical missionaries along with their conscious evangelism, provided a distinct and often life-saving service to Indian women and their families. Missionary women doctors were the pioneers in establishment of medical education for women in India. Missionary physicians ran, often with only a skeleton staff and provided medical services at smaller dispensaries in provincial towns and even in villages.

In the early decades of the twentieth century, the western women physician's efforts led to the establishment of some of the great mission hospitals and medical educational institutes in India. The small-scale training classes for Indian girls were formalized to give certificates and diplomas that equipped them to launch their own dispensaries. Finally, by the middle of the twentieth century, these institutions were compelled to keep pace with the times and match the best standards of medical education and clinical facilities available in the general non-segregated hospitals in India.[18]

St. Stephen's Hospital is a famous Missionary institution established during this period and female missionaries who had no medical training laid the seeds for its establishment. Small dispensaries and training units were taken up by professional medical missionaries and developed into huge establishments. The initial struggles and efforts to reach out, the personal experiences of single female missionaries, the disappointments, success, etc., reflect the immense struggle under gone by these missionary women.

St. Stephen's Hospital, Delhi

One of the earliest Ladies Auxiliaries to be established was of the SPG and women sent out as missionaries were attached to SPG missions in various places in India and their term of service was for three years.

Society for the Propagation of the Gospel was established in India in 1820 and its associate organization Cambridge Mission to Delhi was established in 1877. In 1867, the SPG established the first female medical mission in India, the Delhi Female Medical Mission. Ms. Priscilla Sandys came to Calcutta in 1858 at the age of 16 and got involved in female education work. After her marriage to Rev. Robert Winter, she shifted to Delhi and started working for SPG and CMD. After her arrival to Delhi in 1863, Priscilla Winter found not a single zenana open for instruction in the Punjab and North West Provinces. It is said that she began a system of zenana visitation and teaching[19] and this gave her the opportunity to see the crying need of medical aid for women in the zenana. She started her medical work in 1864.[20] Although she had 'no further medical qualification than a medicine chest', she began to hold an open-air dispensary at the women's bathing *ghats* on the western bank of the holy river Jamuna. Here, she distributed simple remedies and gave advise to all classes of Hindu women, the majority of whom would not go to male physician. Though these women lived secluded in zenanas, they would go down every morning to the river both to make their vows and dip in the sacred stream.[21] This rudimentary dispensary and Mrs. Winter's home nursing of women during epidemics of fever and cholera formed the modest origin of what later became the Delhi Female Medical Mission as well as the St. Stephen's Hospital.

Mrs. Winter worked hard to run the schools and spread the medical aid for women.[22] She called in the help of others through 'White Ladies Association' to run a small dispensary in the city. In 1868, the Winters were visiting England and they began to collect subscriptions for the sending out and support of a lady medical worker, whose work was to attend native ladies in their zenanas, to set on foot a dispensary for women only and to train native women as nurses. There are a great number of respectable but destitute women in Delhi, for whom a means of an honest livelihood will thus be provided.[23]

The Civil Surgeon of Delhi heartily approved the scheme, promised his assistance in supervision, and donated two gold *mohuras* (32 rupees). The Winters raised about 290 pounds and had the assurance of a body of friends interested in the scheme. The first medical missionary was Mrs. Browne who left England in September 1867, but she was dismissed in June 1868. In 1874, a house was rented where patients could be treated in dispensary hours and a woman worker was engaged to manage the dispensary to train nurses and to visit women in their houses. This dispensary evolved into a medical zenana work specializing in midwifery under Miss Engelman, a German. Ironically, Miss Engleman joined the Delhi Mission in 1871 essentially as a teacher-cum-evangelist. When she was posted at Karnal, she picked up interest in medical work and had rudimentary medical training under Dr. Bose, a Christian Bengali doctor. Dr. Bose was also the visiting physician at Delhi dispensary in later years and Miss Engleman lived in the hospital, attending out-patients, teaching nursing students, responding to calls in zenanas, and reading the bible and prayers with students and patients.[24]

With a passion for Christianity embodied in institutions, Miss Engelman and the Cambridge Brothers established St. Stephen's Hospital for Women, in memory of Mrs. Winter who died in 1881 at the age of 39. In 1885, this new hospital with 50-bed capacity, over-looking Queen's Gardens, also known as Company Bagh, was built at Chandni Chowk. The foundation stone was laid by the Dutchess of Connaught on 18th January 1884 and the Hospital was formally opened by Lady Dufferin, the Vicerine, on 31st October 1885. The ground floor was used for the outpatient department; the first floor had rooms for 2doctors, a sister, evangelist, a dispenser, and 10 or 12 nurses. The top floor had a few small wards, which could with difficulty accommodate 20–30 patients.[25] By 1888, she and her Anglo-Indian assistant Alice King lived in a new 30-bed hospital overlooking the Queen's Gardens in Delhi. In 1891, Mr. Winter died and in memory of both Robert and Priscilla Winter, an extension was built to the hospital.

The first full-time doctor in St. Stephen's Hospital was Dr. Jenny Muller who took over as the head of St. Stephen's Hospital in 1891 and worked till 1916. In fact, Jenny Muller came to India to help in the Teacher's training class run by Mrs. Winter, but she got interested in medical work. She later joined Calcutta Medical College and returned as a fully qualified licentiate doctor. It was during her time that the site on which St. Stephen's Hospital now stands was acquired and the present Maternity Wing was built. On 3rd December 1906, the foundation stone of the old hospital was laid by the Countess of Minto at Tis Hazari, overlooking what was then a Police Parade Ground and now the Tis Hazari Courts. On 9th January 1909, the new hospital in Tis Hazari was formally opened. However, in 1908 itself, the doctors' house and out-patient block were finished and opened for work. The Indian staff, consisting of five *dais* (midwives), three dispensers, and seven nurses, together with ten or 12 patients who had to be admitted, were housed, fed and cared for wherever there happened to be vacant spot in the dispensary building.

For the first year or two, the daily average in-patient figure was 25–40; the total for the year being 550–880 odd. The Internal Obstetrical work was low, 20–60 patients in the year; while the External figured at about 208, making a total of 230, of which about 75 needed operative assistance.[26] Dr. Mildred Staley was the first MBBS doctor and she joined St. Stephen's in 1893. She described a typical day of her life at St. Stephen's in the following words:

> Woken at 5 am with *chotahazari*. 'Out' seeing patients by 5.45. Back to take prayers in Urdu by 6.45. Brief ward round, then Dispensary till 11 am. Breakfast. Ward round. Office job till 8 pm. Dinner and more Patients.[27]

Many issues of class, caste, gender, and religion and the local preferences were to be considered and incorporated by the western women doctors in their efforts to build viable medical institutions. In this process, these doctors had to clearly flout prescribed rules of patient care and hospital administration as they had been taught at medical schools in Britain and America. For example, often the abnormal obstetric work was done in the patients' homes as they refused to come to the hospitals and were in critical condition.

> A 'band-gari' (a box with seats back and front on four wheels, with a pair of derelict horses, and the driver on the cool with ropes for reins) would proceed with a portable table, large sterilizers with instruments in them, another large box with trays and medicines tied on to the roof, the anesthetic apparatus and a few smaller bowls and trays carried inside on one's knees, and two hurricane lanterns on the floor, for often the only light in the house would be a vile smoking wick in an earthen saucer of oil. The surgery on most occasions was antiseptic rather than aseptic, and at critical moments anxiety was hair-raising. Even in hospital on visiting days the relations would not hesitate to remove the dressings of a patient who had had a major operation.[28]

Baby welfare clinic was started in connection with Maternity Department in 1933. In the second half of the 1930s, school inspection work in Delhi was also taken up by the Hospital. In 1939, St. Stephen's Hospital had 3 European and 3 Indian doctors, 3 European nursing sisters, 1 Indian nursing sister, 1 pharmacist, 1 Indian staff compounder, 1 Evangelist, 2 Indian bible women, and 1 Secretary-housekeeper. All these were missionaries. Besides these, there were 4trained nurses who were native assistants; 44 nurses and five compounders were in training.[29]

However, the external maternity work had decreased partly due to the Municipal Welfare Centres which had trained midwives, but greatly due to antenatal clinic begun by Dr. Houlton in 1928. Many of the doctors who served at St. Stephen were in demand for other sister missions, Indian Medical Service, colleges, etc. Some of these doctors held very high posts, yet continued to serve at the mission too. In 1913, Dr. Helen Franklin joined the staff and worked in St. Stephen's Hospital until 1920 when she became Vice-Principal and Professor of Surgery in the newly established Lady Hardinge Medical College. She continued to give part-time services to St. Stephen's Hospital. In 1937, she left Lady Hardinge Hospital, took up full-time work in Ranchi and later in St. Stephen's Hospital for a year till her retirement in 1945. Dr. Millicut Webb joined Women's Medical Services three years after coming to India to become CMO. In 1917, Dr. Dorothy Scott joined St. Stephen's succeeding her cousin, Dr. Agnes Scott. She was invited to start a tuberculosis sanatorium near Kasauli in 1927. The same year, Dr. Charlotte Houlton came to India and she went on to become the principal of Lady Hardinge Medical College in 1933. She was a member of the SPG and was involved in the planning of the All India Institute of Medical Sciences. In 1941, she was awarded the Kaiser-i-Hind for her services in India and on her retirement in 1961, was awarded the M.B.E by the British Government between 1919 and 1929 no fewer than 17 new staff arrived from England.

Trainings

Besides trained doctors, there were many nurses who came to serve at St. Stephen's hospital. In 1908, Sister Alice Wilkinson arrived in India from Britain. The Nurses Training School of the hospital was started in her time. She became Nursing Superintendent and was responsible for raising the standard of nursing not only in St. Stephen's Hospital but also in whole of India. Being founder member of the 'Trained Nurses Association of India', she worked as its Secretary until she left India in 1948. She continued working in the S.P.G. House in London until her 90th birthday when she retired and returned to India to spend her last days here. She worked to the last to bring together countless nurses in India and abroad and she died in St. Stephen's Hospital at the age of 92 on 15th May 1967.

Sister Wilkinson was the founder-member of the trained Nurses' Association. She was also involved in the founding of the College of Nursing at Delhi. In 1913, a Board of Missions was formed for examining nurses as several hospitals felt the need of a uniform syllabus and examination. In 1918, the name 'The United Board of Examiners for Mission Hospitals in North India' was adopted. Arrangements were also made for examinations to be held in different languages as its sphere extended from Quetta and Mardan on the North West Frontier to Hazaribagh and Mission hospitals beyond Calcutta. A definite course and examination for dispensers was added and later even non-mission hospitals were affiliated. Sister Wilkinson writes frankly on the training of nurses in the following words:

> I do not think the general standard of the nurses work has been up to its usual mark, due in a great measure to they being new sisters not yet versed in the ways of an Indian hospital nor sufficiently conversant with the fact that Indian nurses are mostly very young and inexperienced girls who have come straight from the school room; and also due to the fact that they do not play the game, and instead of rising to the occasion will often let a sister down. At times I wonder is it worthwhile going on? But the promise shown by one or two senior nurses keep up the hope and preserve.[30]

These trained nurses were not quite satisfied with the Indian nurses and often expressed their displeasure on the training of the native women as nurses. Sister Bury's report of 1915 notes that, the nurses were sent to Karnal for six months to get varied training. They got more operation work, midwifery, and nursing of European patients here in Delhi, and in Karnal, they got out-patient work and the chance of more individual attention and training, which is the advantage of a smaller Hospital.[31] However, some times, the performance of the nurses in the exams was not up to the mark and these nurses were disappointed. Sister E.M. Hughes involved with the nurse training felt it was tough to teach nurses even the basics and she was very upset with the poor performance of the nurses in Mission board Examinations. But she continues to say that

> ...three midwifery candidates we put in for examination are quite a way down in the list of marks, and yet I really do feel they knew their job better than I did when I sat for C.M.B; or any other nurse at that stage in England.[32]

Sister Hughes's following words show how daunting was the task of training Indian women as nurses.

> It is interesting teaching them, but also at times more than depressing and I thought instead of a thorough nursing training to fit you for work in St. Stephen's it would be better to take a course of training in the metropolitan police force, for it does seem at times that there is no nursing to be done beyond constant hammering for cleanliness and order.[33]

The sister in charge of St. Stephen's Hospital in 1934 in her annual report wrote

> The nurses, very good while supervised, were incapable of assuming any responsibility and doctors and sister in turn sat up at night with

any seriously ill patient. Lectures to nurses and dispensers were exceedingly simple. I well remember a hot hour in June spent in trying to get one of the latter to say 'potassium permanganate'

correctly, and when Sister was ill, endeavouring to teach a nurse of two years' standing how to record a temperature chart.[34]

The other point of observation was about the kind of candidates who came for nurse training. Most of the nurses sought hospital as a refuge and means to support the family. Most of them were young women educated in mission schools, married at an early age, widowed or deserted by their husbands. These nurses hardly had any idea of the 'love of nursing' and for most of them it meant just a job. Though there was gradual realization and recognition of nursing as a profession, there was still strong social taboo on nursing as a respectable or ideal profession for educated girls. The advent of an English-trained sister began a new era in nursing and girls, on leaving school, began to think of training before being married. Most of these girls could read and write, so attempts were made at training them on English lines. Besides nursing, these girls were also taught to dispense medicines with a very elementary knowledge of the pharmacopoeia.[35] The trained nurses came to be in great demand.

> St. Luke's hospital in Vengurla some years was asked by one of the Bombay municipal Hospitals to supply some graduate nurses if possible. We were able to spare three. Shortly after their arrival the matron telegraphed 'please send two dozen more'.[36]

For very long time, the nurses were not trusted with any sort of responsibilities.

> A serious operation case or a very ill patient could not be left to the care of the night nurses, the Sister or Doctor had to be called at stated intervals to give a hypodermic injection, and often they had to watch, themselves, by the patient through the night.[37]

Gradually, this attitude changed and the Indian staff were given opportunities of handling responsibilities independently. The Indianization of the hospital staff started at St. Stephen's in 1925 when an Indian nurse staff was put in-charge of an entire ward; doing full sister's duties. Sister Wilkinson writes

> While some who are by no means brilliant make good reliable workers, it is essential to have intelligent well educated girls if they are to be trained to take full charge as sisters, which is the ultimate aim and object of our work of teaching them, so that they may be fitted for the task of helping and teaching their own countrymen.[38]

Most of the sisters had to reason out a lot while delegating the work to Indian nurses. Dr. Morris once wrote

> I admit it is very difficult at times to stand by and see a nurse do badly what you feel you could do much better (one pines to do it but there it is, how shall they learn if we always do it for them and they look on? I do so hope that our venture of faith in this direction will be blessed and that nurses will rise equal to their great responsibility.[39]

Miss Salmon of CMS writes that once she was discussing about the responsibilities of church being handed over to Indians and an Indian doctor said to her 'Do you mean that this is our own, and that we can run it ourselves? I do think it good of you all to stand on one side this way and let us do it' Salmons reacting to this said

> She was not being sarcastic. But the remark hurt for I feel that somehow in the past we had failed and that we had not got across to them that the church was theirs; and that their service was both needed and wanted.[40]

Dai training

Besides nurse training, native *dais* were also given training at St. Stephen's Hospital. Miss Englemann was put in charge of the training of the *dais* by the Delhi Municipality and the first women to be trained were non-Christians, Mohammedans, and low-caste Hindus. At that time, the Punjab Government was giving Rs. 410/- a year for medicines and the Delhi Municipality was contributing Rs. 75/- a month in scholarships for training women as nurses. Later, when *Chamar* Christian women, both wives and widows, entered training, Miss Englemann accompanied them to the patient's houses and acted as a chaperon the whole time.[41]*Dais* in those days were the very lowest class of women. Therefore, when they first began to attend the sick an escort was essential. These pioneers of nursing profession in Delhi were illiterate and Miss. Englemann had to colour the ointments, oils, and medicines in the dispensary to distinguish one from another.

Doctors took the help of the local *dais* to communicate with the native women and often the doctors had to face lot of problems. Dr. Barnaby once said that

> The Dais (midwives) have a really incredible faculty of understanding our halting Urdu, and translating the patients' story into English Urdu. They, the dais, do not speak English but understand the peculiar language spoken by the newly arrived doctor. It is very tiring, of course, when you have to struggle to express in your best Urdu for patient to say, 'I do not understand English'.[42]

The illiterate *dais* were gradually replaced by younger women who could read but who were not capable of passing the United Board of Missions Examination. They were given a 2-year's training chiefly in the wards under the sisters and then take the Punjab Central Midwives' Board Examination for Nurse Dais.[43]

Dispensers

Mission hospitals were also pioneering in the training of girls as dispensers. In the initial years when Dr. Englemann was running the small dispensary, an entirely untrained missionary helped to give out medicines. 'The "doctor" heard the patient's list of complains, looked at her tongue, felt her pulse and called out to the "Dispenser" "Give so and so three fever pills for to-day," etc. The midwife would be told to "Wash the baby's eyes with pink lotion and to put green ointment on its wounds'.[44] In 1919, a separate department was established for training of girls as dispensers. During the time of Miss. Fielding Smith, the hospital dispensary was equipped, and the first staff were trained. In collaboration with colleagues elsewhere, she drew up a syllabus and organized examinations under the auspices of the North India United Board for the Training of Nurses and Dispensers. These enabled girls trained at St. Stephen's Hospital to take a certificate of efficiency and the Hospital came to be recognized for dispensers.[45]

Village camps

Village camps or visits were a very important part of the medical missionaries. Spasmodic attempts at village medical work were made for long time for village visiting. It was possible only in January 1929 to set aside a doctor for itinerating work during the winter. There were a few centres where a room over the Indian Catechist's house or a mud-walled room was used as the temporary dispensary by the visiting doctors. By living with the villagers, these doctors won the confidence of the ignorant and frightened village women, so that when seriously ill and needing proper medical treatment, the doctor could persuade them to come into hospital.[46] Female Medical missionary doctors were objects of great curiosity, especially in the villages; the villagers crowded round them debating loudly with each other. There were instances like when an Indian passer by referred rudely to one missionary as a dancing girl and she jumped out of her carriage and thrashed him with her umbrella.[47] One of missionary doctors described about her first village visit as follows.

> It was my first visit to the village: my companion and I started after breakfast in a 'baili' (ox cart) across 4 miles of rough road. The last half mile being more suitable for camel, we prepared to wade through the sand, the baili following...As we entered the village a Jatni friend invited us into her house and was anxious for me to hold the dispensary

in her courtyard, but that would have absolutely shut out the low caste and sweepers so after a little conversation we moved on, and selected a shady tree to sit under. The usual inquisite crowd collected, some young men gathered round to jeer and make rude remarks. My colleague went off to let the people whom she visits to know of the arrival of the medicines. Sick folk came up very slowly. I had to remind myself that such is the case in some villages. The small crowd of jeering men who would not remove themselves, however politely asked, was largely the cause of women not coming up....we could not get away till the bottles and boxes are empty. Even a buffalo calf with a sore eye was brought for medicine. I am not quite certain of the numbers, but I think 3 patients came into hospital from that one visit.[48]

Dr. Bazely who served at St. Stephen's Hospital was very famous for her village visiting and camping. She had a great sense of humour, loved village people, and was often on loan to other hospitals. When in Delhi, she liked nothing better than camping in the villages. In the later years, she did this in her car known as 'Yellow Biscuit'.[49] She retired officially in 1941 but was still helping at St. Stephen's hospital and elsewhere even after 1946. She was, however, professionally dissatisfied with village visits. She writes

I will honestly confess that camping work does not satisfy me professionally; under the circumstances it is not possible to treat any with the least degree of satisfaction. After a 10 or 12 days tour, I yearn for hospital work. But the latter does, in some measure, depend on the former, and the result being so apparent, no hardships seem too much to endure when clear and insistent is the urgent cry to go out to find the sick and dying.[50]

As to the success of missionaries in terms of number of converts, the medical missionaries often failed. There were very few converts. Dr. Agnes Scott working at St. Elizabeth Hospital at Karnal in her medical reports notes that '.....others have expressed a wish to become Christians, but there is generally some motive behind, an unkind husband or mother-in-law, blindness or incurable lameness or inability to produce the necessary son and heir'.[51] The feelings of the missionaries on the issue of number of conversions can be summed up in the words of Dr. C.L. Houlton, who headed St. Stephen's Hospital. She said,

In our work in the hospital, which is chiefly among non-Christians, results cannot be gauged by the number of conversions that take place. Those are very few and far between; but from time to time evidence is forthcoming that, however faulty and unChrist like our work and lives may be, yet somehow God is working through us and the people we come in contact with do learn something of the love of our Lord and Master.[52]

Indianization of church

The work of missionaries was not impository in nature and many Indians extended their co-operation for the endeavours of the church. A European Missionary who was held in high esteem by Indian students summed up the change in the religious outlook of the Hindu youth during the last few decades of the nineteenth century in the following words.

> In the early days of the Christian Missionary enterprise, the opposition took its stand on the plea, 'Christianity is not true'; gradually the attitude changed to 'Christianity is not new. We have the law and the prophets-our sages have taught us all these truths in the past' but the modern attitude of many enlightened Hindus is 'Christianity is not YOU'.[53]

It was strongly urged by 1910 that missions should associate with them in their work Indian Christians who should have the same position in these missions as the missionaries sent out from western lands. The church establishments were run by Europeans and usually the lower staff were Indian Christians. Even well-qualified Indian professionals were not given independent responsibilities and they served only in subordinate positions. The National Missionary Conference held in Calcutta in 1912 recommended missions 'to place Indians on a footing of complete equality with Europeans' and 'to open for them the highest and most responsible positions in every department of missionary activity'.[54]

Of roughly 622 foreign mission agents in northwest India in 1931, 43 were qualified female doctors and 11 were qualified male doctors. But these foreign practitioners were almost matched in number by Indian doctors, male and female, who numbered 59 by 1935. There were nearly 400 Indian nurses working in Christian medical institutions. Besides the nurses, there were incalculable numbers of hospital and dispensary staff like bible women, compounders and dressers, cooks and laundresses, cleaners and sweepers, some of whom were not even Christians. However, missionaries' objection to the total transfer to Indian Christian doctors was based on the grounds of lack of proper training, mistrust, lack of proselytizing enthusiasm, etc.

Dr. Walter. F. Hume in his article 'Indianization of Medical Missionary Work' said that the devolution of the foreign medical missions depended not only on co-operation but on the development of leadership among Indian doctors. He further went on to say

> Christian Medical Schools and Mission hospitals should equip Indian doctors for independent work by allowing them to carry responsibilities, to do operations, treat serious cases and gain confidence. Unfortunately, Mission hospitals are too much of 'one-man' show and the institutions are too often known by the name of the senior foreign medical missionary.[55]

Gradual change was later seen and one of the first real advancement made was the changing of the original 'Medical Missionary Association of India' where only foreign missionaries with few exception were members, to the 'Christian Medical Association of India' where Indian Christian doctors were equal members.

However, from the point of view of an Indian Christian woman, the mission continued to be an educational and professional bureaucracy, with all the hazards and opportunities of any such bureaucracy including upward mobility and the satisfaction of a profession worth doing. Europeans were doctors but Indians were 'hospital assistants'. Indian Christian women joined as nurses in large numbers. For example, mission hospitals provided the bulk of training for nurses throughout all India and the small Christian community supplied the overwhelming majority (by one estimate, 90 per cent) of all trained nurses. The relationship was inevitably one of subordination. Similarly, in 1928, when the government launched a scheme for village dispensaries with a sub-assistant surgeon and midwife nurse, most of the nurses hired were poor Christian widows of outcaste origins.

While the government doctors did something in medical lines, it did not at all meet the situation and the medical missionaries were able to reach out to these untouched areas. Undeniably, they were very popular among people and were keen to avail their services.

> Instead of the patients arriving, as in old days, often in a moribund condition having been maltreated by the dirty and ignorant indigenous dais, large number now attend the Ante-Natal Clinic all through their pregnancy and come in for their confinements from choice, even if it is not absolutely necessary.[56]

In later years, foreign Medical missionaries and few exceptions came together and formed an association called the 'Medical Association of India'. The greatest advancement was the changing of this association to 'Christian Medical Association of India' in the 1930s and Indian Christian doctors were also members of this association. This body was composed of both men and women medical professionals and was open also to private practitioners who were not connected to any mission hospital.[57] At the Biennial Medical Missionaries Conference at Clifton Spring, it was said

> Their immediate motive is not to convert to Christianity or even to disarm the doubting, opposing minds of the people. They go to cure and prevent and to save just as the Great Physician himself did on earth and would have them do as His disciples.[58]

Conclusion

Female medical missionaries who came to India had the double aim of healing the body and soul of the women. The study in the chapter shows that the work of female medical missionaries had many more related issues like the medical training for women in West, place of women missionaries in the church, Indianization of church, etc. The analysis shows that what started as 'veranda dispensaries' and 'medical chest' treatments have grown into great medical hospitals and medical schools. There was always a sense of mistrust and scepticism in delegation to Indians. Under British rule, women missionaries found space in the public domain and opportunities for achievement denied then at home. However, they had to face continuous tussles and conflicts with their own male church hierarchies and their struggles for equal reorganization, equal remuneration, the rights of single women, and for the autonomy of Christian women's organizations were important manifestations of resistance to patriarchal control. But their lasting contribution was the unintentional creation of a 'feminist' consciousness in local women. Besides these internal ecclesiastical administrative issues, the female medical missionaries had to struggle in foreign countries, sometimes even for survival in the hot climates of eastern lands. Undeniably, these missionary women had worked selflessly for the saving the lives of Indian women. However, the treatment given to Indian doctors was less than desired. Although the missionaries spoke of the contributions made by the Europeans and home societies, the Indian donations and contributions were underplayed. Many local rajas and wealthy men had given land and funds for the construction of dispensaries and hospitals. At the grass root level, it was the Indian Christians who worked as assistants and their contributions also need to be studied further.

There were several layers in the interaction of the medical missionaries in India. It was not merely a colonial or religious enterprise. It's a complex phenomenon expressed itself strongly even when they were living and learning in England. They realized that their education in medicine is incomplete without practicing it, so they took to the colonies as a place where they could experiment and learn. This was much more than their religion motive. Once they reached the colonies, the dynamics took another shape. The colonized even though weak and diseased, were no *tabula rasa* which the foreign missionaries had to contend with.

Notes

1 http://apps.who.int/gho/data/node.main.15?lang=en (accessed on 28 January, 2020).
2 J. C. Pollock, *Shadows Fall Apart*, London: Hodder and Stoughton, 1958, p. 34.
3 Cox Jeffery, 'Audience and Exclusion at the Margins of Imperial History', *Women's History Review*, Vol. 3, No. 4, 1994, pp. 501–514.
4 Rosemary Fitzgerald, 'A Peculiar And Exceptional Measure: The Call For Women Medical Missionaries For India In the Late Nineteenth Century', in Robert A Bickers and Rosemary Seton,(eds.) *Missionary Encounters*, Surrey: Curzon Press, 1996; Kumari Jayawardena, *The White Woman's Other Burden*, New York: Routledge, 1995; Leslie Flemming (ed.), *Women's Work for Women*, Boulder: Westview Press, 1990; Leslie Flemming, 'A New Humanity: American Missionary Ideals for Women in North India, 1870–1930', in Nupur Chaudhuri and Margaret Strobel (eds.), *Western Women and Imperialism*, Bloomington:

Indiana University Press, 1992; Ruth Brouwer Compton, 'New Women for God: Canadian Presbyterian Women and Indian Missions,1876–1914', *Social History of Canada*, 44, Toronto: University of Toronto Press, 1990; JharnaGourly, *Florence Nightingale and the Health of the Raj*, London, Routledge, 2018.

5 Jeffery Cox, *Imperial Fault Lines*, Stanford: Stanford University Press, 2002, p. 15.

6 Between 1809–1825, CMS began work at Chunar, Agra, Benaras, Meerut and Gorakhpur in United Provinces, at Simla and Kotgarh in 1840, and at Amritsar in Punjab in 1852.

7 Leslie Flemming,'New Models,New Roles: U.S Presbyterian women Missionaries and Social Change' in North India, 1870–1910' in Leslie Flemming (ed.), *Women's Work for Women*, Boulder: Westview Press, 1990, p. 42.

8 Deborah Kirkwood, 'Protestant Missionary Women: Wives and Spinsters', in Fiona Bowie *et al* (eds.), *Women and Missions*, Rhode Island: Berg, Providence, 1993, p. 27.

9 Rosemary Fitzgerald, 'A Peculiar and Exceptional Measure: The Call for Women medical Missionaries for India in the Late Nineteenth Century', in Robert A. Bickers and Rosemary Seton (eds.), *Missionary Encounters*, Surrey: Curzon Press, 1996, p. 178.

10 Jane Haggis, "Good Wives and Mothers' or 'Dedicated Workers?' Contradictions of Domesticity in the 'Mission of Sisterhood', Travancore, South India', in Kalpana Ram and Margaret Jolly (eds.), *Maternities and Modernities*, Cambridge: CUP, 1998, p. 106.

11 *International Review of Missions*, Vol. 13, 1924, p. 257.

12 Kumari Jayawardena, The *White Woman's Other Burden*, op. cit., p. 25.

13 Jane Haggis, 'Good Wives and Mothers' or 'Dedicated Workers?' Contradictions of Domesticity in the 'Mission of Sisterhood', Travancore, South India', in Kalpana Ram, and Margaret Jolly (eds.), op. cit., p. 106.

14 Leslie Flemming, 'A New Humanity: American Missionary Ideals for Women in North India, 1870–930', in Nupur Chaudhuri, and Margaret Strobel (eds.), op. cit., p. 193.

15 *The Ministry of Healing in India: Handbook of the Christian Medical Association of India*. Mysore, 1932, p. 27.

16 Margaret Balfour and Ruth Young, *The Work of Medical Women in India*, London: Humphrey Milford, 1929, p.14.

17 Irene. H. Barnes, *Between Life and Death: The Story of C.E.Z.M.S. Medical Missions in India, China and Ceylon*, London: Marshal Brothers, 1901, p. 9.

18 Maina Chawla Singh, 'Gender, Medicine and Empire: Early Initiatives in Institution Building and Professionalization (1890s–1940s)', in Biwamoy Pati and Shakti Kak (eds.), *Exploring Gender Equations*, New Delhi: NMML, 2005, pp. 96–97.

19 Geraldine. H. Forbes, 'In Search of Pure Heathen: Missionary women in Nineteenth Century India', *EPW*, Vol. 21, No. 17, 1986, pp. WS2-8.

20 F.J. Western, *Early History of the Cambridge Mission to Delhi in Connection with SPG*, private publication, July 1950, p. 102.

21 Rosemary Fitzgerald, 'A Peculiar and Exceptional Measure: The Call for Women Medical Missionaries for India in the Late Nineteenth Century', in Robert A Bickers and Rosemary Seton (eds.), op. cit., p. 188.

22 In one letter Mrs. Winter wrote "I have to study the language, teach in schools and zenanans, nurse the sick, visit Indian Christians and write begging letters." Her husband once wrote "She has almost lost the use of her right handwriting begging letters" as quoted in Ruth Roseveare, '*Delhi- Community of St. Stephen's 1886–1986*', Reepham, Norwich: Privately Published, 1986, p. 11.

23 Western, *Early History of the Cambridge Mission to Delhi in Connection with SPG*, op. cit., p. 102.

24 Ruth Roseveare, *Delhi- Community of St. Stephen's 1886–1986*, op. cit., p. 35.
25 *DELHI*, Vol. XII, No.9, 1930, p. 172.
26 Ibid., p. 35.
27 Ruth Roseveare, op. cit., p. 36.
28 *DELHI*, Vol. XIV, 1934, p. 35.
29 *Annual Report of the SPG and CMD and South Punjab*, 1939, p. 12.
30 *Annual Report of the SPG and CMD and South Punjab*, 1924.
31 *Delhi Mission News*, Vol. VIII, No. 2, 1916, p. 22.
32 *DELHI*, Vol. XIII, No 7, 1932, p.10.
33 Ibid., p.23.
34 *DELHI*, Vol. XIV, No.2, 1934, p. 35.
35 *DELHI*, Vol. XII, No. 9, 1930, pp. 172–173.
36 R.H.H. Goheen, 'Medical Missions in India', *International Review of Mission*, Vol. 19, 1930, p. 215.
37 Ibid., p. 173.
38 *Annual Report of SPG and CMD and South Punjab* 1927, p. 25.
39 *Annual Report of SPG and CMD and South Punjab* 1929, p. 27.
40 'Towards the India of Tomorrow', *Annual Report of Church Missionary Society*, 1944, p. 29.
41 Ibid., p. 171.
42 *DELHI*, Vol. XII., No. 3, 1928, p. 54.
43 *DELHI*, Vol. XII, No. 9, 1930, pp. 173–174.
44 Ibid., p. 172.
45 *DELHI*, Vol. XVI, No. 7, 1941, p. 41.
46 *DELHI*, Vol. XII, No. 9, 1930, p. 174.
47 Margaret Mac Millan, *Women of Raj*, London: Thames & Hudson, 1988, p. 212.
48 *DELHI*, Vol. XII, No. 6, 1929, p. 114.
49 Ruth Roseveare, op. cit., p. 69.
50 *Delhi Mission News*, Vol. VIII, No. 9, 1918, p. 118.
51 *Delhi Mission News*, Vol. VII, No.10, 1915, p. 120.
52 *Annual Report of SPG and CMD and South Punjab*, 1932.
53 C. A. Cumarswamy, 'The Indian Christian Church and the Spirit of Nationality', *International Review of Missions*, Vol. 13, 1924, p. 64.
54 John McKenzie, 'Church and Mission in India', *International Review of Missions*, 1920, Vol.9, p.76.
55 Hume. F. Walter 'Indianization of Medical Missionary Work', *Missionary Herald*, Vol. CXXIV, No. 11, 1928, p. 424.
56 *DELHi*, Vol. XII, no.9, January 1930, p.174.
57 Hume F. Walter, op. cit., p. 424.
58 *Missionary Herald*, Vol. CXXXII, No. 5, 1935, p. 271.

Appendix

List of publications of Professor Deepak Kumar

Books

1. *Science and the Raj, 1857–1905*, Delhi: Oxford University Press, 1995 (Paperback edition published in 1997, Reprinted in 2000)
   Science and the Raj: A Study of British India; Revised second edition by Oxford University Press in 2006; Paperback in 2006; Reprinted in 2011 and 2015.
   Earlier Hindi version published by Granthsilpi, Delhi, 1998; Urdu version published by NCPUL, Delhi, 2009; Bengali version by Sujan Publication, Kolkata, 2007; Another edition in Bengali from K. P. Bagchi, Kolkata, 2021.
2. *The Trishanku Nation: Memory, Self and Society in Contemporary India*, New Delhi: Oxford University Press, 2016; Revised second edition in 2021; Hindi version published by Rajkamal Prakashan, Delhi, 2019.
3. *'Culture' of Science and the Making of India*, Delhi: Primus, 2022 (forthcoming).
4. *Atam Khabar: Sanskriti, Samaj aur Hum* Delhi: Aakar, 2022 (forthcoming).
5. *Science and Society in Modern India,*. (Currently under preparation).

Edited books

1. *Science and Empire: Essays in Indian Context*, Delhi: Anamika Prakashan, 1990 (Paperback edition in 1995).
2. *Technology and the Raj: Western Technology and Technical Transfers to India 1700–1947*, New Delhi: Sage, 1995 (Co-edited with Roy Macleod).
3. *Disease and Medicine in India: A Historical Overview*, New Delhi: Tulika, 2001 (Paperback in 2002).

4. *The British Empire and the Natural World: Environmental Encounters in South Asia*, New Delhi: Oxford University Press, 2011 (Co-edited with Vinita Damodaran and Rohan D'Souza).

5. *HashiyekaVritatnta* (Narratives of the Margins in Hindi), Panchkula: Adhar Prakashan, 2011 (Co-edited with Devendra Chaubey).

6. *Medical Encounter in British India*, New Delhi: Oxford University Press, 2013 (Co-edited with Raj Sekhar Basu).

7. *Education in Colonial India: Historical Insights*, New Delhi: Manohar, 2013 (Co-edited with Nandita Khadria, Joseph Bara and Radha Gayathri). Reprinted by Manohar in 2020.

8. *Tilling the Land: Agricultural Knowledge and Practices in Colonial India*, Delhi: Primus, 2016 (Co-edited with Bipasha Raha).

Articles in referred journals

1. 'Patterns of Colonial Science', *Indian Journal of History of Science*, Vol. 15, No. 1, 1980, pp. 105–113.

2. 'Economic Compulsions and the Geological Survey of India', *Indian Journal of History of Science*, Vol. 17, No. 2, 1982, pp. 298–300.

3. 'Science, Resources and the Raj', *Indian Historical Review*, Vol. X, Nos. 1–2, 1984, pp. 66–89.

4. 'Science in Education: A Study in Victorian India', *Indian Journal of History of Science*, Vol. 19, No. 3, 1984, pp. 253–260.

5. 'Racial Discrimination and Science in 19th Century India', *The Indian Economic and Social History Review*, Vol. XIX, No. 1, 1983, pp. 63–82.

6. 'Pre-Colonial Science and Technology: Lessons from 18th Century India', *Journal of Japan-Netherlands Research Institute*, March, 1992.

7. 'Calcutta: The Emergence of a Science City', *Indian Journal of History of Science*, Vol. 29, No. 1, 1994, pp. 1–7.

8. 'Culture of Science and Colonial Culture, India 1820–1920', *British Journal of History of Science*, Vol. 29, No. 2, 1996, pp. 195–209.

9. 'Gandhi on Technology', *Gandhi Marg*, January-March 1997, pp. 428–437.

10. 'Colony under a Microscope: The Life and Works of W. Haffkine', *Science, Technology and Society*, Vol. 4, No. 2, 1999, pp. 239–271.

11. 'Reconstructing India: Disunity in Science and Technology for Development Discourse, 1900–1947', *OSIRIS*, Vol. 15, No. 1, 2000, pp. 1201–1217.

12. 'Science and Society in Colonial India: Exploring an Agenda', *Social Scientist*, Vol. 28, Nos. 5–6, 2000, pp. 24–46. (Presidential Address, Modern India Section, Indian History Congress, Calicut, 1999)

13. 'Developing a History of Science and Technology in South Asia', *Economic and Political Weekly*, Vol. 38, No. 23, 2003, pp. 2248–2251.

14. 'Emergence of Scientocracy: Snippets from Colonial India, *Economic and Political Weekly*, Vol. 39, No. 35, 2004, pp. 3893–3898.

15. 'Medical Encounter: Concepts and Practices in India, A Historical Outline', *Archiv Internationales d'Historie des Sciences*, Vol. 55, No. 155, 2005, pp. 357–366.
16. 'An Imperial Apology', Review Article, *The Indian Historical Review*, Vol. XXXIV, No. 2, 2007, pp. 189–196.
17. 'Bioethics, medicine and society – a philosophical inquiry', *Current Science*, Vol. 97, No. 8, 2009, pp. 1128–1136 (Co-author V. K. Yadavendu).
18. 'Probing History of Public Health and Medicine in India: Encounter at Multiple Sites', *The Indian Historical Review*, Vol. 37, No. 2, 2010, pp. 259–273.
19. 'Reason, Science and Religion: Gleanings from the Colonial Past', *Studies in People's History*, Vol. 1, No. 2, 2014, pp. 181–198.
20. 'Science Administration: A Historical Outline', *Social Scientist*, Vol. 43, Nos. 1-2, 2015, pp. 31–42.
21. 'HISTEM and the Making of Modern India: Some Questions and Explanations, *Indian Journal of History of Science*, Vol. 50, No. 4, 2015, pp. 616–628.
22. 'Techno-scientific Education and the Development Discourse: India (1900–1947), *The Indian Historical Review*, Vol. 45, No. 2, 2018, pp. 1–14.
23. 'Science Institutions in Colonial India: Some Snippets, Some Lessons', *Indian Journal of History of Science*, Vol. 53, No. 4, 2018, pp. 23–31.
24. 'Medical History: British India, the Dutch Indies and Beyond', *Journal of the Asiatic Society*, Vol. LXII, No. 4, 2020, pp. 229–246.
25. 'Colonialism and knowledge transformation: A study of Victorian India', *Studies in People's History*, Vol. 8, No. 1, 2021, pp. 92–105.

Book chapters

1. 'Science in Agriculture: A Study in Victorian India', in A. Rahman (ed.), *Science in Indian Tradition: A Historical Perspective*, New Delhi: NISTADS, 1984.
2. 'The Evolution of Colonial Science in India: Natural History and the East India Company', in J. MacKenzie (ed.), *Imperialism and the Natural World*, Manchester: Manchester University Press, 1990.
3. 'Colonial Science: A Look at Indian Experiences' in Deepak Kumar (ed.), *Science and Empire*, Delhi: Anamika Prakashan, 1991.
4. 'Problems of Science Administration: A Study of the Survey Organizations', in P. Petitjean et al. (eds.), *Science and Empires*, Dordrecht: Kluwer Academic Publishers, 1992.
5. 'The Concept of Colonial Science: A Review', in Jean Dhombres et al (eds.), *Symposia Survey Papers: IXI International Congress of History of Science*, Zaragoza, 1993.

6. 'Technology, Education and Colonialism: A Study of Victorian India, 1835–1880', in Yamada Keiji (ed.), *Transfer of Technology between Europe and Asia*, Tokyo, 1995.

7. 'Natives in Colonial Parlance', in Indu Banga and Jaidev (eds.), *Cultural Reorientation in Modern India*, Simla: IIAS, 1996.

8. 'Medical Encounters in British India', in A. Cunningham and B. Andrews (eds.), *Western Medicine as Contested Knowledge*, Manchester: Manchester University Press, 1997.

9. 'Educational Ideas of the Indian Men of Science: Bengal 1850–1920', in S. Bhattacharya (ed.), *Contested Terrain: Perspectives on Education*, New Delhi: Orient Longman, 1997.

10. 'Techno-Scientific Knowledge in an Age of Decline: A Study of Eighteenth Century India', in Roy Porter (ed.), *Science in Eighteenth Century*, Cambridge: Cambridge University Press, 2003.

11. 'Health and Medicine in British India and Dutch Indies: A Comparative Study', in Joseph S. Alter (ed.), *Asian Medicine and Globalization*, Philadelphia: University of Pennsylvania Press, 2005.

12. 'Perceptions of Public Health: A Study in British India', in Amiya K. Bagchi and Krishna Soman (eds.), *Maladies, Preventives and Curatives: Debates in Public Health in India*, New Delhi: Tulika, 2005.

13. 'HISTEM in India: An Introduction', in Arun Bandopadhyay (ed.), *Science and Society in India*, New Delhi: Manohar, 2009.

14. 'Investigating the Histories of Technology and Medicine in South Asia', *Proceedings of the Indian Association for Asian and Pacific Studies*, Kolkata, 2006.

15. 'Colony and Science: A Study of British India', in J. B. Dasgupta (ed.), *Science, Technology, Imperialism and War*, New Delhi: Pearson Longman, 2007.

16. 'Foreign Philanthropy and Medical Research in India', in K. L. Tuteja and S. Pathania (eds.), *Histories Diversities: Society, Politics and Culture*, New Delhi: Manohar, 2009.

17. 'Scientific Surveys in British India', in Uma Dasgupta (ed.), *Science and Modern India: An Institutional History*, New Delhi: Pearson Longman, 2010.

18. 'Indian National Congress, Science and Society', in Aditya Mukherjee (ed.), *A Centenary History of the Indian National Congress*, Vol. V, New Delhi: Academic Foundation, 2011.

19. 'Botanical Explorations and the East India Company: Revisiting Plant Colonialism', in Vinita Damodaran, Anna Winterbottom and Alan Lester (eds.), *The East India Company and the Natural World*, Basingstoke, Palgrave Macmillan, 2014.

20. 'Tagore's Pedagogy and Rural Reconstruction', in Michael Mann (ed.), *Shantiniketan Hellerau: New Education in the 'Pedagogic Provinces' of India and Germany*, Heidelberg: Draupadi Verlag, 2015.

21. 'Religion, Polity and Media: A Look at Contemporary India', in Detlef Briesen, Sigrid Baringhorst and Arvind Das (eds.), *Religion, Politics and Media: German and Indian Perspectives*, New Delhi: Palm Leaf, 2015.
22. 'Science in Modern Bengal', in Sabyasachi Bhattacharya (ed.), *A Comprehensive History of Modern Bengal, 1700–1950*, Vol. 3, Delhi: Primus, 2020.
23. 'India', in Hugh Richard Slotten, Ronald L Numbers and David N Livingstone (eds.), *The Cambridge History of Science*, Vol. 8: Modern Science in National, Transnational and Global Context, Cambridge: CUP, 2020.
24. 'Scientific Knowledge and Society in India', in Syed Ejaz Hussain and Sanjay Garg (eds.), *Alternative Arguments: Essays in Honour of Surendra Gopal*, Delhi: Primus, 2020.
25. 'Education in Modern India', in Farida Khan (ed.), *Oxford Encyclopaedia on Higher Education*, Oxford: Oxford University Press, 2022 (Forthcoming).

Select Bibliography

This is not a comprehensive bibliography for HISTEM. The literature is, at present, so vast that a separate monograph is needed to cover all the sub-themes – science, technology, environment, and medicine. However, this is essential reading, at least, to the early-career researchers. This can be a ready reference tool for senior scholars and colleagues as well. For the logistic reason, I mention only books published after 1950, there are several significant journal articles and book chapters on the theme, which can be found in the endnotes of the individual chapter.

Adas, Michael, *Machines as the Measure of Men: Science, Technology and Ideologies of Western Dominance*, Ithaca: Cornell University Press, 1989.

Agarwal, Arun, *Greener Pastures: Politics, Markets, and Community among a Migrant Pastoral People*, New Delhi: OUP, 1999.

Agarwal, Arun, *Environmentality: Technologies of Government and the Making of Subjects*, New Delhi: OUP, 2005.

Ahluwalia, Sanjam, *Reproductive Restraints: Birth Control in India 1877–1947*, New Delhi: Permanent Black, 2008.

Ahmed, Sahara, *Woods, Mines and Minds: Politics of Survival in Jalpaiguri and the Jungle Mahals, 1860–1970*, New Delhi: Primus, 2019.

Alavi, Seema, *Islam and Healing: Loss and Recovery of an Indo-Muslim Medical Tradition*, Basingstoke: Palgrave Macmillan, 2008.

Allender, Tim, *Learning Femininity in Colonial India, 1820–1932*, Manchester: MUP, 2016.

Alter, Joseph S., *Asian Medicine and Globalization*, Philadelphia: University of Pennsylvania, 2005.

Amrith, Sunil S., *Crossing the Bay of Bengal: The Furies of Nature and the Fortunes of Migrants*, Cambridge, MA: Harvard University Press, 2013.

Anderson, Clare, *Subaltern Lives: Biographies of Colonialism in the Indian Ocean World, 1790–1920*, Cambridge: CUP, 2012.

Anderson, Robert S., *Nucleus and Nation: Scientists, International Networks and Power in India*, Chicago and London: The University of Chicago Press, 2010.

Anderson, Warwick, *Colonial Pathologies: American Tropical Medicine, Race and Hygiene in the Philippines*, Durham, NC: Duke University Press, 2006.

Armstrong, David, *Political Anatomy of the Body: Medical Knowledge in Britain in the 20th Century*, Cambridge: CUP, 1983.

Arnold, David (ed.), *Imperial Medicine and Indigenous Societies*, New Delhi, OUP, 1989.

Arnold, David, *Colonizing the Body: State Medicine and Epidemic Disease in Nineteenth Century India*, Berkeley: University of California Press, 1993.

Arnold, David, *Warm Climates and Western Medicine: The Emergence of Tropical Medicine, 1500–1900*, Amsterdam and Atlanta, GA: Rodopi, 1996.

Arnold, David, *The New Cambridge History of India, Vol. III:5, Science, Technology and Medicine in Colonial India*, Cambridge: CUP, 2000.

Arnold, David,*The Tropics and the Traveling Gaze: India, Landscape, and Science 1800–1856*, New Delhi: Permanent Black, 2005.

Arnold, David, *Everyday Technology: Machines and the Making of India's Modernity*, Chicago and London, University of Chicago Press, 2013.

Arnold, David, *Toxic Histories: Poison and Pollution in Modern India*, Cambridge: CUP, 2016.

Arnold, David and Ramachandra Guha (eds.), *Nature, Culture, Imperialism: Essays on the Environmental History of South Asia*, New Delhi: OUP, 1995.

Attewell, Guy, *Refiguring Unani Tibb: Plural Healing in Late Colonial India*, New Delhi: Orient Blackswan, 2007.

Axelby, Richard, Savithri Preetha Nair and Andrew S Cook, *Science and the Changing Environment in India, 1780–1920: A Guide to Sources in the India Office Records*, London: The British Library, 2010.

Baber, Zaheer, *The Science of Empire: Scientific Knowledge, Civilization and Colonial Rule in India*, New Delhi: OUP, 1998.

Bagchi, Amiya Kumar, *Private Investment in India 1900–1939*, Cambridge: CUP, 1972.

Bagchi, Amiya Kumar and Krishna, Soman, (eds.), *Maladies, Preventives and Curatives: Debates in Public Health in India*, New Delhi: Tulika, 2005.

Bala, Poonam, *Imperialism and Medicine in Bengal: A Socio-Historical Perspective*, New Delhi: Sage, 1991.

Bala, Poonam (ed.), *Contesting Colonial Authority: Medicine and Indigenous Responses in Nineteenth and Twentieth-Century India*, Lanham, MD: Lexington Books, 2012.

Bala, Poonam (ed.), *Medicine and Colonialism: Historical Perspectives in India and South Africa*, London: Pickering &Chatto, 2014.

Ballhatchet, Kenneth, *Race, Sex and Class under the Raj: Imperial Attitudes and Policies and Their Critics, 1793–1905*, London: Weidenfeld and Nicolson, 1980.

Bandopadhyay, Arun (ed.), *Science and Society in India, c. 1750–2000*, New Delhi: Manohar, 2010.

Bandopadhyay, Arun (ed.), *Nature, Knowledge and Development: Critical Essays on the Environmental History of India*, New Delhi: Primus, 2016.

Banerjee, Madhulika, *Power Knowledge, Medicine: Ayurvedic Pharmaceuticals at Home and in the World*, Hyderabad: Orient Blackswan, 2009.

Barman, Rupkumar, *Practice of Folk Medicine in Sub-Himalayan Bengal: A Study on the Folk Medicinal Practices of the Rajbanshis in Historical Perspective*, New Delhi: INSA and Abhijeet Publications, 2019.

Basalla, George, *The Evolution of Technology*, Cambridge: CUP, 1988.

Bassett, Ross, *The Technological Indian*, Cambridge, Massachusetts and London: Harvard University Press, 2016.

Basu, Aparna, *The Growth of Education and Political Development in India, 1898–1920*, New Delhi: OUP, 1974.

Basu, Malika, *History of Indigenous Pharmaceutical Companies in Colonial Calcutta*, New Delhi: Manohar, 2021.

Bayly, Christopher A., *Empire and Information: Intelligence Gathering and Social Communication in India, 1780–1870*, Cambridge: CUP, 1996.

Beattie, James, *Empire and Environmental Anxiety: Health, Science, Art and Conservation in South Asia and Australia, 1800–1920*, Basingstoke and New York: Palgrave Macmillan, 2011.

Berger, Rachel, *Ayurveda Made Modern: Political Histories of Indigenous Medicine, 1900–1955*, Basingstoke: Palgrave Macmillan, 2013.

Bernal, J. D., *Science in History*, London: Watts, Second Edition, 1957.

Bhattacharya, Amit, *Swadeshi Enterprise in Bengal: The First Phase 1880–1920*, Kolkata: Readers Service, 2ndEdition, 2008 (first published 1986).

Bhattacharya, Nandini, *Contagion and Enclaves: Tropical Medicine in Colonial India*, Liverpool: Liverpool University Press, 2012.

Bhattacharya, Sabyasachi and Pietro Redondi (eds.),*Techniques to Technology: A French Historiography of Technology*, New Delhi: Orient Longman, 1990.

Bhattacharya, Sanjay, *Expunging Variola: The Control and Eradication of Small Pox in India, 1947–1977*, Hyderabad: Orient Longman, 2006.

Bhattacharya, Sanjay, Mark Harrison, and Michael Worboys, *Fractured States: Small Pox, Public Health and Vaccination Policy in British India, 1800–1947*, New Delhi: Orient Longman, 2005.

Bhattacharyya, Debjani, *Empire and Ecology in the Bengal Delta: The Making of Calcutta*, Cambridge: CUP, 2018.

Bijker, Wiebe E, Thomas P. Hughes and Trevor J. Pinch (eds.), *The Social Construction of Technological Systems: New Directions in the Sociology and History of Technology*, Massachusetts: MIT Press, 1987.

Bijker, Wiebe E. and John Law (eds.), *Shaping Technology/ Building Societies: Studies in Socio-technical Change*, Cambridge, MA: MIT Press, 1992.

Biswas, Arun Kumar, *Gleanings of the Past and the Science Movement in the Diaries of Dr Mahendralal and Amritlal Sircar*, Calcutta: The Asiatic Society, 2000.

Biswas, Arun Kumar, *Collected Works of Mahendralal Sircar, Eugene Lafont and the Science Movement (1860–1910)*, Calcutta: Asiatic Society, 2001.

Biswas, Arun Kumar (ed.), *History, Science and Society in the Indian Context*, Calcutta: Asiatic Society, 2001.

Bonea, Amelia, *The News of Empire: Telegraphy, Journalism, and the Politics of Reporting in Colonial India, c. 1830–1900*, New Delhi: OUP, 2016.

Borthwick, Meredith, *The Changing Role of Women in Bengal, 1849–1905*, Princeton: Princeton University Press, 1984.

Bose, D. M., S. N. Sen and B. V. Subbarayappa (eds.), *A Concise History of Science in India*, New Delhi: INSA, 1971.

Bose, Pradip Kumar (ed.), *Health and Society in Bengal: A Selection from the Late 19th Century Bengali Periodicals*, New Delhi: Sage, 2006.

Brimnes, Niels, *Languished Hopes: Tuberculosis, the State and International Assistance in Twentieth-century India*, New Delhi: Orient Blackswan, 2016.

Cardwell, Donald Stephen Lowell, *Technology, Science and History*, London: Heinemann, 1972.

Cederlof, Gunnel, *Landscapes and the Law: Environmental Politics, Regional Histories, and Contests over Nature*, New Delhi: Permanent Black, 2008.

Cederlof, Gunnel, and K. Sivaramakrishnan (eds.), *Ecological Nationalisms, Nature, Livelihoods and Identities in South Asia*, New Delhi: Permanent Black, 2005.

Chakrabarti, Pratik, *Western Science in Modern India: Metropolitan Methods, Colonial Practices*, Ranikhet: Permanent Black, 2004.

Chakrabarti, Pratik, *Materials and Medicine: Trade, Conquest and Therapeutics in the Eighteenth Century*, Manchester: MUP, 2010.

Chakrabarti, Pratik, *Bacteriology in British India: Laboratory Medicine and the Tropics*, Rochester, NY: University of Rochester Press, 2012.

Chakrabarti, Pratik, *Medicine and Empire: 1600–1960*, Basingstoke: Palgrave Macmillan, 2014.

Chakrabarti, Ranjan (ed.), *Situating Environmental History*, New Delhi: Manohar, 2007.

Chakrabarti, Ranjan (ed.), *Critical Themes in Environmental History of India*, New Delhi: Sage, 2020.

Chakrabarty, Dipesh, *Provincializing Europe: Postcolonial Thought and Historical Difference*, Princeton: Princeton University Press, 2000.

Chatterjee, Santimay, M. K. Dasgupta and Amitabha Ghosh (eds.), *Studies in History of Sciences*, Calcutta: The Asiatic Society, 1997.

Chatterjee, Srilata, *Western Medicine and Colonial Society: Hospitals of Calcutta, c. 1757–1860*, New Delhi: Primus, 2017.

Chattopadhyay, Debiprasad, *History of Science and Technology in Ancient India: The Beginning*, Calcutta: Firma KLM, 1996.

Chattopadhyay, Debiprasad, *Lokayata: A Study in Ancient Indian Materialism*, New Delhi: People's Publishing House, 6thReprint, 2012 (First Published1959).

Chattopadhyay, Debiprasad, *Science and Society in Ancient India*, Kolkata: K. P. Bagchi, 2ndReprint, 2014 (First Published 1977).

Chaudhuri, B. B., and Arun Bandopadhyay (eds.), *Tribes, Forest and Social Formation in Indian History*, New Delhi: Manohar, 2004.

Cherry, Steven, *Medical Services and the Hospital in Britain, 1860–1939*, Cambridge: CUP, 1996.

Chowdhury, Indira, *Growing the Tree of Science: Homi Bhaba and the Tata Institute of Fundamental Science*, New Delhi: OUP, 2016.

Cipolla, Carlo M., *Miasmas and Disease: Public Health and the Environment in the Pre-Industrial Age*, New Haven and London: Yale University Press, 1992.

Cohn, Bernard S., *An Anthropologist among the Historians and Other Essays*, New Delhi: OUP, 1987.

Coleman, Leo, *Electrification as Political Rituals in New Delhi: A Moral Technology*, New Delhi: Speaking Tiger, 2017.

Cook, Harold J., *Matters of Exchange: Commerce, Medicine, and Science in the Dutch Golden Age*, New Haven: Yale University Press, 2007.

Coopersmith, Jonathan, *The Electrification of Russia, 1880–1926*, Ithaca and London: Cornell University Press, 1992.

Cowan, Ruth Schwartz, *A Social History of American Technology*, New York and Oxford: OUP, 1997.

Crook, Nigel, *The Transmission of Knowledge in South Asia*, New Delhi: OUP, 1996.

Crosby, A. W., *Ecological Imperialism: The Biological Expansion of Europe, 900–1900*, Cambridge and New York: CUP, 1986.

D'Antonio, Patricia, Julie A. Fairman and Jean C. Whelan (eds.), *Routledge Handbook on the Global History of Nursing*, London and New York: Routledge, 2013.

D'Souza, Rohan, *Drowned and Dammed: Colonial Capitalism, and Flood Control in Eastern India*, New Delhi: OUP, 2006.

D'Souza, Rohan (ed.), *Environment, Technology and Development: Critical and Subversive Essays*, New Delhi: Orient Blackswan, 2012.

Damodaran, Vinita, Anna Winterbottom and Alan Lester (eds.), *The East Indian Company and the Natural World*, Basingstoke: Palgrave Macmillan, 2015.

Das, Debjani, *Houses of Madness: Insanity and Asylums of Bengal in Nineteenth Century India*, New Delhi: OUP, 2015.

Das, Shinjini, *Vernacular Medicine in Colonial India: Family, Market and Homeopathy*, Cambridge: CUP, 2019.

Dasgupta, Subrata, *Jagadish Chandra Bose and the Indian Response to Western Science*, Ranikhet: Permanent Black, 1999.

Daston, Lorraine and Peter Galison, *Objectivity*, New York: Zone Books, 2010.

Datta, Partho, *Planning the City: Urbanization and Reform in Calcutta c. 1800–c. 1940*, New Delhi: Tulika, 2012.

Drayton, R. H., *Nature's Government: Science, Imperial Britain, and the 'Improvement' of the World*, New Haven: Yale University Press, 2000.

Dunlap, Thomas R., *Nature and the English Diaspora: Environment and History in the United States, Canada, Australia, and New Zealand*, Cambridge and New York: CUP, 1999.

Dutta, Achintya Kumar, *Trauma in Public Health: Tuberculosis in Twentieth-century India*, Kolkata: K P Bagchi& Co, 2018.

Ede, Andrew, *Technology and Society: A World History*, Cambridge: CUP, 2019.

Edgerton, David, *The Shock of the Old: Technology and Global History since 1900*, London: Profile, 2008.

Elvin, Mark, *The Retreat of the Elephants: An Environmental History of China*, New Haven: Yale University Press, 2004.

Ernst, Waltraud, *Mad Tales from the Raj: The European Insane in British India, 1800–1858*, London: Routledge, 1991.

Ernst, Waltraud (ed.), *Plural Medicine, Tradition and Modernity, 1800–2000*, London and New York: Routledge, 2002.

Fan, Fa-ti, *British Naturalists in Qing China: Science, Empire, and Cultural Encounter*, Cambridge, MA: Harvard University Press, 2004.

Fisher, Michael H., *An Environmental History of India: From Earliest Times to the Twenty-First Century*, Cambridge: CUP, 2018.

Forbes, Geraldine, *Women in Colonial India: Essays on Politics, Medicine and Historiography*, New Delhi: Chronicles Books, 2005.

Forbes, Geraldine and Tapan Raychaudhuri (eds.), *The Memoirs of Dr. Haimabati Sen: From Child Widow to Lady Doctor*, New Delhi: Roli Books, 2000

Fox, Robert (ed.), *Technological Change: Methods and Themes in the History of Technology*, Amsterdam: Harwood Academic Publisher, 1996.

Fox, Robert and Anne Guagnini (eds.), *Education, technology and industrial performance in Europe, 1850–1939*, Cambridge: CUP, 1993.

Gadgil, Madhav, *Ecological Journeys: The Science and Politics of Conservation in India*, New Delhi: Permanent Black, 2002.

Gadgil, Madhav and Ramachandra Guha, *This Fissured Land: An Ecological History of India*, New Delhi: OUP, 1992.

Gadgil, Madhav and Ramachandra Guha, *Ecology and Equity, the Use and Abuse of Nature in Contemporary India*, New Delhi: OUP, 1995.

Gascoigne, John, *Science in the Service of Empire: Joseph Banks, the British State and the uses of Science in the Age of Revolution*, Cambridge and New York: CUP, 1998.

Golinski, Jan, *Making Natural Knowledge: Constructivism and the History of Science*, Chicago: University of Chicago Press, 2005.

Gooday, Graeme, *Domesticating Electricity: Technology, Uncertainty and Gender, 1880–1914*, London: Pickering &Chatto, 2008.

Goswami, Manu, *Producing India: From Colonial Economy to National Space*, New Delhi: Permanent Black, 2004.

Gourlay, Jharna, *Florence Nightingale and the Health of the Raj*, Aldershot: Ashgate, 2003.

Gourlay, Jharna, *Piety, Profession and Sisterhood: Medical Women and Female Medical Education in Nineteenth Century India*, Kolkata: K. P. Bagchi, 2017.

Grove, Richard H., *Green Imperialism: Colonial Expansion, Tropical Island Edens, and the Origins of Environmentalism, 1600–1860*, Cambridge: CUP, 1995.

Grove, Richard H., *Ecology, Climate, and Empire: Colonialism and Global Environmental History, 1400–1940*, Cambridge: White Horse Press, 1997.

Grove, Richard H., Vinita Damodaran and Satpal Sangwan (eds.), *Nature and the Orient: The Environmental History of South and Southeast Asia*, New Delhi: OUP, 1998.

Guha, Ambalika, *Colonial Modernities: Midwifery in Bengal, c. 1860–1947*, London and New York: Routledge, 2018.

Guha, Ramachandra, *The Unquiet Woods: Ecological Change and Peasant Resistance in the Himalayas*, New Delhi: OUP, 1989.

Guha, Ramachandra, *How Much Should a Person Consume? Thinking through the Environment*, New Delhi: Permanent Black, 2006.

Guha, Ramachandra, *Environmentalism: A Global History*, Gurgaon: Allen Lane, 2014.

Guha, Sumit, *Ecology and Ethnicity in India, c. 1200–1991*, Cambridge: CUP, 1999.

Guha, Sumit, *Health and Population in South Asia: From Earliest Times to the Present*, New Delhi: Permanent Black, 2001.

Gunergun, Feza and Dhruv Raina (eds.), *Science Between Europe and Asia: Historical Studies on the Transmission, Adoption and Adaptation of Knowledge* Dordrecht and Heidelberg: Springer, 2011.

Habib, Irfan, *Technology in Medieval India c. 650–1750*, New Delhi: Tulika, 2008.

Habib, Irfan, *Man and Environment: The Ecological History of India*, New Delhi: Tulika, 2010.

Habib, S. Irfan and Dhruv Raina (eds.), *Social History of Science in Colonial India*, New Delhi: OUP, 2007.

Hahn, Barbara, *Technology in the Industrial Revolution*, Cambridge: CUP, 2020.

Hardiman, David, *Missionaries and their Medicine: A Christian Modernity for Tribal India*, Manchester and New York: MUP, 2014.

Hardiman, David and Projit Bihari Mukhherji (eds.), *Medical Marginality in South Asia: Situating Subaltern Therapeutics*, London and New York: Routledge, 2012.

Harding, Sandra, *Is Science Multicultural? Postcolonialisms, Feminisms, and Epistemologies*, Bloomington: Indiana University Press, 1998.

Harrison, Mark, *Public Health in British India: Anglo-Indian Preventive Medicine, 1859–1914*, Cambridge: CUP, 1994.

Harrison, Mark, *Climates and Constitutions: Health, Race, Environment and British Imperialism in India 1600–1850*, Oxford: OUP, 2002.

Harrison, Mark, *Medicine in an Age of Commerce and Empire: Britain and its Tropical Colonies*, Oxford: OUP, 2010.

Harrison, Mark, Margaret Jones and Helen Sweet (eds.), *From Western Medicine to Global Medicine: The Hospital Beyond the West*, New Delhi: Orient Blackswan, 2009.

Hassan, Narin, *Diagnosing Empire: Women, Medical Knowledge, and Colonial Mobility*, Surrey and Burlington: Ashgate, 2011.

Headrick, Daniel R., *Tools of Empire: Technology and European Imperialism in the Nineteenth Century*, New York: OUP, 1981.

Headrick, Daniel R., *The Tentacles of Progress: Technology Transfer in the Age of Imperialism, 1850–1940*, New York: OUP, 1988.

Healey, Madelaine, *Indian Sisters: A History of Nursing and the State, 1907–2007*, London, New York and New Delhi: Routledge, 2013.

Herbert, E. W., *Flora's Empire: British Gardens in India*, Philadelphia: University of Pennsylvania Press, 2011.

Hill, Christopher V., *River of Sorrow: Environment and Social Control Riparian North India, 1770–1994*, Ann Arbor, MI: Association for Asian Studies, 1997.

Hodges, Sarah (ed.), *Reproductive Health in India: History, Politics and Controversies*, Hyderabad: Orient Blackswan, 2006.

Huff, Toby E., *Intellectual Curiosity and the Scientific Revolution: A Global Perspective*, New York: CUP, 2011.

Hughes, Thomas P., *Networks of Power: Electrification in Western Society, 1880–1930*, Baltimore and London: Johns Hopkins University Press, 1983.

Inkster, Ian, *Science and Technology in History: An Approach to Industrial Development*, London: Palgrave Macmillan, 1991.

Jain, Sanjeev and Alok Sarin (eds.), *The Psychological Impact of the Partition of India*, New Delhi: Sage, 2018.

Jayawardena, Kumari, *The White Woman's Other Burden: Western Women and South Asia during British Colonial Rule*, New York and London: Routledge, 1995.

Jeffery, Roger, *The Politics of Health in India*, Berkeley: University of California Press, 1988.

Jones, Margaret, *Health Policy in Britain's Model Colony: Ceylon (1900–1948)*, New Delhi: Orient Blackswan, 2005.

Jones, Margaret, *The Hospital System and Health Care: Sri Lanka 1815–1960*, New Delhi: Orient Blackswan, 2009.

Kale, Sunila, *Electrifying India: Regional Political Economic of Development*, Stanford and California: Stanford University Press, 2014.

Kapila, Shruti (ed.), *An Intellectual History for India*, Cambridge and Delhi: CUP, 2010.

Kerr, Ian J., *Building the Railways of the Raj, 1850–1900*, New Delhi: OUP, 1995.

Kerr, Ian J. (ed.), *Railways in Modern India*, New Delhi: OUP, 2001.

Kerr, Ian J. (ed.), *27 Down: New Departures in Indian Railway Studies*,Hyderabad: Orient Longman, 2007.

Kerr, Ian J., *Engines of Change: The Railroads That Made India*, New Delhi: Orient Blackswan, 2012.

Kopf, David, *British Orientalism and the Bengal Renaissance: The Dynamics of Indian Modernisation 1773–1835*, Berkeley: University of California Press, 1969.

Kour, Kawal Deep, *A History of Intoxication: Opium in Assam*, New Delhi: Manohar, 2019.

Kragh, Helge, *An introduction to the historiography of science*, Cambridge: CUP, 1987.

Kranzberg, Melvin (ed.), *Technological Education—Technological Style*, San Francisco: San Francisco Press, 1986.

Kumar, Anil, *Medicine and the Raj: British Medical Policy in India, 1835–1911*, New Delhi: Sage, 1998.

Kumar, Deepak (ed.), *Science and Empire: Essays in Indian Context*, New Delhi: Anamika Prakashan, 1991.

Kumar, Deepak, *Science and the Raj: A Study of British India*, New Delhi: OUP, 2ndEdition, 2006 (1st Edition 1995).

Kumar, Deepak, *The Trishanku Nation: Memory, Self and Society in Contemporary India*, New Delhi: OUP, 2016.

Kumar, Deepak, Joseph Bara, Nandita Khadria, and Ch. Radha Gayathri (eds.), *Education in Colonial India: Historical Insights*, New Delhi: Manohar, 2013.

Kumar, Deepak and Rajsekhar Basu (eds.), *Medical Encounters in British India*, New Delhi: OUP, 2013.

Kumar, Deepak, Vinita Damodaran and Rohan D'Souza (eds.), *The British Empire and the Natural World: Environmental Encounters in South Asia*, New Delhi: OUP, 2011.

Kumar, Deepak, and Bipasha Raha (eds.), *Tilling the Land: Agricultural Knowledge and Practices in Colonial India*, New Delhi: Primus, 2016.

Kumar, Neelam (ed.), *Women and Science in India: A Reader*, New Delhi: OUP, 2009.

Kumar, Neelam (ed.), *Gender and Science: Studies Across Culture*, New Delhi: Foundation Books, 2012.

Kumar, Prakash, *Indigo Plantations and Science in Colonial India*, New Delhi: CUP, 2012.

Kumar, Sudit Krishna and Suvobrata Sarkar (eds.), *Contextualizing the Body: An Indian Experience*, New Delhi: Manohar, 2021.

Lahiri, Choudhury, Deep Kanta, *Telegraphic Imperialism: Crisis and Panic in the Indian Empire, c. 1830–1920*, Basingstoke: Palgrave Macmillan, 2010.

Lambert, David, and Alan Lester (ed.), *Colonial Lives across the British Empire: Imperial Careering in the Long Nineteenth Century*, Cambridge: CUP, 2011.

Landes, David S., *The Unbound Prometheus: Technological Change and Industrial Development in Western Europe from 1750 to the Present*, Cambridge: CUP, 2ndEdition, 2003.

Latour, Bruno, *Science in Action: How to Follow Scientists and Engineers Through Society*, Cambridge, MA: Harvard University Press, 1987.

Latour, Bruno, *The Pasteurization of France*, Cambridge, MA: Harvard University Press, 1988.

Leslie, Charles M. (ed.), *Asian Medical Systems: A Comparative Study*, Vol. 3, New Delhi: Motilal Bararsidass, 1998.

Lingam, Lakshmi (ed.), *Understanding Women's Health Issues: A Reader*, New Delhi: Kali for Women, 1998.

Lourdusamy, J., *Science and National Consciousness in Bengal 1870–1930*, New Delhi: Orient Longman, 2004.

Macleod, Roy and Deepak Kumar (eds.), *Technology and the Raj: Western Technology and Technical Transfers to India 1700–1947*, New Delhi: Sage, 1995.

Madhwi, *Health, Medicine and Migration: The Formation of Indentured Labour, c. 1834–1920*, Delhi: Primus, 2020.

Mann, Michael, *South Asia's Modern History: Thematic Perspectives*, London and New York: Routledge, 2015.

Mann, Michael, *Wiring the Nation: Telecommunication, Newspaper-Reportage, and Nation Building in British India, 1850–1930*, New Delhi: OUP, 2017.

McNeil, Ian, (ed.), *An Encyclopaedia of the History of Technology*, London and New York: Routledge, 1990.

Mendonca, Angela, Ana Cunha and Ranjan Chakrabarti (eds.), *Natural Resources, Sustainability and Humanity: A Comprehensive View*, Dordrecht: Springer, 2012.

Menke, Henk, Jane Buckingham, Farzana Gounder, Ashutosh Kumar and Maurits S. Hassankhan (eds.), *Social Aspects of Health, Medicine and Disease in Colonial and Post-Colonial Era*, New Delhi: Manohar, 2020.

Miller, David Philip, and Peter HannsReill (eds.), *Visions of Empire: Voyages, Botany, and Representations of Nature*, Cambridge and New York: CUP, 1996.

Mills, James and Satadru Sen (eds.), *Confronting the Body: The Politics of Physicality in Colonial and Post-colonial India*, London: Anthem Press, 2004.

Misra, B. B., *The Indian Middle Class: Their Growth in Modern Times*, New Delhi: OUP, 1978.

Mital, K. V., *History of the Thomason College of Engineering (1847–1949)*, Roorkee: The University of Roorkee, 1986.

Mitchell, Timothy, *Carbon Democracy: Political Power in the Age of Oil*, London and New York: Verso, 2011.

Moscucci, Ornella, *The Science of Women: Gynaecology and Gender in England 1800–1929*, Cambridge: CUP, 1990.

Mukherjee, Haridas and Uma Mukherjee, *The Origins of the National Education Movement*, Calcutta: NCE, Bengal, 2ndRevised Edition, 2000.

Mukherjee, Jenia (ed.), *Sustainable Urbanization in India: Challenges and Opportunities*, Singapore: Springer, 2018.

Mukherjee, Sujata, *Gender, Medicine, and Society in Colonial India: Women's Health Care in Nineteenth and Early Twentieth Century Bengal*, New Delhi: OUP, 2017.

Mukherji, Projit Bihari, *Doctoring Traditions: Ayurveda, Small Technologies, and Braided Sciences*, Chicago and London: The University of Chicago Press, 2016.

Mukherji, Projit Bihari, *Nationalizing the Body: The Medical Market Print and Daktari Medicine*, London: Anthem Press, 2009.

Mukhopadhyay, Aparajita, *Imperial Technology and 'Native' Agency: A Social History of Railways in Colonial India, 1850–1920*, London and New York: Routledge, 2018.

Naik, Pramod V., *Meghnad Saha: His Life in Science and Politics*, Gewerbestrasse: Springer, 2017.

Nandy, Ashis, *The Intimate Enemy: Loss and Recovery of Self under Colonialism*, New Delhi: OUP, 1983.

Nandy, Ashis (ed.), *Science, Hegemony and Violence: A Requiem for Modernity*,New Delhi: OUP, 1990.

Nath, Pratyay, *Climate of Conquest: War, Environment and Empire in Mughal North India*, New Delhi: OUP, 2019.

Nye, David E., *Electrifying America: Social Meanings of a New Technology, 1880– 1940*, Cambridge: The MIT Press, 1990.

Pacey, Arnold, *Technology in World Civilization: A Thousand-Year History*, Cambridge and Massachusetts: The MIT Press, 1990.

Palit, Chittabrata, *Scientific Bengal: Science, Technology, Medicine and Environment under the Raj*, New Delhi: Kalpaz, 2006.

Palit, Chittabrata, *Science and Nationalism in Bengal (1876–1947)*, Kolkata: Institute of Historical Studies, 2ndEdition, 2016 (First Published 2004).

Palit, Chittabrata and Aparajita Dhar (eds.), *Medical History of India: Discipline, Disease, Death*, New Delhi: Kunal Books, 2019.

Palit, Chittabrata and Achintya Kumar Dutta (eds.), *History of Medicine in India: The Medical Encounter*, New Delhi: Kalpaz, 2005.

Pande, Ishita, *Medicine, Race and Liberalism in British Bengal: Symptoms of Empire*, London and New York, Routledge, 2010.

Panikkar, K. N., *Culture, Ideology, Hegemony: Intellectuals and Social Consciousness in Colonial India*, New Delhi: Tulika, 2001.

Pati, Biswamoy and Mark Harrison (eds.), *Health, Medicine and Empire: Perspectives on Colonial India*, New Delhi: Orient Longman, 2001.

Pati, Biswamoy and Mark Harrison (eds.), *The Social History of Health and Medicine in Colonial India*, Oxon: Routledge, 2009.

Pati, Biswamoy and Mark Harrison (eds.), *Society, Medicine and Politics in Colonial India*, London and New York: Routledge, 2018.

Pattnaik, Binay Kumar (ed.), *Sociology of Science and Technology in India*, New Delhi: Sage, 2014.

Phalkey, Jahnavi, *Atomic State: Big Science in Twentieth Century India*, Ranikhet: Permanent Black, 2013.

Polu, Sandhya L., *Infectious Disease in India, 1892–1940: Policy-Making and the Perception of Risk*, Basingstoke: Palgrave Macmillan, 2012.

Porter, Roy, *Disease, Medicine, and Society in England, 1550–1860*, Cambridge and New York: CUP, 1995.

Porter, Roy (ed.), *The Cambridge Illustrated History of Medicine*, Cambridge and New York: CUP, 1996.

Porter, Roy (ed.), *The Cambridge History of Science*, Volume 4: Eighteenth Century Science, Cambridge: CUP, 2017.

Prakash, Gyan, *Another Reason: Science and the Imagination of Modern India*, Princeton: Princeton University Press, 1999.

Prasad, Ritika, *Tracks of Change: Railways and Everyday Life in Colonial India*, New Delhi: CUP, 2015.

Pyenson, Lewis, *Empire of Reason: Exact Sciences in Indonesia, 1840–1940*, Leiden and New York: E. J. Brill, 1989.

Rafferty, Anne Marie, Jane Robinson and Ruth Elkan (eds.), *Nursing history and the Politics of Welfare*, London and New York: Routledge, 1997.

Raha, Bipasha, *The Plough and the Pen: Peasantry, Agriculture and the Literati in Colonial Bengal*, New Delhi: Manohar, 2012.

Raha, Bipasha, *Living a Dream: Rabindranath Tagore and Rural Resuscitation*, New Delhi: Manohar, 2014.

Rahman, Abdur, *Intellectual Colonisation: Science and Technology in West-East Relations*, New Delhi: Vikas Publishing, 1983.

Raina, Dhruv, *Images and Contexts: The Historiography of Science and Modernity in India*, New Delhi: OUP, 2003.

Raina, Dhruv, *Needham's Indian Network: The Search for a Home for the History of Science in India (1950–1970)*, New Delhi: Yoda Press, 2015.

Raina, Dhruv and S. Irfan Habib, *Domesticating Modern Science: A Social History of Science and Culture in Colonial India*, New Delhi: Tulika, 2004.

Raj, Kapil, *Relocating Modern Science: Circulation and the Construction of Scientific Knowledge in South Asia and Europe Seventeenth to Nineteenth Century*, Basingstoke: Palgrave Macmillan, 2007.

Rajpal, Shilpi, *Curing Madness? A Social and Cultural history of Insanity in Colonial North India, 1800–1950s*, New Delhi: OUP, 2021.

Ramanna, Mridula, *Western Medicine and Public Health in Colonial Bombay, 1845–1895*, New Delhi: Orient Longman, 2002.

Ramanna, Mridula, *Health Care in Bombay Presidency, 1896–1930*, New Delhi: Primus, 2012.

Ramanna, Mridula, *Facets of Public Health in Early Twentieth-Century Bombay*, New Delhi: Primus, 2020.

Ramasubba, Radhika, *Public Health and Medical Research in India: Their Origins under the Impact of British Colonial Policy*, Stockholm: SAREC, 1982.

Ramnath, Aparajith, *The Birth of an Indian Profession: Engineers, Industry and the State, 1900–1947*, New Delhi: OUP, 2017.

Rangarajan, Mahesh, *Fencing the Forests: Conservation and Ecological Change in India's Central Provinces, 1860–1914*, New Delhi: OUP, 1996.

Rangarajan, Mahesh (ed.), *Environmental Issues in India: A Reader*, New Delhi: Pearson, 2007.

Rangarajan, Mahesh, *Nature and the Nation: Essays on Environmental History*, New Delhi: Permanent Black, 2015.

Rangarajan, Mahesh and K. Sivaramakrishnan (eds.), *India's Environmental History*, New Delhi: Permanent Black, 2012.

Rao, Mohan, *From Population Control to Reproductive Health: Malthusian Arithmetic*, New Delhi: Sage, 2004.

Ray, Kabita, *History of Public Health: Colonial Bengal, 1921–1947*, Calcutta: K. P. Bagchi, 1998.

Ray, Utsa, *Culinary Culture in Colonial India: A Cosmopolitan Platter and the Middle-Class*, New Delhi: CUP, 2015.

Raychaudhuri, Tapan, *Europe Reconsidered: Perceptions of the West in Nineteenth Century Bengal*, New Delhi: OUP, 1988.

Reid, Anthony, *Southeast Asia in the Age of Commerce, 1450–1680*, New Haven: Yale University Press, 1988.

Reiser, Stanley, *Technological Medicine: The Changing World of Doctor and Patients*, New York: CUP, 2009.

Rotter, Andrew J., *Empires of the Senses: Bodily Encounters in Imperial India and the Philippines*, New York: OUP, 2019.

Roy, Rohan Deb, *Malarial Subjects: Empire, Medicine and Nonhumans in British India, 1820–1909*, Cambridge: CUP, 2017.

Roy, Rohan Deb, and Guy N. A. Attewell (eds.), *Locating the Medical: Explorations in South Asian History*, New Delhi: OUP, 2018.

Roy, Tirthankar, *Artisans and Industrialization: Indian Weaving in the Twentieth Century*, New Delhi: OUP, 1993.

Roy, Tirthankar, *The Economic History of India, 1857–1947*, New Delhi: OUP, 2000.

Roy, Tirthankar, *The Crafts and Capitalism, Handloom Weaving Industry in Colonial India*, London and New York: Routledge, 2020.

Rycroft, Daniel J., and Sangeeta Dasgupta (eds.), *The Politics of Belonging in India: Becoming Adivasi*, London and New York: Routledge, 2011.

Saberwal, Vasant, *Pastoral Politics, Shepherds, Bureaucrats and Conservation in the Western Himalaya*, New Delhi: OUP, 1999.

Sahu, Bhairabi Prasad (ed.), *Iron and Social Change in Early India*, New Delhi: OUP, 2006.

Saikia, Arupjyoti, *Forests and Ecological History of Assam, 1826–2000*, New Delhi: OUP, 2011.

Saikia, Arupjyoti, *The Unquiet River: A Biography of Brahmaputra*, New Delhi: OUP, 2019.

Samanta, Arabinda, *Malarial Fever in Colonial Bengal, 1820–1939: Social History of an Epidemic*, Calcutta: Firma KLM, 2002.

Samanta, Arabinda, *Living with Epidemic in Colonial Bengal, 1818–1945*, New Delhi: Manohar, 2017.

Samanta, Samiparna, *Meat, Mercy, Morality: Animals and Humanitarianism in Colonial Bengal, 1850–1920*, New Delhi: OUP, 2021.

Sangwan, Satpal, *Science, Technology and Colonization: An Indian Experience 1757–1857*, New Delhi: Anamika, 1991.

Sarkar, Benoy Kumar, *Education for Industrialization: An Analysis of the Forty Years' Work of Jadavpur College of Engineering and Technology, 1905–1945*, Kolkata: The National Council of Education, 2ndEdition, 2017 (First Published 1946).

Sarkar, Smritikumar, *Technology and Rural Change in Eastern India 1830–1980*, New Delhi: OUP, 2014.

Sarkar, Sumit, *The Swadeshi Movement in Bengal*, Ranikhet: Permanent Black, New Edition, 2010 (First Published 1973).

Sarkar, Sutapa Chatterjee, *The Sundarbans: Folk Deities, Monsters and Mortals*, New Delhi: Social Science Press, 2010.

Sarkar, Suvobrata, *The Quest for Technical Knowledge: Bengal in the Nineteenth Century*, New Delhi: Manohar, 2012.

Sarkar, Suvobrata, *Let there be Light: Engineering, Entrepreneurship, and Electricity in Colonial Bengal, 1880–1945*, New Delhi: CUP, 2020.

Schiebinger, Londa and Claudia Swan (eds.), *Colonial Botany: Science, Commerce, and Politics in the Early Modern World*, Philadelphia: University of Pennsylvania Press, 2005.

Sebastian, Joseph, *Cochin Forests and the British Techno-ecological Imperialism in India*, New Delhi: Primus, 2016.

Sehgal, Narendra K., Satpal Sangwan and Subodh Mahanti (eds.), *Uncharted Terrains: Essays on Science Popularisation in Pre-Independence India*, New Delhi: Vigyan Prasar, 2000.

Sehrawat, Samiksha, *Colonial Medical Care in North India: Gender, State, and Society c. 1840–1920*, New Delhi: OUP, 2013.

Sen, S. N., *Scientific and Technical Education in India*, New Delhi: INSA, 1991.

Sen, Satadru, *Benoy Kumar Sarkar: Restoring the nation to the world*, New Delhi, London and New York: Routledge, 2015.

Sengoopta, Chandak, *The Rays Before Satyajit: Creativity and Modernity in Colonial India*, New Delhi: OUP, 2016.

Shapin, Steven and Simon Schaffer, *Leviathan and the Air-pumps: Hobbes, Boyle, and the Experimental Life*, Princeton and Oxford: Princeton University Press, 2011.

Sharma, Madhuri, *Indigenous and Western Medicine in Colonial India*, New Delhi: Foundation Books, 2012.

Shiva, Vandana, *Staying Alive: Women, Ecology and Development*, London: Zed Press, 1989.

Shiva, Vandana, *Ecology and the Politics of Survival, Conflicts over Natural Resources in India*, New Delhi: Sage, 1991.

Singh, Chetan, *Natural Premise: Ecology and Peasant Life in the Western Himalaya, 1800–1950*, New Delhi: OUP, 1988.

Singh, Maina Chawla, *Gender, Religion, and "Heathen Lands": American Missionary Women in South Asia, 1860s–1940s*, New York: Garland, 2000.

Sinha, Jagdish N., *Science, War and Imperialism: India in the Second World War*, Leiden & Boston: Brill, 2008.

Sinha, Nitin, *Communication and Colonialism in Eastern India: Bihar, 1760s–1880s*, London, New York and New Delhi: Anthem Press, 2013.

Sinha, Sandip, *Public Health Policy and the Indian Public: Bengal 1850–1920*, Calcutta: Vision Publications, 1998.

Sivaramakrishnan, Kavita, *Modern Forests: State-making and Environmental Change in Colonial Eastern India*, New Delhi: OUP, 1999.

Sivaramakrishnan, Kavita, *Old Poisons, New Bottles: Recasting Indigenous Medicine in Colonial Punjab (1850–1945)*, Hyderabad: Orient Longman, 2006.

Sivasundaram, Sujit, *Nature and the Godly Empire: Science and Evangelical Mission in the Pacific, 1795–1850*, Cambridge: CUP, 2005.

Sivasundaram, Sujit, *Islanded: Britain, Sri Lanka, and the Bounds of an Indian Ocean Colony*, Chicago: University of Chicago Press, 2013.

Skaria, Ajay, *Hybrid Histories: Forests, Frontiers, and Wildness in Western India*, New Delhi: OUP, 1999.

Slotten, Hugh Richard, Ronald L Numbers and David N Livingstone (eds.), *The Cambridge History of Science*, Vol. 8: Modern Science in National, Transnational and Global Context, Cambridge: CUP, 2020.

Subbarayappa, H. V., *In Pursuit of Excellence: A History of the Indian Institute of Science*, New Delhi: Tata Mc Graw-Hill Publishing Company, 1992.

Subrahmanyam, Sanjay, *Europe's Asia: Words, People, Empires, 1500–1800*, Cambridge.: Harvard University Press, 2017.

Subrahmanyan, Lalita, *Women Scientists in the Third World: The Indian Experience*, New Delhi: Sage, 1998.

Subramanian, Ajantha, *The Caste of Merit: Engineering Education in India*, Cambridge, Massachusetts and London: Harvard University Press, 2019.

Sur, Abha, *Dispersed Radiance: Caste, Gender and Modern Science in India*, New Delhi: Navayana Publishers, 2012.

Tyabji, Nasir, *Colonialism, Chemical Technology and Industry in Southern India, 1880–1937*, New Delhi: OUP, 1995.

Tyabji, Nasir, *Industrialization and Innovation: The Indian Experience*, New Delhi: Sage, 2000.

Tyabji, Nasir, *Forging Capitalism in Nehru's India: Neocolonialism and the State, c. 1940–1970*, New Delhi: OUP, 2015.

Varughese, Shiju Sam, *Contested Knowledge: Science, Media and Democracy in Kerala*, New Delhi: OUP, 2017.

Visvanathan, Shiv, *Organizing for Science: The Making of an Industrial Research Laboratory*, New Delhi: OUP, 1985.

Wickramasinghe, Nira, *Metallic Modern: Everyday Machines in Colonial Sri Lanka*, Oxford: Berghahn Books, 2014.

Winterbottom, Anna, *Hybrid Knowledge in the Early East India Company World*, Basingstoke: Palgrave Macmillan, 2016.

Worster, Donald, *Nature's Economy: A History of Ecological Ideas*, Cambridge: CUP, 1977.

Wujastyk, Dominik, *The Roots of Ayurveda: Selections from Sanskrit Medical Writings*, New Delhi: Penguin, 1998.

Zachariah, Benjamin, *Developing India: An Intellectual and Social History*, New Delhi: OUP, 2005.

Index